Crystallization of Nucleic Acids and Proteins

The Practical Approach Series

SERIES EDITORS

D. RICKWOOD
Department of Biology, University of Essex
Wivenhoe Park, Colchester, Essex CO4 3SQ, UK

B. D. HAMES
Department of Biochemistry and Molecular Biology, University of Leeds
Leeds LS2 9JT, UK

Affinity Chromatography
Animal Cell Culture
Animal Virus Pathogenesis
Antibodies I and II
Biochemical Toxicology
Biological Membranes
Biosensors
Carbohydrate Analysis
Cell Growth and Division
Cellular Calcium
Cellular Neurobiology
Centrifugation (2nd edition)
Clinical Immunology
Computers in Microbiology
Cytokines
Directed Mutagenesis
DNA Cloning I, II, and III
Drosophila
Electron Microscopy in Biology
Electron Microscopy in Molecular Biology
Essential Molecular Biology I and II
Fermentation
Flow Cytometry
Gel Electrophoresis of Nucleic Acids (2nd edition)
Gel Electrophoresis of Proteins (2nd edition)
Genome Analysis
HPLC of Macromolecules
HPLC of Small Molecules
Human Cytogenetics
Human Genetic Diseases
Immobilised Cells and Enzymes
Iodinated Density Gradient Media
Light Microscopy in Biology
Liposomes
Lymphocytes
Lymphokines and Interferons
Mammalian Cell Biotechnology
Mammalian Development
Medical Bacteriology
Medical Mycology
Microcomputers in Biology
Microcomputers in Physiology
Mitochondria

Molecular Neurobiology
Mutagenicity Testing
Neurochemistry
Nucleic Acid and Protein Sequence Analysis
Nucleic Acids Hybridisation
Nucleic Acids Sequencing
Oligonucleotide Synthesis
Oligonucleotides and Analogues
PCR
Peptide Hormone Action
Peptide Hormone Secretion
Photosynthesis: Energy Transduction
Plant Cell Culture
Plant Molecular Biology
Plasmids
Polymerase Chain Reaction
Post-implantation Mammalian Embryos
Prostaglandins and Related Substances
Protein Architecture
Protein Function
Protein Purification Applications
Protein Purification Methods
Protein Sequencing
Protein Structure
Proteolytic Enzymes
Radioisotopes in Biology
Receptor Biochemistry
Receptor–Effector Coupling
Ribosomes and Protein Synthesis
Solid Phase Peptide Synthesis
Spectrophotometry and Spectrofluorimetry
Steroid Hormones
Teratocarcinomas and Embryonic Stem Cells
Transcription and Translation
Virology
Yeast

Crystallization of Nucleic Acids and Proteins
A Practical Approach

Edited by

A. DUCRUIX and R. GIEGÉ

CNRS, Gif-sur-Yvette, France
and
CNRS, Strasbourg, France

Oxford University Press, Walton Street, Oxford OX2 6DP

Oxford is a trade mark of Oxford University Press

Published in the United States
by Oxford University Press, New York

British Library Cataloguing in Publication Data
A cataloguing record for this book is available from the British Library

Library of Congress Cataloging in Publication Data
Crystallization of nucleic acids and proteins: a practical approach/
edited by A. Ducruix and R. Giegé.
(The Practical approach series)
Includes bibliographical references and index.
1. Proteins–Analysis. 2. Nucleic acids–Analysis.
3. Crystallization–Methodology. I. Ducruix, A. II. Giegé, R.
III. Series.
QP552.C79 1991 547.7'5046—dc20 91–21164
ISBN 0–19–963245–6
ISBN 0–19–963246–4 (pbk.)

Typeset by Cambrian Typesetters, Frimley, Surrey
Printed in Great Britain by
Information Press Ltd, Eynsham, Oxford

Preface

WITH the development of biotechnological techniques and macromolecular engineering there is an increasing need to elucidate the three-dimensional structure of proteins, nucleic acids, and multi-macromolecular assemblies by X-ray methods. To achieve this aim, crystals diffracting at high resolution are needed. The major aim of this book in the *Practical Approach* series is to present the methods employed to produce crystals of biological macromolecules. As usual in the series, detailed laboratory protocols are given throughout the book. However, we have not given protocols just as 'recipes', instead we have presented the methods with reference to the theoretical concepts and principles underlying them. In fact, one of the aims of this book was to combat the falacious idea that crystal growth of biological macromolecules is more an 'art' than a science. This may sometimes be true from a practical point of view, but it is certainly incorrect in principle. Therefore, emphasis has been given to the physical parameters involved in crystallization and to our knowledge of the crystal growth of small molecules, as well as to the particular properties of biological macromolecules.

This book is intended to be read by a wide range of scientists. Firstly, by the crystallographers who have to solve three-dimensional structures of macromolecules. Secondly, by all molecular biologists who have access to macromolecules but often do not know how to handle them for crystallization, and who may consider crystallization as an esoteric 'art' because of a lack of basic knowledge about the crystallization process. Thirdly, by the physicochemists and physicists who for other reasons consider biology an esoteric science. It is our wish that this practical manual will contribute to a better understanding of crystallogenesis by these scientists and to the improved perception of the biological requirements that have to be taken into account for physical studies. Finally, the book should be a laboratory guide for all students and beginners, helping them to avoid making mistakes when entering the field of crystal preparation.

Chapter 1 is an introduction to crystallogenesis of biological macromolecules. It includes a brief historical survey of the subject and introduces the general principles and major achievements of this discipline. The preparation of biological macromolecules and the concept of 'crystallography-grade purity' are developed in Chapter 2. Statistical methods to screen for crystallization are described in Chapter 3 with a presentation of the theory and practical advice (and a computer program) for protocol design. One of the goals of this book is to give to crystal growers of biological macromolecules the conceptual and methodological tools needed to perform such control. This is examined in Chapter 4 which includes a description of the classical crystallization methods together with workshop examples. Because it is sometimes difficult to reproduce appropriate nucleation conditions, Chapter 5 is devoted to seeding procedures with preformed crystalline material, including micro-, macro-, and cross-seeding with numerous examples.

Crystallization in gels, which is an old technique revisited for biological macromolecules, is described with practical considerations in Chapter 6. Related methods such as crystal growth under microgravity and supergravity are included in this chapter. The special cases of nucleic acids (and their complexes with proteins as well as nucleo-protein assemblies) and membrane proteins are covered in two individual chapters (7 and 8). In the latter case, the purification and the role of detergents are explained in great detail. The concept of solubility and the methods used to establish phase diagrams are discussed in Chapter 9 in a quantitative approach to crystallogenesis which allows for a correlation between the variation of one parameter and the ability of a given biological macromolecule to crystallize. Chapter 10 deals with physical methods of protein crystallization; the use of light scattering to probe crystallization conditions is advocated with a description of the materials used. Chapter 11 discusses the soaking of crystals of biological macromolecules for resolving the structure (heavy-atom derivatives), or for diffusing inhibitors, activators, or cofactors (eventually photoactivable).

Chapter 12 provides an introduction to X-ray crystallography that is geared towards biochemists wanting to characterize crystals by themselves rather than explaining how to solve a structure. Robotics, although not widely used among academic laboratories at the moment, is fully presented in Chapter 13 by pioneers of the technique. It is expected to be used by all laboratories in the near future. Finally, Chapter 14 covers a new field which the editors believe to be very important, namely the preparation of selenomethionyl protein crystals. Procotols are given for expression, purification, and crystallization of selenomethionyl proteins, for which several structures have been elucidated by the multiple anomalous dispersion (MAD) method.

It is a great pleasure to acknowledge our gratitude to a number of friends and colleagues. First, our colleagues from Gif-sur-Yvette and Strasbourg deserve particular thanks for having participated over the years in the development of our studies on crystallogenesis; their enthusiasm was essential and gave us the impetus for the preparation of a book covering this field. However, without the invaluable help of many friends from both sides of the Atlantic who agreed to cover specialized topics this venture would not have been possible. We would like to warmly thank all of them for having rapidly prepared updated manuscripts in a rapidly moving field. We would also like to thank all those who communicated results prior to publication. Special thanks should be given to Bernard Lorber and Madeleine Ries-Kautt who assisted us in the preparation of the manuscripts. The French Centre National de la Recherche Scientifique (CNRS) and Centre National d'Etudes Spatiales (CNES) are acknowledged for their permanent support in developing biological crystallogenesis and their interest for the physico-chemical aspects of the field.

Gif-sur-Yvette and Strasbourg A. D.
February 1990 R. G.

Contents

Contributors

C. W. CARTER, JR.
Department of Biochemistry and Biophysics, CB 7260, University of North Carolina at Chapel Hill, Chapel Hill, NC 27599–7260, USA.

P. CHEN
Department of Molecular Biology, The Scripps Research Institute, 10666 North Torrey Pines Road, La Jolla, CA 92037, USA.

A. C. DOCK-BREGEON
Institut de Biologie Moléculaire et Cellulaire du CNRS, 15 rue René Descartes, F-67084 Strasbourg Cedex, France.

S. DOUBLIÉ
Department of Biochemistry and Biophysics, CB 7260, University of North Carolina at Chapel Hill, Chapel Hill, NC 27599–7260, USA.

A. DUCRUIX
Institut de Chimie des Substances Naturelles du CNRS, avenue de la Terrasse, F-91198 Gif-sur-Yvette Cedex, France.

R. GIEGÉ
Institut de Biologie Moléculaire et Cellulaire du CNRS, 15 rue René Descartes, F-67084 Strasbourg Cedex, France.

F. LEFAUCHEUX
Laboratoire de Minéralogie-Cristallographie, Universités Pierre et Marie Curie (Paris VI) et Paris VII, 4 place Jussieu, F-75252 Paris Cedex 05, France.

B. LORBER
Institut de Biologie Moléculaire et Cellulaire du CNRS, 15 rue René Descartes, F-67084 Strasbourg Cedex, France.

V. MIKOL
Sandoz Ltd., Preclinical Research, CH-4002 Basel, Switzerland.

D. MORAS
Institut de Biologie Moléculaire et Cellulaire du CNRS, 15 rue René Descartes, F-67084 Strasbourg Cedex, France.

M. A. PEROZZO
Laboratory for the Structure of Matter, Code 6030, Naval Research Laboratory, Washington DC 20375–5000, USA.

K. PROVOST
Laboratoire de Minérologie-Cristallographie, Universites Pierre et Marie Curie (Paris VI) et Paris VII, 4 place Jussieu, F-75252 Paris Cedex 05, France.

F. REISS-HUSSON
UPR 407, CNRS, avenue de la Terrasse, F-91198 Gif-sur-Yvette Cedex, France.

M. RIES-KAUTT
Institut de Chimie des Substances Naturelles du CNRS, avenue de la Terrasse, F-91198 Gif-sur-Yvette Cedex, France.

M. C. ROBERT
Laboratoire de Minérologie-Cristallographie, Universités Pierre et Marie Curie (Paris VI) et Paris VII, 4 place Jussieu, F-75252 Paris Cedex 05, France.

L. SAWYER
Department of Biochemistry, University of Edinburgh, George Square, Edinburgh EH8 9XD, UK.

E. A. STURA
Department of Molecular Biology, The Scripps Research Institute, 10666 North Torrey Pines Road, La Jolla, CA 92037, USA.

M. A. TURNER
Department of Biochemistry, University of Edinburgh, George Square, Edinburgh EH8 9XD, UK.

K. B. WARD
Laboratory for the Structure of Matter, Code 6030, Naval Research Laboratory, Washington DC 20375–5000, USA.

I. A. WILSON
Department of Molecular Biology, The Scripps Research Institute, 10666 North Torrey Pines Road, La Jolla, CA 92037, USA.

W. A. ZUK
Laboratory for the Structure of Matter, Code 6030, Naval Research Laboratory, Washington DC 20375–5000, USA.

Abbreviations

ACA	American Crystallography Association
AcO	acetate
ACS	American Chemical Society
ANS	8-anilino-1-naphtalenesulfonic acid
BD	benzoyl DEAE
BTP	Bis-Tris-Propane
CHAPS	3-[(3-cholamidopropyl)-dimethyl-amino]-1-propane sulfonate
CHAPSO	3-[(3-cholamidopropyl)-dimethyl-amino)-2-hydroxy-1-propane sulfonate
CM	carboxy-methyl
CMC	critical micellar concentration
C_nDAO	*n*-alkyl-dimethylamineoxides
C_nEm	*n*-alkyl-oligoethylene glycol-monoethers
C_nG	*n*-alkyl-β-glucosides
C_nM	*n*-alkyl-maltosides
DEAE	diethyl-amino-ethyl
DIFP	diisopropylfluorophosphate
DNA	deoxyribonucleic acid
DTE, DTT	dithioerythritol, dithiothreitol
EDTA	ethylene diamine tetraacetic acid
EGTA	ethylene glycol bis(2-aminoethylether)-*N*,*N*,*N′*,*N′*-tetraacetic acid
EM	electron microscopy
FPLC	fast protein liquid chromatography
Hepes	*N*,2-hydroxyethylpiperazine-*N′*-ethane sulfonic acid
HEW	hen egg white
HIC	hydrophobic interaction chromatography
HPLC	high performance liquid chromatography
IEF	isoelectric focusing
LB medium	Luria-Bertami medium
LHC	light harvesting complex
MAD	multiwavelength anomalous dispersion
Mes	2-(*N*-morpholino)ethane sulphonic acid
MIR	multiple isomorphous replacement
MPD	2-methyl 2-4-pentane diol
PBC	periodic bond chain
PEG	polyethylene glycol
pI	isoelectric point
p.p.m.	part per million
PMSF	phenylmethylsulfonyl fluoride
RNA	ribonucleic acid
R_f	relative chromatographic migration
RPC	reverse phase chromatography
r.p.m.	revolution per minute

SDS	sodium dodecyl sulphate
SDS–PAGE	sodium dodecyl sulphate polyacrylamide gel electrophoresis
TEOS	tetraethoxysilane
TEMED	*N*,*N*,*N*′,*N*′ tetramethylenethylenediamine
TLC	thin layer chromatography
TMOS	tetramethoxysilane
Tris	tris(hydroxymethyl)methylamine
tRNA	transfer ribonucleic acid

1

An introduction to the crystallogenesis of biological macromolecules

R. GIEGÉ and A. DUCRUIX

Il y a là des mystères, qui préparent à l'avenir d'immenses travaux et appellent dès aujourd'hui les plus sérieuses méditations de la science

Pasteur, 1860, in *Leçons de Chimie.*

1. Introduction

The word 'crystal' is derived from the Greek root 'krustallos' meaning 'clear ice'. Like ice, crystals are chemically well defined, and many of them are of transparent and glittering appearance, like quartz, which was for a long time the archetype. Often they are beautiful geometrical solids with regular faces and sharp edges, which probably explains why crystallinity, even in the figurative meaning, is taken as a symbol of perfection and purity. From the physical point of view, crystals are regular three-dimensional arrays of atoms, ions, molecules, or molecular assemblies. Ideal crystals can be imagined as infinite and perfect arrays in which the building blocks (the asymmetric units) are arranged according to well-defined symmetries (forming the 230 space groups) into unit cells that are repeated in the three-dimensions by translations. Experimental crystals, however, have finite dimensions. An implicit consequence is that a macroscopic fragment from a crystal is still a crystal, because the orderly arrangement of molecules within such a fragment still extends at long distances. The practical consequence is that crystal fragments can be used as seeds (see Chapter 5). In laboratory-grown crystals the periodicity is never perfect, due to different kinds of local disorders or long-range imperfections like dislocations. Also, these crystals are often of polycrystalline nature. The external forms of crystals are always manifestations of their internal structures and symmetries, even if in some cases these symmetries may be hidden at the macroscopic level, due to differential growth kinetics of the crystal faces. The periodicity in crystal architecture is also reflected in their macroscopic physical properties. The most straightforward example is given by the ability of crystals to diffract X-rays or neutrons, the phenomenon underlying structural chemistry (for an introductory text see ref. 1), and the major aim of this book is to present the methods employed to produce crystals of biological macromolecules needed for diffraction studies. Other properties of invaluable practical applications should not be overlooked either, as in the case of optical and electronic properties which are all the basis of non-linear optics and modern electronics (for an introduction to physical properties of molecular crystals see ref. 2). Crystals furnish one of the most beautiful examples of order and symmetry in nature and it is not surprising that their study fascinates scientists (3).

What characterizes biological macromolecular crystals from small-molecular crystals? In terms of morphology, one finds with macromolecular crystals the same diversity as for small molecular crystals (*Figure 1*). In terms of crystal size, however, biological macromolecular crystals are rather small, with volumes rarely exceeding 10 mm^3, and thus they have to be examined under a binocular microscope. Except for special usages, such as neutron diffraction, this is not too severe a limitation. Among the most striking differences between the two families of crystals are the poor mechanical properties and the high content of solvent of macromolecular crystals. These crystals are always extremely fragile and are very sensitive to external conditions. This property can be used as a preliminary identification test: protein crystals are brittle or will crush when touched with the tip of a needle, while salt crystals that can sometimes develop in macromolecule crystallization experiments will resist this treatment. This fragility is a consequence both of the weak interactions between macromolecules within crystal lattices and of the high solvent content (from 20% to more than 80%) in these crystals (see Chapter 12). For that reason, macromolecular crystals have to be kept in a solvent-saturated environment, otherwise dehydration will lead to crystal cracking and destruction. The high solvent content, however, has useful consequences because solvent channels permit diffusion of small molecules, a property used for the preparation of isomorphous heavy-atom derivatives needed to solve the structures (see Chapters 11 and 12). Further, crystal structures can be considered as native structures, as is indeed directly verified in some cases by the occurrence of enzymatic reactions within crystal lattices upon diffusion of the appropriate ligands (4, 5). Other characteristic properties of macromolecular crystals are their rather weak optical birefringence under polarized light: colours may be intense for large crystals but less bright than for salt crystals (isotropic cubic crystals or amorphous material will not be birefringent). Also, because the building blocks composing macromolecules are enantiomers (L-amino acids in proteins—except in the case of some natural peptides—and D-sugars in nucleic acids) macromolecules will not crystallize in space groups with inversion symmetries. Accordingly, out of the 230 possible space groups, macromolecules only crystallize in the 65

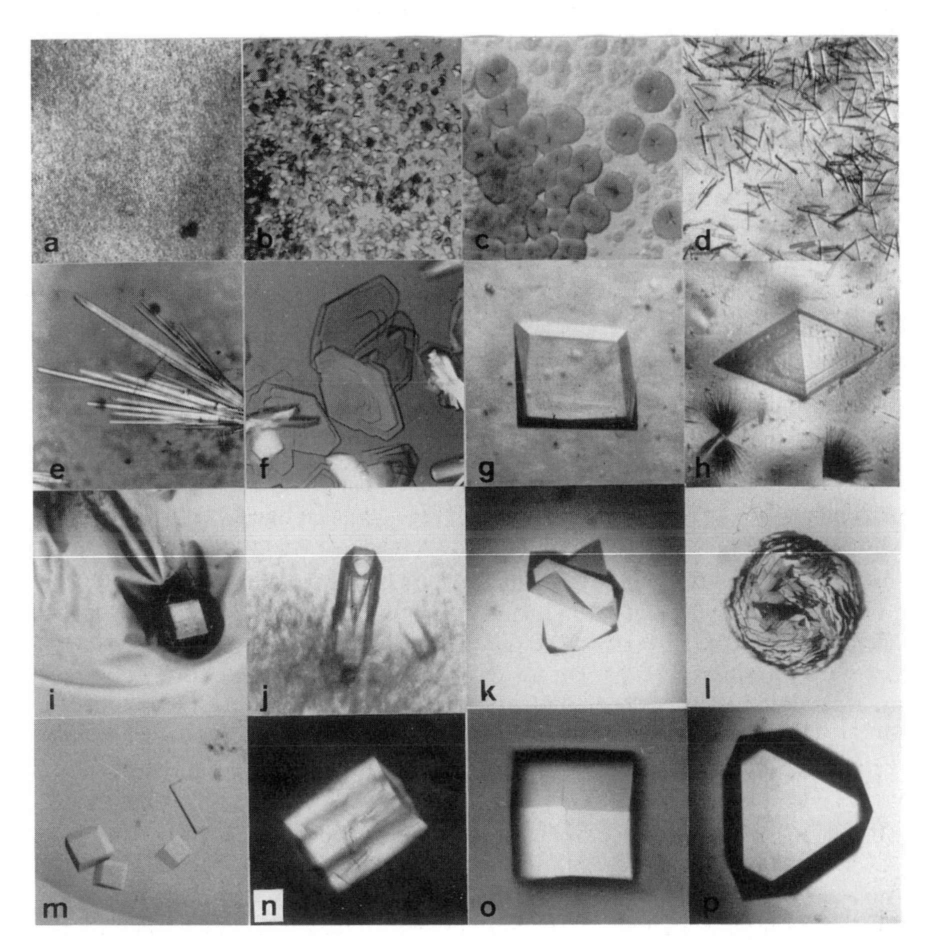

Figure 1. From precipitates to perfect crystals of biological macromolecules. (a) precipitate of HEW lysozyme; (b) yeast aspartyl-tRNA synthetase microcrystals; (c) spherulites of the complex between aspartyl-tRNA synthetase and $tRNA^{Asp}$; (d) short needles of tissue inhibitor protein of metalloprotease; (e) aspartyl-tRNA synthetase long needle-like crystals; (f) yeast initiator $tRNA^{Met}$ thin plates with growth defects; (g) plate shaped crystals of *Hypoderma lineatum* collagenase; (h) tetragonal crystals of aspartyl-tRNA synthetase showing growth defects together with brush-like needle bunches; (i) example of 'skin' of denatured protein around a HEW lysozyme crystal; (j) crystal of mellitin with a hollow extremity; (k, l) twinned and twinned-embedded crystals of collagenase and HEW lysozyme; (m, n, o, p) perfect three-dimensional crystals; (m) polymorphism in a same crytallization drop showing cubic and orthorhombic crystal habits of aspartyl-tRNA synthetase/$tRNA^{Asp}$ complex; (n) crystals of $tRNA^{Asp}$ with cracks due to ageing; (o) crystals of HEW lysozyme with 4-fold symmetry; (p) crystal of *H. lineatum* collegenase.

space groups without such inversions (6). Macromolecular crystals are also characterized by large unit cells with dimensions that can reach up to 1000 Å for virus crystals (7). From a practical point of view, it is important to remember that crystal morphology is not synonymous with crystal quality. Therefore the final diagnostic of the suitability of a crystal for structural studies will always be the quality of the diffraction pattern which reveals its internal order, as is reflected at first glance by the so-called 'resolution' parameter (see Chapter 12).

Crystal growth, which is a very old activity (8), has always intrigued mankind, and many philosophers and scientists have compared it with the biological process of reproduction, and it has even been speculated that the duplication of genetic material would occur through crystallization-like mechanisms (9). Nowadays, the theoretical and practical frames of crystallogenesis are well established for small molecules, but not for macromolecules, although it can be anticipated that many principles underlying the growth of small molecule crystals will apply for that of macromolecules (10, 11). Until recently, crystallization of macromolecules was rather empirical, and because of its unpredictability and frequent irreproducibility, it has long been considered an 'art' rather than a science. It is only in the last decade that a real need has emerged to better understand and to rationalize the crystallization of biological macromolecules.

2. Crystallization and biology: a historical background

2.1 Before X-rays

It is often forgotten that some advances in biochemistry and molecular biology have their origin in crystallization data and that empirical crystal growth of biological materials is as old as biochemistry. The first reports on protein crystals were published more than a century ago when Funke, Hünefeld, Lehman, Teichman, and others crystallized haemoglobin from the blood of various invertebrates and vertebrates, and it was prophesied that the study of crystals would shed light on the exact nature of proteinic substances (12–15). This was followed by the crystallization of hen-egg albumin and a series of plant proteins (reviewed in 14, 15). The beauty of crystals certainly fascinated the physiological chemists in these early days, since an atlas with extensive descriptions of the morphologies of haemoglobin crystals was published in 1909 (13). In 1926 Sumner reported the crystallization of urease from jack beans (16), soon followed by Northrop who crystallized pepsine and a series of other proteolytic enzymes (17). Besides being a method of purification, crystallization experiments established the view that pure enzymes are proteins, a fact not obvious to all at that time (18). Another scientific achievement arose in 1935 when Stanley crystallized

Tobacco Mosaic Virus. Influenced by Northrop's conception of enzymes, and using methods developed for proteins, he prepared the virus in a pure crystalline state (he believed the virus was an autocatalytic protein) and showed that it retains its infectivity after several recrystallizations (reviewed in 9). The importance and the implications for biology of these discoveries was recognized rapidly and in 1946 the Nobel Prize for Chemistry was awarded to Sumner, Northrop, and Stanley.

2.2 Crystallogenesis and structural biology

The use of crystals in structural biology goes back to 1934, when Bernal and Crowfoot (D. Hodgkin) produced the first X-ray diffraction pattern of a protein, that of crystalline pepsin (19). Since then representatives of most families of macromolecules have been crystallized (20), but using mainly empirical methods and without rational control of the growth mechanisms. This can be well understood because in the early days of structural biology the interest of scientists was mainly directed at the development of the X-ray methods needed to resolve the structures rather than at a rationalization of the crystallization procedures. Therefore, only limited efforts were expended in understanding or improving crystallization procedures. At present X-ray methods are well established (6, 21). The overall scheme of the various steps of the resolution of a three-dimensional structure is summarized in *Figure 2*. But production of suitable crystals diffracting at high resolution often remains the bottleneck in structure determination projects. With the rapid development of biotechnologies and the unlimited potential of macromolecular engineering (which requires structural knowledge for site-directed mutagenesis experiments or for design of factitious macromolecules) there is now an increasing need for macromolecular crystals diffracting at high resolution, not only proteins but also nucleic acids and multimacromolecular assemblies. Thanks to the biotechnological tools it is now possible to obtain rather easily the amounts of pure macromolecules (in most cases several mg) needed to start a crystallization project. In this introductory chapter we present the general trends of the science of crystal growth in biology; more extensive discussions and the practical details will be presented in following chapters.

The first major breakthrough towards better and easier crystallizations was the development, in the 1960s, of crystallization micromethods (e.g. dialysis and vapour phase diffusion) (see Chapter 4). It was promoted by structural projects on macromolecules reluctant to crystallize easily and available in limited amounts (22). Further significant improvements came from the discovery of specific properties of additives to be included in crystallization solvents, such as the polyamines (23, 24) and the non-ionic detergents (25–27) which gave the clue for crystallizing nucleic acids (see Chapter 7) or membrane proteins (see Chapter 8).

More recently, the perception of the importance of purity for growing

1. Purification of macromolecules
from wild-type, engineered, or overproducing organisms
(possibility of *in vitro* synthesis for nucleic acids and small peptides)

2. Crystallization
by *de novo* crystallization or seeding techniques

3. Data measurements
characterization of space group and diffraction resolution; measurements of diffraction intensities by diffractometry or photographic methods or using an electronic area detector or an image plate (possible use of neutron and frequently of tunable X-ray synchrotron radiation)

4. Phase determination
using methods based on isomorphous replacement (preparation of heavy-atom derivatives), anomalous scattering, molecular replacement and non-crystallographic symmetry, or direct calculations (e.g. from maximum entropy)

5. Electron density map computation and interpretation
interpretation of mini-maps; model building on computer graphic displays

6. Model refinement
least-square refinements; restrained refinements; ...

Figure 2. Steps involved in the resolution of the three-dimensional structure of a biological macromolecule. For more details concerning step 1 see Chapter 2, for step 2 see elsewhere in this book, and for steps 3–6 see reference 6, Chapter 12, and Chapter 14 for the preparation of selenomethionyl protein crystals used for anomalous scattering phase determination.

better crystals (28) was an important achievement in the field (see Chapter 2), and the adequate choice of the biological material was an important determinant for the success of many crystallizations. With the ribosome, for instance, the preparation of homogeneous particles from halophilic or thermophilic bacteria, instead of from mesophilic bacteria, considerably improved crystal quality (29, 30). Also, the ease of synthesizing oligonucleotides with automated methods, or the development of genetic engineering technologies for overexpression of proteins, explains the increasing number of crystallized DNA fragments or rare proteins.

Today, crystal growth research is stimulated by macromolecular engineering requirements, but also to some extent by space-science projects (crystallization under microgravity conditions, see Chapter 6) and its descriptive stage is moving towards a new, more quantitative discipline, biocrystallogenesis, which includes biology, biochemistry, physics, and engineering related aspects. For specialized literature see refs 31–34, and for general reviews 35–38.

3. General principles

3.1 A multiparametric process

Biocrystallization, like any crystallization, is a multiparametric process involving the three classical steps of nucleation, growth, and cessation of growth. What makes crystal growth of biological macromolecules different is, firstly, the much larger number of parameters than those involved in small molecule crystal growth (*Table 1*) and, secondly, the peculiar physico-

Table 1. Parameters affecting the crystallization (and/or the solubility) of macromolecules[a]

Intrinsic physico-chemical parameters:
- Supersaturation (concentration of macromolecules and precipitants)
- Temperature, pH (fluctuations of these parameters)
- Time (rates of equilibration and of growth)
- Ionic strength and purity of chemicals (nature of precipitant, buffer, additives)
- Diffusion and convection (gels, microgravity)
- Volume and geometry of samples and set-ups (surface of crystallization chambers)
- Solid particles, wall and interface effects (homogeneous versus heterogeneous nucleation, epitaxy, . . .)
- Density and viscosity effects (differences between crystal and mother liquor)
- Pressure, electric, and magnetic fields
- Vibrations and sound (acoustic waves)
- Sequence of events (experimentalist versus robot)

Biochemical and biophysical parameters:
- Sensitivity of conformations to physical parameters (temperature, pH, ionic strength, solvents, . . .)
- Binding of ligands (substrates, cofactors, metal ions, other ions, . . .)
- Specific additives (reducing agents, non-ionic detergents, polyamines, . . .)
- Related with properties of macromolecules (oxydation, hydrophilicity versus hydrophobicity, polyelectrolyte nature of nucleic acids, . . .)
- Ageing of samples (red-ox effects, denaturations or degradations)

Biological parameters:
- Rarity of most biological macromolecules
- Biological sources and physiological state of organisms or cells (thermophiles versus halophiles or mesophiles, growing versus stationary phase, . . .)
- Bacterial contaminants

Purity of macromolecules:
- Macromolecular contaminants (odd macromolecules or small molecules)
- Sequence (micro) heterogeneities (fragmentation by proteases or nucleases—fragmented macromolecules may better crystallize, partial or heterogeneous post-translational modifications, . . .)
- Conformational (micro) heterogeneities (flexible domains, oligomer, and conformer equilibria, aggregations, denaturations, . . .)
- Batch effects (two batches are not identical!)

[a] Although all these parameters have not been screened systematically, especially for the crystallization of a given macromolecule, all of them have been evaluated individually in isolated cases.

chemical properties of these compounds. For instance, their optimal stability in aqueous media is restricted to a rather narrow temperature and pH range. But the main difference from small molecules crystal growth is the conformational flexibility and chemical versatility of macromolecules, and their consequent greater sensitivity to external conditions. This complexity is the main reason why systematic investigations were not undertaken earlier. Furthermore, the importance of some parameters, such as the geometry of crystallization vessels or the biological origin of macromolecules, had not been recognized. It is only in the last few years that the hierarchy of parameters has been perceived. A practical consequence of this new perception was the development of statistical methods to screen crystallization conditions (see Chapter 3). For a rational design of growth conditions, however, physical and biological parameters have to be controlled. One of the aims of this book is to give to crystal growers of biological macromolecules the conceptual and methodological tools needed to achieve such control (see Chapters 4 to 10).

3.2 Purity

Because macromolecules are extracted from complex biological mixtures, purification plays an extremely important role in crystallogenesis. Purity, however, is not an absolute requirement since crystals of macromolecules can sometimes be obtained from mixtures. But such crystals are mostly small or grow as polycrystalline masses, are not well shaped, and are of bad diffraction quality, and thus cannot be used for diffraction studies. However, crystallization of macromolecules from mixtures may be used as a tool for purification (39). For the purpose of X-ray crystallography, high-quality monocrystals of appreciable size (0.2 mm at least for the dimension of a face) are needed. It is our belief that poor purity is the most common cause of unsuccessful crystallization, and for crystallogenesis the purity requirements of macromolecules have to be higher than in other fields of molecular biology. Purity has to be of 'crystallography grade': the macromolecules not only have to be pure in terms of lack of contaminants, they have also to be conformationally 'pure' (28). Denatured macromolecules, or macromolecules with structural microheterogeneities, adversely affect crystal growth more than do unrelated molecules, especially when structural heterogeneities concern domains involved in crystal packing. On the other hand presence of microquantities of proteases (or nucleases) can alter the structure of the macromolecules during storage or the rather long time needed for crystallization. As a consequence, when starting a crystallization project one has to be primarily concerned with purification methodologies and to take all precautions against protease and nuclease action. To have reproducible results, the physiological state of cells should be controlled, because protease (or nuclease) levels may vary as well as the balance between cellular components. As a general rule, batches of macromolecules should not be mixed and crystallization experiments should be conducted on fresh material so that ageing phenomena are limited. For more details see Chapter 2.

In summary, we emphasize the importance of macromolecular purity in biological crystallogenesis, and in case of unsuccessful experiments we recommend firstly improvement or reconsideration of the purification procedure of the molecules of interest.

3.3 Solubilities and supersaturation

To grow crystals of any compound, molecules have to be brought in a supersaturated, thermodynamically unstable state, which may develop in a crystalline or amorphous phase when it returns to equilibrium. Supersaturation can be achieved by slow evaporation of the solvent or by varying parameters (listed in *Table 1*). The recent use of pressure as a parameter is of note (40). From this it follows that knowledge of macromolecular solubilities is a prerequisite for controlling crystallization conditions. However, the theoretical background underlying solubility is still controversial, especially regarding salt effects (40), so that solubility data almost always originates from experimental determinations. Recently, quantitative methods permitting such determinations on small protein samples were described (42, 43). The main output was the experimental demonstration of the complexity of solubility behaviours, emphasizing the importance of phase-diagram determinations for a rational design of crystal growth (see Chapter 9).

As to the nature of the salt used to reach supersaturation, one can wonder why ammonium sulphate is so frequently chosen by crystal growers (20). This usage is in fact incidental and results from the practices of biochemists for salting-out proteins. Indeed many other salts can be employed, but their effectiveness for inducing crystallization is variable (43). The practical consequence is that protein supersaturation can be reached (or changed) in a large concentration range of protein and salt, provided the adequate salts are used.

3.4 Nucleation, growth, and cessation of growth

Because proteins and nucleic acids require defined pH and ionic strength for stability and function, biomacromolecule crystals have to be grown from chemically rather complex aqueous solutions. Crystallization starts by a nucleation phase (i.e. the formation of the first ordered aggregates) which is followed by a growth phase. Nucleation conditions are sometimes difficult to reproduce, and thus seeding procedures with preformed crystalline material should not be overlooked as in many cases they represent the only method to obtain reproducible results (44, and Chapter 5). It should be noticed that nucleation requires a greater supersaturation than growth, and that crystallization rates increase when supersaturation increases. Thus nucleation and growth should be uncoupled, which is almost never done consciously but which occurs sometimes under uncontrolled laboratory conditions. From a practical point of view, interface or wall effects as well as shape and volume of drops can affect nucleation or growth, and consequently the geometry of crystallization chambers or drops has to be defined. For additional informations see Chapter 10.

Cessation of growth can have several causes. Apart from trivial ones, like depletion of the macromolecules from the crystallizing media, it can result from growth defects, poisoning of the faces, or ageing of the molecules. Better control growth conditions, in particular of the flow of molecules around the crystals, may in some cases overcome the drawbacks as was shown in microgravity experiments (ref. 45, and Chapter 6).

3.5 Packing

With biological macromolecules, crystal quality may be correlated with the packing of the molecules within the crystalline lattices, and external crystal morphology with internal structure. As shown by the Periodic Bond Chain (PBC) method, direct protein–protein contacts are essential in determining packing and morphology (46). Forces involved in packing of macromolecules may be considered as weak as compared to those maintaining the cohesion of small-molecule crystals. They involve salt-bridges, hydrogen bonds, van der Waals, dipole–dipole, and stacking interactions (47–49). It must also be borne in mind that the weak cohesion of macromolecular crystals results from the fact that only a small part of macromolecular surfaces participates in intermolecular contacts, the remainder being in contact with the solvent (exceptions may be found for small proteins). This explains the commonly observed polymorphism of biological macromolecular crystals.

4. Major achievements

Some landmarks obtained by crystal growers of biological macromolecules are given in *Table 2*. Representative of most families of proteins, nucleic acids, and even of multimolecular assemblies, have been crystallized [see the Biological Macromolecule Crystallization Database from NIST, Gaithersburg, MD, USA (20) which was recently updated (78)]. According to this database, more than 1726 crystal forms corresponding to 876 macromolecular species have been crystallized (these numbers correspond to the entries up to May 1989). While the number of new crystallizations remained low and more or less constant between 1940 and 1965, it has grown very rapidly ever since in a quasi-exponential fashion (*Figure 3*), and indeed more than half of all crystallizations compiled to date have been published after 1983 (78). The early successes concerned mainly proteins, but more recently there are increasing experiments on nucleic acids, nucleo–protein or protein–protein

Table 2. Landmarks in the crystallization of macromolecules or macromolecular assemblies

Classes / Families	Selected examples	References[a]
Proteins:		
Antibiotics proteins:	Colicin A fragment (1986)	(50)
Antifreeze peptides:	Alanine-rich AFP (1986)	(51)
Antibodies:	IgG Kol (1974)	(52)
Blood proteins:	Haemoglobin (1853)	(**12**, **13**)
Chaperone proteins:	PAPD protein (1989)	(53)
Enzymes:	Urease (1926)	(16)
Glycoproteins:	Neuraminidase (1983)	(54)
Hormones:	Insulin (1966)	(55)
Immune response mediators:	Interleukin (1987)	(56)
Membrane proteins:	Porin from *E. coli* (1983)	(57)
Periplasmic binding proteins:	Leu/Ile/Val binding protein (1983)	(58)
Structural proteins:	TMV disk protein (1974)	(59)
Sweet taste proteins:	Thaumatin I (1985)	(60)
Toxins:	Erabutoxin (1971)	(61)
Artificial peptides	α-Helical peptide (1986)	(62)
Fragmented proteins:	Methionyl-tRNA synthetase (1971)	(63)
Overproduced and engineered proteins:	T4 lysozyme variants (1988)	(64)
Selenomethionine containing proteins:	engineered RNAse H (1990)	(65)
Nucleic acids:		
Oligodeoxyribonucleotides:	various duplexes (1988)	(**49**)
Oligoribonucleotides:	octamer duplex (1988)	(66)
Oligonucleotides hybrids:	decamer RNA/DNA hybrid (1982)	(67)
Transfer RNAs:	$tRNA^{Met}$, $tRNA^{Phe}$ (1968)	(**24**, 68)
Complexes, assemblies, and particles:		
DNA/Drug complexes:	DNA oligomers/Triostatin (1988)	(**49**)
Enzyme/RNA complexes:	$tRNA^{Asp}$/AspRS (1980)	(69)
Enzyme/DNA complexes:	EcoR1/DNA fragment (1984)	(70)
Enzyme/saccharide complexes:	Concanavalin A complex (1989)	(71)
Protein/Protein complexes:	Antigen/Antibody (1988)	(**72**)
Protein/DNA complexes:	various Repressor/DNA fragments (1988)	(73, 74)
	Nucleosome (1984)	(75)
Photosynthetic centres:	from *R. viridis* (1982)	(25)
Ribosomes:	30S or 50S particle (1988)	(**29**)
	entire 70S ribosome (1989)	(30)
Viruses:	tobacco mosaic virus (1935)	(9)
	picornaviruses, e.g. poliovirus (1985)	(76)
	isometric plant and insect viruses (1988)	(**77**)

[a] References in bold characters concern reviews. The year of publication is given for the reference which is citated; if two references are cited the most ancient one is given.

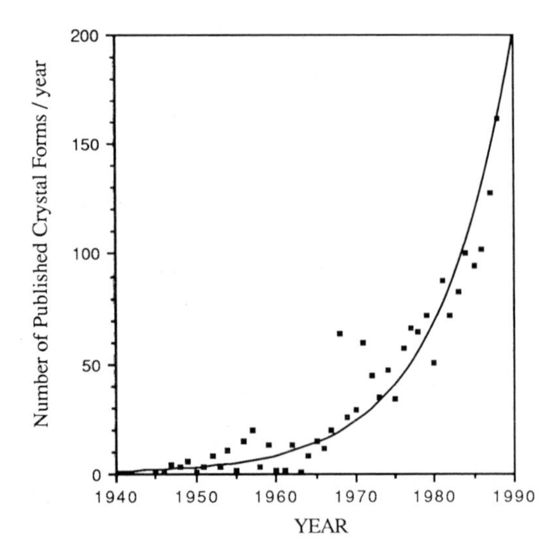

Figure 3. The increase of reported macromolecular crystal forms from 1940 to 1988. The data have been fitted with a simple exponential function (adapted from Roussell *et al.* 1990).

complexes, as well as on particles of higher complexity. It is likely that this tendency will be pursued in future, together with the crystallization of engineered macromolecules. From the practical point of view it is worth mentioning the increased use of PEG as the precipitant (in 40% of the compiled crystallization assays), concomitant with a decrease in the use of salts. Among the salts ammonium sulphate continues to be the most frequently followed by phosphate and citrate. Organic solvents were used from the beginning, but only in about 20% of the experiments (78). Noticeable is the use of MPD, which proved to be often successful in recent experiments.

4.1 Proteins

Representatives of the protein family were crystallized a hundred years ago (12, 13). To date, monocrystals suitable for X-ray diffraction studies have been obtained for most subclasses of proteins, including enzymes, hormones, toxins, structural proteins, and antibodies (at present about 400 X-ray structures have been solved and are listed in the Brookhaven Data Bank). Among them mainly soluble proteins are found, but recently successes were also obtained for hydrophobic membrane proteins (see Chapter 8) and for glycoproteins (54). Proteins of all molecular sizes and oligomeric structures have been crystallized: from the very small monomeric plant seed crambin, a 46 amino-acid long polypeptide (79), to the large *E. coli* aspartate transcarbamoylase holoenzyme (M_r 310 000) composed of two catalytic and three regulatory subunits (80).

There is, however, no correlation between the size of proteins and the quality of the crystals they can yield. Instead, a correlation seems to exist between the structural stability of proteins in solution and crystal quality, as well as to the ability of these proteins to crystallize. For instance, crystals of the extremely stable crambin diffract to better than 0.9 Å resolution (79). Conversely, proteins known to undergo structural rearrangements during their functioning often yield crystals with internal disorder, and sometimes, even if the overall resolution is high, their crystalline structure reveals the existance of disordered domains, as was found in tyrosyl-tRNA synthetase (81).

Worthy of note is the increasing number of proteins or protein variants obtained by genetic engineering (e.g. 64), or even the number of completely artificial proteins (62), that have been crystallized in recent years. This is explained by the relative ease of designing and overproducing variants in sufficient amounts, as well as the relative ease of crystallizing these molecules as soon as crystallization conditions of the parental molecules are known. Genetic engineering methods can also assist X-ray crystallography and help to overcome its second bottleneck, the preparation of isomorphous heavy-atom derivatives needed to solve the phase problem. This can be done by designing protein variants with new heavy-atom acceptor sites such as cysteine residues, a possibility explored on lysozyme (82) and already used to solve new structures (83). An alternative method derives phases by a multiwavelength anomalous diffraction (MAD) analysis using data from selonomethioninyl protein crystals (84) (see Chapter 14).

4.2 Nucleic acids

Because they are rather small and have compact structures, tRNAs were the first nucleic acids to be crystallized (see Chapter 7), but only a few species yielded crystal forms suitable for X-ray analysis (24). Among other natural RNAs, only ribosomal 5S RNAs could be crystallized, but up to now no high-resolution diffracting crystals could be grown (85). This difficulty may be linked to an intrinsic flexibility of RNA molecules and especially to their chemical fragility (28). With DNA the situation is different. Although large DNA molecules are likely to be unsuitable for packing into regular crystalline arrangements, except as fibres, smaller fragments of oligomers (2–12 mers or more) do crystallize readily in form of monocrystals (49). This development of DNA crystallography was linked to that of DNA chemistry, and in particular to the automated synthesis of the required DNA fragments and their purification by HPLC.

4.3 Complexes, assemblies, and particles

Despite their great size, certain viruses were among the first biological compounds to be crystallized. This is because these multimacromolecular assemblies, in particular spherical viruses, possess highly symmetrical structures favouring their ordering in crystalline lattices (76, 77). Ribosomes, the backbone of protein synthesis, which are also multimacromolecular particles composed of three RNAs and numerous proteins, were soon obtained in crystalline forms, but it is only recently that crystals diffracting at medium resolution were grown from materials originating from halophilic or extreme thermophilic organisms (29, 30). As to reconstitued complexes formed by two interacting macromolecules, the history of their crystallization follows the progress of the biochemical methods needed to obtain highly homogeneous preparations of the individual components. This is in particular the case for complexes between proteins and fragments of DNA, the crystallization of which was dependent upon the availability of DNA oligomers of required length and sequence, or in that of antigen/antibody complexes of the availability of monoclonal antibodies. On the other hand, knowledge of the physico-chemical conditions of complex formation was essential for the crystallization of tRNA/aminoacyl-tRNA synthetase complexes (86–88 and Chapter 7) and solubilization of membrane proteins by adapted non-ionic detergents for that of photosynthetic centres (25–27, and Chapter 8).

5. From empiricism to rationality

5.1 Towards a better understanding of parameters

To date, the major parameters underlying crystallization of macromolecules have been recognized (see *Table 1*), and if their respective contributions in the crystallization process are not known with certainty, the theoretical frame needed for explaining their role is well established (10, 11). Also, the experimental tools exist that are needed for measuring the contributions of these parameters. This is the case for solubility and aggregation-state measurements as well as for monitoring pH and temperature (see Chapters 9, 10, and 14), or even the effect of microgravity (see Chapter 6). Correlations between the variation of a parameter and the ability of a given macromolecule to crystallize are expected to be found. Finally, and perhaps most important, was the recognition of the importance of purity. Thus again we emphasize that experimenters should primarily devote their efforts to starting crystallization attempts with molecules of the highest biochemical quality.

5.2 Towards an active control of crystal growth

Only a few attempts have been published describing the active control of crystallization experiments. In general, the history of experiments is not well known, because crystal growers do not monitor parameters. This is especially the case for temperature, which is almost never known with accuracy, even if experiments are conducted in thermostated cabinets (this may be advantageous because many microconditions may be screened, although at the cost of reproducibility). Also, the kinetics of events are practically never monitored. In the following chapters the first attempts to control macromolecular crystal growth will be discussed and the experimental tools to reach such a goal will be described (e.g. for the kinetics of evaporation in vapour diffusion crystallization, see Chapters 4 and 10; for active video and temperature control and automation of assays and robotics, see Chapter 13). Although these different aspects are all in their infancy, and required instrumentation often does not exist, or exists only as prototypes, we believe that user-friendly methodologies will be developed soon, and that laboratories may be equipped with the adequate instrumentation.

5.3 The future of biological crystallogenesis

As mentioned before, genetic methods will provide access to molecules present at very low amounts in cells and help to solve problems linked to structural or conformational heterogeneities of proteins reluctant to crystallize. In particular it is expected that engineering of active variants containing compact cores should permit easier crystallizations of proteins with flexible domains. Macromolecular engineering will certainly find many applications in the RNA world, where ribozymes, pseudo-knots, and regulatory structures in messenger RNAs, have not yet been crystallized.

For the biologist, studying crystal growth should be correlated with biological problems, and crystallization projects on macromolecular complexes, on membrane proteins, and especially on engineered proteins, will certainly be developed. For the physicist, growing large monocrystals can be a goal in itself, and one might speculate that exploration of optical, electrical, mechanical, and other physical properties of crystalline arrays made from biological macromolecules or assemblies can lead to novel frontiers in the material sciences of tomorrow. Finally, for the chemist, usage of chemical and molecular biology tools could lead in future to the design of molecular devices mimicking macromolecular crystals, as was discussed for artificial self-assembling nucleic acids (89).

In conclusion, the rational approach for prompt crystallizations will demand a synergy between biochemically- and physically-directed research and usage of automated methods for the control of nucleation and growth, as well as the rapid preparation of high-quality monocrystals.

References

1. Pickworth Glusker, J. and Trueblood, K. N. (1985). *Crystal structure analysis, a primer*. Oxford University Press, New York.
2. Wright, J. D. (1987). *Molecular crystals*. Cambridge University Press.
3. Lima-de-Faria, J. (ed.) (1990). *Historical atlas of crystallography*. Kluwer, Dordrecht.

4. Ottonello, S., Mozzarella, A., Rossi, G. L., Carotti, D., and Riva, F. (1983). *Eur. J. Biochem.*, **133**, 47.
5. Hajdu, J., Acharaya, K. R., Stuart, D. I., Barford, D., and Johnson, L. N. (1988). *Trends Biochem. Sci.*, **13**, 104.
6. Blundell, T. L. and Johnson, L. M. (1976). *Protein crystallography.* Academic Press, New York.
7. Usha, R., Johnson, J. E., Moras, D., Thierry, J.-C., Fourme, R., and Kahn, R. (1984). *J. Appl. Cryst.*, **17**, 147.
8. Bohm, J. (1985). *Acta Phys. Hungarica*, **57**, 161.
9. Kay, L. L. (1986). *ISIS/J. History Sci. Soc.*, **77**, 450.
10. Feigelson, R. S. (1988). *J. Crystal Growth*, **90**, 1.
11. Boistelle, R. and Astier, J. P. (1988). *J. Crystal Growth*, **90**, 14.
12. Lehman, C. G. (1853). *Lehrbuch der physiologische Chemie.* Leipzig.
13. Reichert, E. T. and Brown, A. P. (1909). *The differentiation and specificity of corresponding proteins and other vital substances in relation to biological classification and evolution: the crystallography of hemoglobins.* Carnegie Institution, Washington DC.
14. Debru, C. (1983). *L'esprit des protéines: histoire et philosophie biochimiques.* Hermann, Paris.
15. McPherson, A. (1991). *J. Crystal Growth*, **110**, 1.
16. Sumner, J. B. (1926). *J. Biol. Chem.*, **69**, 435.
17. Northrop, J. H., Kunitz, M., and Herriot, R. M. (1948). *Crystalline enzymes.* Columbia University Press, New York.
18. Dounce, A. L. and Allen, P. Z. (1988). *Trends Biochem. Sci.*, **13**, 317.
19. Bernal, J. D. and Crowfoot, D. (1934). *Nature*, **133**, 794.
20. Gilliland, G. L. (1988). *J. Crystal Growth*, **90**, 51.
21. Hirs, C. H. W., Timasheff, S. N., and Wyckoff, H. W. (ed.) (1985). *Methods in Enzymology*, **114**, 1.
22. McPherson, A. (1982). *Preparation and analysis of protein crystals.* Wiley, New York.
23. Kim, S. H. and Rich, A. (1968). *Science*, **162**, 1381.
24. Dock, A.-C., Lorber, B., Moras, D., Pixa, G., Thierry, J.-C., and Giegé, R. (1984). *Biochimie*, **66**, 179.
25. Michel, H. (1982). *J. Mol. Biol.*, **158**, 567.
26. Kühlbrandt, W. (1988). *Quat. Rev. Biophys.*, **21**, 429.
27. Arnoux, B., Ducruix, A., Reiss-Husson, F., Lutz, M., Norris, J., Schiffer, M., and Chang, C. H. (1989). *FEBS Lett.*, **258**, 47.
28. Giegé, R., Dock, A.-C., Kern, D., Lorber, B., Thierry, J.-C., and Moras, D. (1986). *J. Crystal Growth*, **76**, 554.
29. Yonath, A., Frolow, F., Shoham, M., Müssig, J., Makowski, I., Glotz, C., *et al.* (1988). *J. Crystal Growth*, **90**, 231.
30. Trakhanov, S., Yusupov, M., Shirikov, V., Garber, M., Kitschler, A., Ruff, M., *et al.* (1989). *J. Mol. Biol.*, **209**, 327.
31. Feigelson, R. S. (ed.) (1986). Proc. 1th. Int. Conf. Protein Crystal Growth, Stanford, CA, USA, 1985. *J. Crystal Growth*, **76**, 529.
32. Giegé, R., Ducruix, A., Fontecilla-Camps, J., Feigelson, R. S., Kern, R., and McPherson, A. (ed.) (1988). Proc. 2nd Int. Conf. Crystal Growth of Biological Macromolecules, Bischenberg, France, 1987. *J. Crystal Growth*, **90**, 1.

33. Ward, K. and Gilliland, G. (ed.) (1991). Proc. 3nd Int. Conf. Crystal Growth of Biological Macromolecules, Washington, DC, USA, 1989. *J. Crystal Growth*, **110**, 1.
34. Carter, C. W. Jr. (ed.) (1990). *Methods, A Companion to Methods in Enzymology*, **1**, 1.
35. Giegé, R. and Mikol, V. (1989). *Trends in Biotechnology*, **7**, 277.
36. Ollis, D. and White, S. (1990). *Methods in Enzymology*, **182**, 646.
37. McPherson, A. (1990). *Eur. J. Biochem.*, **189**, 1.
38. Wood, S. P. (1990). In *Protein purification applications: a practical approach* (ed. E. L. V. Harris and S. Angal), pp. 45–58. IRL, Oxford.
39. Jakoby, W. B. (1971). *Methods in Enzymology*, **22**, 248.
40. Visuri, K., Kaipainen, E., Kivimäki, J., Niemi, H., Leissla, M., and Palosaari, S. (1990). *Biotechnology*, 547.
41. VonHippel, P. H. and Schleich, T. (1969). In *Structure and stability of biological macromolecules* (ed. S. N. Timasheff and G. D. Fasman), Vol. 2, pp. 417–574. Dekker, New York.
42. Mikol, V. and Giegé, R. (1989). *J. Crystal Growth*, **97**, 324.
43. Ries-Kautt, M. M. and Ducruix, A. F. (1989). *J. Biol. Chem.*, **264**, 745.
44. Thaller, C., Weaver, L. H., Eichele, G., Wilson, E., Karlson, R., and Jansonius, J. N. (1981). *J. Mol. Biol.*, **147**, 465.
45. DeLucas, L. J. and Bugg, C. E. (1987). *Trends in Biotechnology*, **5**, 188.
46. Frey, M., Genevesio-Taverne, J.-C., and Fontecilla-Camps, J. C. (1988). *J. Crystal Growth*, **90**, 245.
47. Bergdoll, M. and Moras, D. (1988). *J. Crystal Growth*, **90**, 283.
48. Salemme, F. R., Genieser, L., Finzel, B. C., Hilmer, R. M., and Wendolosky, J. J. (1988). *J. Crystal Growth*, **90**, 273.
49. Wang, A. H. J. and Teng, M. K. (1988). *J. Crystal Growth*, **90**, 295.
50. Tucker, A. D., Pattus, S., and Tsernoglou, D. (1986). *J. Mol. Biol.*, **190**, 133.
51. Yang, D. S. C., Sax, M., Chakrabartty, A., and Hew, C. L. (1988). *Nature*, **333**, 232.
52. Palm, W. and Colman, P. M. (1974). *J. Mol. Biol.*, **82**, 587.
53. Holmgren, A. and Braenden, C. I. (1989). *Nature*, **342**, 248.
54. Varghese, J. N., Laver, W. G., and Colman, P. M. (1983). *Nature*, **303**, 35.
55. Harding, M. M., Hodgkin, D. C., Kennedy, A. F., O'Connor, A., and Weitzmann, P. D. J. (1966). *J. Mol. Biol.*, **16**, 212.
56. Schär, H. P., Priestle, J. P., and Grütter, M. (1987). *J. Biol. Chem.*, **262**, 13724.
57. Garavito, R. M., Jenkins, J., Jansonius, J. N., Karlsson, R., and Rosenbusch, J. P. (1983). *J. Mol. Biol.*, **164**, 313.
58. Saper, M. A. and Quiocho, F. A. (1983). *J. Biol. Chem.*, **258**, 11057.
59. Finch, J. T., Lebermann, R., Chang, Y. S., and Klug, A. (1966). *Nature*, **212**, 349.
60. de Vos, A. M., Hatada, M. H., van der Wel, H., Krabbendam, H., Peerdelman, A. F., and Kim, S. H. (1985). *Proc. Natl. Acad. Sci. USA*, **82**, 1406.
61. Low, B. W., Patter, R., Jackson, R. B., Tamiya, N., and Sato, S. (1971). *J. Biol. Chem.*, **246**, 4366.
62. Eisenberg, D., Wilcox, W., Eshita, S. M., Pryciak, P. M., Ho. S. P., and DeGrado, W. F. (1986). *Proteins*, **1**, 16.

63. Waller, J.-P., Risler, J.-L., Monteilhet, C., and Zelwer, C. (1971). *FEBS Lett.*, **16**, 186.
64. Brennan, R. G., Wozniak, J., Faber, R., and Matthews, B. W. (1988). *J. Crystal Growth*, **90**, 160.
65. Yang, W., Hendrickson, W. A., Kalman, F. T., and Conch, R. J. (1990). *J. Biol. Chem.*, **265**, 13553.
66. Dock-Bregeon, A.-C., Chevrier, B., Podjarny, A., Moras, D., deBear, J. S., Gough, G. R., Gilham, P. T., and Johnson, J. E. (1988). *Nature*, **335**, 375.
67. Wang, A. H. J., Fujii, S., Van Boom, J. H., Van der Marel, G. A., Van Boeckel, S. A. A., and Rich, A. (1982). *Nature*, **299**, 601.
68. Hampel, A., Labananskas, M., Conners, P. G., Kirkegard, L., RajBhandary, U. L., Sigler, P. B. *et al.* (1968). *Science*, **162**, 1384.
69. Giegé, R., Lorber, B., Ebel, J.-P., Moras, D., and Thierry, J.-C. (1980). *C.R. Acad. Sci. Paris, D-2*, **291**, 393.
70. Frederick, C. A., Grable, J., Melia, M., Samudzi, C., Jen-Jacobson, L., Wang, B., *et al.* (1984). *Nature*, **309**, 327.
71. Derewenda, Z., Yariv, J., Heliwell, J. R., Kalb(Gilboa), A. J., Dodson, E. J., Papiz, M. Z., *et al.* (1989). *EMBO J.*, **8**, 2189.
72. Boulot, G., Guillon, V., Marriuza, R. A., Poljiak, R. J., Riottot, M. M., Souchon, *et al.* (1988). *J. Crystal Growth*, **90**, 213.
72. Joachimiak, A. (1988). *J. Crystal Growth*, **90**, 201.
74. Aggarwal, A. K., Rodgers, D. W., Drottar, M., Ptashne, M., and Harrison, S. C. (1988). *Science*, **242**, 899.
75. Richmond, T. J., Finch, J. T., Rushton, B., Rhodes, D., and Klug, A. (1984). *Nature*, **311**, 532.
76. Hogle, J. M., Chow, M., and Filman, D. J. (1985). *Science*, **229**, 1358.
77. Sehnke, P. C., Harrington, M., Hosur, M. V., Li, Y., Usha, R., Tucker, R. C., *et al.* (1988). *J. Crystal Growth*, **90**, 222.
78. Roussell, A., Serre, L., Frey, M., and Fontecilla-Camps, J. C. (1990). *J. Crystal Growth*, **106**, 405.
79. Hendrickson, W. A. and Teeter, M. M. (1981). *Nature*, **290**, 107.
80. Kantrowitz, E. R. and Lipscomb, W. N. (1990). *Trends Biochem. Sci.*, **15**, 53.
81. Brick, P., Bhat, T. N., and Blow, D. M. (1989). *J. Mol. Biol.*, **208**, 83.
82. Dao-Pin, S., Alber, T., Bell, J. F., Weaver, L. H., and Matthews, B. W. (1988). *Protein Eng.*, **1**, 115.
83. Stock, A. M., Mottonen, J. M., Stock., J. B., and Schutt, C. E. (1989). *Nature*, **337**, 745.
84. Hendrickson, W. A., Horton, J. R., and LeMaster, D. M. (1990). *EMBO J.*, **9**, 1665.
85. Lorenz, S., Betzel, C., Raderschall, E., Dauter, Z., Wilson, K. S., and Erdmann, V. A. (1991). *J. Mol. Biol.* **219**, 399.
86. Lorber, B., Giegé, R., Ebel, J.-P., Berthet, C., Thierry, J.-C., and Moras, D. (1983). *J. Biol. Chem.*, **258**, 8429.
87. Giegé, R., Lorber, B., Ebel, J.-P., Moras, D., Thierry, J.-C., Jacrot, B., *et al.* (1982). *Biochimie*, **54**, 357.
88. Ruff, M., Cavarelli, J., Mikol, V., Lorber, B., Mitschler, A., Giegé, R., *et al.* (1988). *J. Mol. Biol.*, **201**, 235.
89. Robinson, B. H. and Seeman, N. C. (1987). *Protein Engineering*, **1**, 295.

2

Preparation and handling of biological macromolecules for crystallization

B. LORBER and R. GIEGÉ

1. Introduction

In the crystallization of biological macromolecules the quality and quantity of the required material is important. Although experimenters may have no choice, difficulties in crystal growth may be linked to the nature or source of the biological material. Purification, stabilization, storage, and handling of macromolecules are therefore essential steps prior to crystallization. As a general rule purity and homogeneity are regarded as conditions *sine qua non*; however, quality of the macromolecules will depend upon the way they are prepared. Although some structures were solved with less than 1 mg of material (for example, ref. 1), a few milligrams should be available for the first crystallization trials. Once crystals can be produced which are suitable for X-ray analysis, additional material will be needed to improve their quality and size and to prepare heavy-atom derivatives. For these reasons it is essential that isolation procedures are able to supply enough fresh macromolecules of reproducible quality.

The aim of this chapter is to give a brief overview of biochemical methods used to prepare and characterize biological macromolecules. Practical aspects concerning manipulation and qualitative analysis of macromolecules intended for crystallization experiments will be emphasized. The case of nucleic acids and membrane proteins is described in more detail in Chapters 7 and 8. Finally, methods for the rapid characterization of the macromolecular content of crystals will be presented as well.

2. The biological material

2.1 Sources of macromolecules

Many specific biological functions are sustained by classes of proteins and nucleic acids universally present in living organisms and therefore the source of the macromolecules may seem unimportant. In fact, better crystallization conditions or diffracting crystal habits are frequently found by switching from one organism to another. This is because variability in sequences between heterologous species may lead to different conformations, and consequently to different crystallization behaviours. Also, because of the high solvent content (50–80%) (2, 3) and the existence of relatively few contacts in macromolecular crystal lattices, differences in crystal quality or habit may result from addition or suppression of intermolecular contacts (4). In practice, proteins isolated from eukaryotes are frequently more difficult to crystallize than their prokaryotic counterparts. Often their degree of structural complexity is higher. They can possess additional domains that may contribute to less compact and/or more flexible structures. Post-translational modifications are often responsible for structural or conformational micro-heterogeneity. Proteins isolated from thermophilic micro-organisms are more stable at higher temperatures than those from other cells and therefore may be more amenable to crystallization. Better crystallizations of aminoacyl-tRNA synthetases and ribosomes isolated from extreme thermophiles illustrate this well (see Chapters 1 and 7). Proteins from halophilic micro-organisms (5) are alternative candidates because they have their optimal stability in the presence of high salt concentrations close to those used to reach supersaturation. Finally, 'freshness' of the starting material and physiological state of cells may be important. Indeed, certain proteins from unicellular organisms are isolated in their native state only when cells are in exponential or pre-stationary growth phase (for example, refs 6–8). Also, since catabolic processes are predominant in tissues of dead organisms, such materials should not be stored before use, unless they have been frozen immediately *post mortem*.

2.2 Macromolecules produced in host cells or *in vitro*

In the past, those macromolecules that were most abundant, easy to isolate, and most stable, were the first studied and crystallized. Today, many researchers deal with biological molecules that are only present in trace amounts, and thus preparation of the quantities needed for crystallization assays is a limiting step even when using micromethods (see Chapter 4). In a number of cases, this problem can be circumvented owing to the advancement of genetic engineering methods which make it possible to clone and over-express genes in bacterial or eukaryotic cells (for examples, ref. 9). To use these methods, proteins must be isolated and a small part of their sequence known (10). In engineered cells, over-produced proteins can accumulate to levels ranging up to a quarter of total proteins (typically about a hundred milligrams of protein can be isolated from less than a hundred grams of bacterial cells), but a high intracellular concentration of certain proteins may lead to the formation of inclusion bodies of denaturated, aggregated, or pseudo-crystallized material which require the use of adapted isolation procedures (11). Therefore, over-production levels should be optimized to obtain functionally active proteins in their natural conformation. The separation of foreign macromolecules from endogenous ones can be difficult when no specific biological assay is available (12). Finally, the presence of ligands or chaperone proteins may be indispensable to maintain the native conformation of certain proteins (13). All the above factors have to be kept in mind when planning a purification strategy.

One of the advantages of recombinant DNA technology is the possibility of expressing the same macromolecule in different cell lines. Maturation enzymes responsible for co- or post-translational modifications in host cells may not work with the same efficiency on recombinant macromolecules and different end-products may be obtained (for example, glycosylation occurs to various extents in eukaryotic cells but it is not known to occur in prokaryotes). Consequently, different modification patterns will lead to structural variants (10, 14) and this may be amplified when modification enzymes are limiting in the presence of over-produced proteins. In future, continuous cell-free translation systems may represent an alternative tool to produce natural or factitious proteins for crystallization (15).

A similar situation is encountered when genes for nucleic acids are transcribed in host cells or *in vitro*. For tRNAs produced *in vivo*, the modification of bases may be partial (16). The tRNAs transcribed *in vitro* by polymerases (from bacteriophages SP_6 or T_7) do not possess modified bases since enzymes responsible for post-transcriptional maturation are absent (17), but these transcripts can be heterogeneous in length because polymerases do not terminate transcription at a unique position (18) (see Chapter 7).

2.3 Design of modified macromolecules

Using site-directed mutagenesis, unlimited changes can be introduced in the sequence of genes coding for polypeptide chains or RNAs (8, 9). Such alterations can have various consequences, including conformational changes, decreases in stability, loss of activity, incomplete folding, changes in solubility, variations in the extent of post-synthetic modifications, or unforeseen degradation (14). Design of protein mutants may include

- substitution of amino acid residues to increase stability or solubility (10, 19),
- deletion to remove parts of the polypeptide chain forming flexible domains (see also Section 2.4),
- fusion with sequences recognizable by immobilized ligands in affinity chromatography (20).

The production of mutants more suitable for the preparation of heavy-atom derivatives is another potential application of genetic engineering: this was first shown in the case of phage T4 lysozyme (21), but also in that of seleno-methionine-containing proteins for anomalous scattering studies (see Chapter 14).

3. Isolation and storage of pure macromolecules

The topic of macromolecule purification has been reviewed extensively (refs 22–24 for proteins, and 25–28 for nucleic acids). Here, only a brief overview will be given. Readers are also encouraged to consult manufacturers or suppliers catalogues and application notes containing updated technical information.

3.1 Preparative isolation methods

3.1.1 Generals

Purification procedures differ widely because individual macromolecules have peculiar properties. For that reason it is difficult to give a general scheme to facilitate the purification of an unknown macromolecule, each having to be considered as unique. Development, improvement, and optimization of purification protocols are mainly achieved by trial-and-error approaches. In crude extracts, macromolecules may be protected by interaction with ligands or other molecules which will probably be eliminated during purification. Consequently, the sequence of events in purification protocols is important. In any case macromolecules must be separated as quickly as possible from harmful compounds that can damage them (e.g. hydrolytic enzymes). Under no circumstances should these harmful compounds be enriched or co-fractionated. Major methods used in purification processes are listed in *Table 1* together with the appropriate equipment.

Two major stages common to most purification protocols are the preparation of a cellular extract (except for secreted macromolecules) and the subsequent fractionation of its components. Intracellular macromolecules have to be released from cells or tissues using physical, chemical, or enzymic disruption methods. Hydrophobic macromolecules must be solubilized and extracts have to be clarified by centrifugation or ultrafiltration. Extracellular compounds, i.e. those secreted by the cells in the culture medium, and macromolecules synthesized *in vitro*, may be recovered either by ultra-filtration, centrifugation, flocculation, or liquid–liquid partitioning.

3.1.2 Proteins

Isolation procedures for proteins usually involve a gross fractionation done by one or several precipitations induced either by addition of salts (e.g. ammonium sulphate), organic solvents (e.g. acetone), organic polymers (e.g. PEG), or by physical treatments such as pH or temperature changes, in order to decrease solubility or denature unwanted macromolecules. Fractionation between two liquid phases and selective precipitation (e.g. of nucleic acids by

Table 1. Methods and equipment useful for the purification of biological macromolecules

- **Cell culture**
  Fermentors, culture plates, thermostated cabinets
  High capacity centrifuges or filtration devices for cell recovery
- **Cell disruption**
  Mechanical disruption devices (grinders, glass bead mills, French press)
  Chemical or biochemical treatments (e.g. permeation of cells by enzymes, phenol treatment for recovery of small RNAs)
  Others (e.g. sonication, freezing/thawing)
- **Centrifugation**
  Centrifuges (low speed to eliminate cell debris or recover precipitates and high speed to fractionate sub-cellular components)
- **Dialysis and ultrafiltration**
  Dialysis tubings, hollow fibers or membranes (various porosities and sizes)
  Concentrators (various capacities from 50 μl to several liters, high flow rate membranes with low macromolecule-binding and various cut-offs)
- **Chromatography** (use preferentially metal-free systems)
  Low pressure chromatography or HPLC columns and matrices
  Other equipment including pumps, programmer, on-line absorbance detector, fraction collector, recorder
- **Preparative electrophoresis or isoelectric focusing**
  Electrophoresis apparatus for large rod or slab gels
  Preparative liquid IEF apparatus (column or horizontal rotating cell)
  Power supplies
- **Detection, characterization, and quantitation**
  Spectrophotometer, fluorimeter
  pH meter, conductimeter, refractometer (for monitoring solutions and chromatographic elutions)
  Liquid scintillation counter (for radioactivity detection)
  Analytical electrophoresis and IEF equipment.

protamine) are additional methods. The next steps involve more resolutive separation methods, generally a combination of column chromatographies. These are based on separation by charge (adsorption, anion or cation exchange chromatography, chromatofocusing), hydrophobicity (hydrophobic interaction or reverse-phase chromatography), size (exclusion chromatography or ultracentrifugation), peculiar structural features (e.g. affinity chromatography on matrices substituted by heparine, antibodies, metal ions, or sulphydryl containing compounds), or activity (affinity chromatography based on catalytic site or receptor/ligand recognition, biomimetic chromatography). HPLC yields higher resolution than the standard technique because of the monodispersity and small size of spherical matrix particles (28). Free-flow electrophoresis (29) and preparative IEF [in gels (30) or in a Rotofor® cell commercialized by BioRad] are alternative separation methods. Monitoring of specific activities during the purification procedure allows detection of unsatisfactory steps in which macromolecules are lost or inactivated.

Guidelines for effective protein purification may be summarized as follows:

- work in the cold room (i.e. at 4 °C) with chilled equipment and solutions if the protein is unstable at higher temperatures;
- use precipitation steps to speed up fractionation;
- limit the number of chromatographic steps to not more than three or four, if possible;
- avoid the use of time-consuming assays for the characterization of macromolecules;
- use quick intermediate treatments (e.g. dialysis and concentration) that are non-denaturing and efficient;
- prevent inactivation, denaturation, and degradation by adding protectors, stabilizing agents, and/or protease inhibitors (31, 32; see also Sections 2.3 and 4).

3.1.3 Nucleic acids

The purification of nucleic acids requires appropriate methods. For tRNAs, the first step is separation from proteins by phenol extraction, usually followed by a gross fractionation by counter-current distribution, or chromatography on benzoyl-DEAE-cellulose or BD-Sepharose (25). Further purification is based on anion-exchange, adsorption, reverse-phase, hydrophobic interaction, or affinity chromatographies. Intermediary treatments include precipitation by ethanol, dialysis, and concentration by evaporation under vacuum. Improvements have been made by using HPLC on various matrices (e.g. ion exchange, reverse-phase, adsorption, hydrophobic interaction, or mixed-mode matrices) (33). Recently, chromatography on a matrix made of silica substituted by short aliphatic chains has proven useful (27). An alternative approach is the benzoylation of DEAE groups bound to resin or silica-based matrices (Lorber *et al.* unpublished). Oligonucleotides used for crystallization are obtained by automated chemical synthesis on solid-phase supports. Since final products are often contaminated by abortive sequences, a purification step (usually by reverse-phase HPLC) is needed. Ribosomal 5S RNA which was used in crystallization attempts was separated from other ribosomal components by molecular sieving (34). For further details see Chapter 7.

3.2 Specific preparation methods

Specific methods can help to improve the quality of samples to be crystallized. In some instances re-chromatography of macromolecules just before the assays favours their crystallization. This may be due to either removal of minor contaminants (e.g. degradation or ageing products), of aggregates appearing during storage, or of small molecules (e.g. additives like glycerol). Resolution techniques like HIC (applicable to both proteins and nucleic acids at a 1–100 mg scale) can bring macromolecules directly into the precipitant for crystallization (either salt, MPD, or PEG) (35, 36). In other cases, adsorption chromatography over a hydroxyapatite column was determinative for crystallization of proteins. Microscale concentration devices are useful to remove small-size contaminants and to exchange buffers prior to crystallization (see Chapter 4). Additional advices are given in Section 4.5.

3.3 Stabilization and storage of pure macromolecules

Biological macromolecules are frequently fragile, sensitive to variations of their environment, and can easily lose their native structure or activity. Once they have been removed from their natural medium, they must therefore be placed in solvents having properties close to those of the cellular medium, to maintain their native conformation intact and active.

Improper storage conditions may spoil the precious macromolecules obtained after long and hard work. Buffers whose pK is only weakly affected by temperature (37, 38) must be chosen so that samples stored at −20 °C and assayed at room or higher temperatures are not subjected to important pH variations. Ionic strength should also be controlled since macromolecules may require a minimal salt concentration to stay soluble, although all ions may not be compatible with their native structure or activity. In most cases denaturation can be minimized by avoiding denaturing treatments (39), pH or temperature extremes, as well as contact with organic solvents, chaotropic agents, or oxidants. Proteins containing cysteine residues usually require a reducing agent (e.g. DTE, DTT, 2-mercaptoethanol, or glutathione). Finally, proteins should not be stored as diluted solutions because they may adsorb on to the walls of glass or plastic containers.

Glycerol at high concentration (e.g. 50–60% v/v) is a good stabilizing agent for storage of proteins. It has the advantage of staying liquid at −20 °C so that denaturation by ice formation can be minimized (its high viscosity also reduces diffusion by about two orders of magnitude). Storage as a suspension in ammonium sulphate solution is also recommendable but sometimes less convenient. Ligands can help to increase protein stability. Most proteins to be crystallized should not be dried or lyophilized, the latter process removing bound water molecules that belong to the macromolecule solvatation shell. A bactericidal agent (e.g. Na azide, ethyl-mercuri-thiosalicylate—*highly toxic*) may be added for storage. Na azide may also be an additive in crystallization attempts, as well as the antifungal and volatile thymol which may be added in reservoirs during crystallizations by vapour diffusion methods (see Section 4.3 in Chapter 4).

Nucleic acids may be stored either as alcoholic precipitates or dry, but should never stay in the presence of phenol which leads to alkaline-type hydrolysis. tRNA molecules should be stored at rather acidic pH (4.5–6.0) in the presence of Mg^{2+} (26) (see Chapter 7).

4. Characterization and handling of macromolecules

Numerous analytical tools are available to detect, characterize, and quantitate macromolecules. This section deals with general methods of particular interest for crystal growers and gives practical advice for handling pure macromolecules (for additional details see refs 39 and 40).

4.1 Analytical biochemical methods

Gel electrophoresis (PAGE) is the most convenient method to quickly visualize the macromolecular content of a sample. *Protocol 1* gives a simplified procedure for the preparation of denaturing gels used for the analysis of proteins and *Protocol 2* for that of small nucleic acids.

Protocol 1. SDS-polyacrylamide gel electrophoresis of proteins (M_r range 10 000–200 000 (adapted from 41)

1. Prepare two solutions to make gels having a total polyacrylamide concentration T (% w/v), and a cross-linker concentration C (% w/w).

Chemicals	Separating gel (T = 10%, C = 2.7%)	Stacking gel (T = 4%, C = 2.7%)
• 1 M Tris–HCl solution pH 8.8[a]	19.2 ml	–
• 1 M Tris–HCl solution pH 6.8[a]	–	19.2 ml
• Acrylamide (*toxic*!)	5.66 g	2.05
• *N,N'*methylenebisacrylamide (*toxic*!)	150 mg	55 mg
• Distilled water	up to 50 ml	up to 50 ml

2. Dissolve and filter the solutions over 0.22 μm pore size membranes.
3. To 10 ml of above solutions add 100 μl of 10% w/v SDS solution, 10 μl TEMED, and 100 μl of fresh 5% w/v ammonium peroxodisulphate. Mix and pour in the mould. Polymerization occurs in about 30 min.
4. Denature proteins before electrophoresis by boiling samples in 30 mM Tris–HCl pH 6.8, 120 mM glycine, 25% v/v glycerol, 1.25% v/v 2-mercaptoethanol, 1.25% w/v SDS detergent, and a trace of bromophenol blue.
5. Run electrophoresis in the appropriate set-ups[b]. Electrophoresis buffer solution contains (for 1 litre): 3 g Tris-base, 14 g glycine, and 1 g SDS.
6. Precipitate proteins after migration by soaking the gel in a solution containing 10% v/v acetic acid and 25% v/v ethanol. Stain proteins in the

Protocol 1. *Continued*

same solution containing 0.05% w/v Coomassie Blue R-250 'dye'[c] or using silver (42).

[a] Another PAGE technique utilizes phosphate buffer instead of Tris (43, 44).
[b] Set-ups (for cylindric or vertical slab gel) and ready-to-use gels are commercially available.
[c] Micromethods can detect submicrogram amounts of protein in routine experiments.

Protocol 2. Gel electrophoresis of nucleic acids (up to 150mers)

1. Prepare two solutions to make gels having a total acrylamide concentration T (% w/v), a cross-linker concentration C (% w/w), and containing 8 M urea.

Chemicals	Separating gel (T = 8%, C = 5%)	Stacking gel (T = 4%, C = 5%)
• Acrylamide (*toxic*!)	3.04 g	1.52g
• N,N′methylenebisacrylamide (*toxic*!)	160 mg	80 mg
• Urea	20 g	20 g
• Distilled water	up to 38 ml	up to 38 ml

2. Dissolve and deionize for 30 min in the presence of immobilized chelatant (e.g. Chelex from BioRad). Filter on 0.22 μm filter. Add 2 ml electrophoresis buffer (20-fold concentrated buffer contains 216 g Tris-base, 110 g boric acid, and 18.6 g EDTA for 1 litre, final pH is 8.3).
3. To 10 ml final solution add 10 μl TEMED and 100 μl of fresh 5% w/v ammonium peroxodisulphate solution. Mix and pour in the mould[a]. Polymerization occurs in about 30 min.
4. Prepare nucleic acid samples for electrophoresis in a solution containing 8 M urea, 20% w/v saccharose (or 20% v/v glycerol), and a trace of bromophenol blue.
5. Detect nucleic acids after electrophoresis by staining[b] in the dark with 'Stains all' dye. Soak gels in a solution containing 30 mg 'Stains all' powder (Eastman Kodak or Aldrich), 100 ml dimethylformamide, and distilled water up to 1 litre. Destain in the light.

[a] Microgels can be prepared in the same set-ups as gels for proteins.
[b] Silver stain techniques for protein are suitable to visualize nucleic acids (see *Protocol 1*).

Electrophoresis allows estimation of the apparent size of proteins (39, 43) and nucleic acids (45). IEF in polyacrylamide or agarose gels separates proteins according to their charge and gives a good estimate of their isoelectric point. In free or immobilized pH gradients it is a very resolutive method that can reveal differences in isoelectric points smaller that 0.01 pH unit (46). An example for the preparation of IEF gels is given in *Protocol 3*.

Protocol 3. Isoelectric focusing of proteins[a] in polyacrylamide gels containing free ampholytes

1. Prepare a solution with a total polyacrylamide concentration T = 5% w/v, a cross-linker concentration C = 2.7% w/w, and an ampholyte concentration of 2% w/v. For 20 ml solution, mix:
   - Acrylamide (*toxic*!) — 975 mg
   - N,N′methylenebisacrylamide (*toxic*!) — 26 mg
   - Glycerol — 2 ml
   - Ampholyte solution (commercial solutions are usually 40% w/v) — 1 ml
   - Distilled water — up to 19.8 ml.
2. Dissolve and filter on 0.22 μm pore size membranes. Before use add 20 μl Temed and 200 μl fresh 5% w/v ammonium peroxodisulphate solution. Homogenize and pour in the mould. Polymerization occurs in 30 min at room temperature[b,c].
3. Remove upper glass plate of horizontal gels. Soak paper wicks with 10 mM glutamic acid (anode) or 20 mM NaOH (cathode) for contact between gel and electrodes. For rod gels, electrolytes are directly in contact with the gel.
4. Protein samples should have a low ionic strength and contain some glycerol to increase viscosity. Deposit samples on horizontal gels or on top of rod gels.
5. Run electrofocusing under constant power on a cooling plate maintained at 4–10 °C.
6. After migration, eliminate well ampholytes by repeated washes in the acetic acid/ethanol solution before staining of proteins (see *Protocol 1*).

[a] IEF is not applicable to nucleic acids. [b] Ready-to-use IEF gels are commercially available. [c] IEF can be performed under denaturing conditions by adding 8.0–9.5 M urea in the gel solutions. Other IEF techniques utilize immobilized pH gradients (no cathodic drift) (46) or agarose gels for proteins with M_r over 200 000 (47).

Electrophoretic titration of proteins in a pH gradient may be used to confirm their purity (48). Capillary electrophoresis is an emerging technique well adapted for rapid purity and homogeneity analysis. HPLC can be useful to separate small amounts of macromolecules for further analysis (28). Sequence analysis is a good control, more sensitive than analysis of amino acid composition (49). *Figure 1* shows an example of the sensitivity and resolution of standard analytical techniques (see also Section 4).

The choice of appropriate methods to quantitate macromolecules is dictated by the amount and concentration of samples and the required degree of specificity. Spectrophotometry is direct and non-destructive and can be very accurate when extinction coefficients are known. The theoretical molar extinction coefficient ε of pure proteins can be calculated from tryptophan and tyrosine content using $\varepsilon_{280nm} = 5690\, n_x + 1280\, n_y$, where 5690 and 1280 are the molar absorption coefficients at 280 nm of tryptophan and tyrosine, and n_x and n_y the numbers of tryptophan and tyrosine residues in polypeptide chains, respectively (50). Thus protein concentrations are obtained from:

$$c\ (\text{mg/ml}) = A_{280nm} \times \varepsilon_{280nm}/M_r.$$

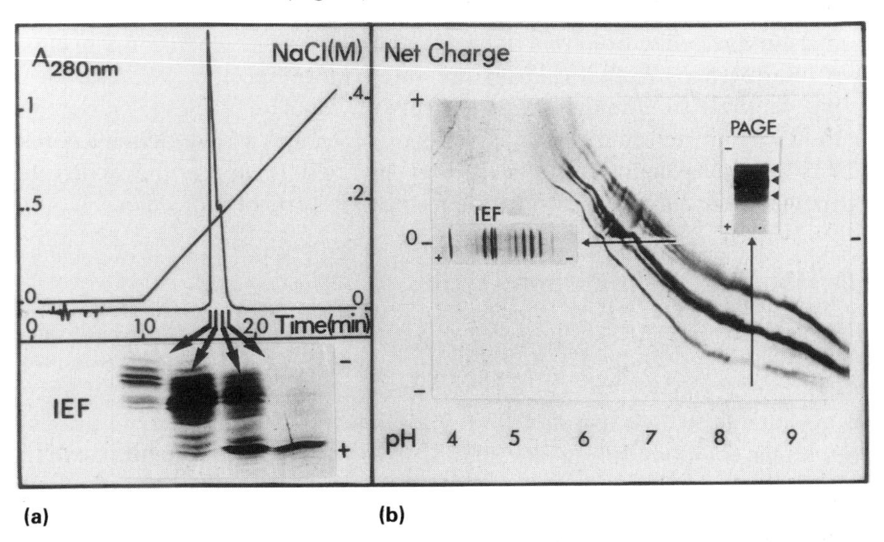

Figure 1. Comparison of the resolution of various analytical techniques. (left) IEF versus HPLC. Aspartyl-tRNA synthetase (250 μg, pure according to standard ion-exchange chromatography) was fractionated by anion exchange HPLC on an Mono Q column (i.d. 5 mm × length 5 cm, v = 1 ml) (Pharmacia-LKB Biotechnology, Inc.) in 50 mM Tris–HCl buffer pH 7.5 and was eluted at 0.5 ml/min with increasing NaCl concentration. IEF was performed on aliquots (3 μg protein) of the fractions. Polyacrylamide gel was 10 × 10 cm^2 (thickness 0.5 mm) and contained 2% w/v ampholytes (pH range 4–7). Staining with Coomassie Blue reveals several protein populations differing by charge. (right) Electrophoretic titration versus IEF or PAGE. Pure aspartyl-tRNA synthetase (25 μg) was subjected to electrophoresis perpendicularily to a pH gradient in a polyacrylamide gel as above but with ampholytes having pKs in the pH range 3–9. Focusing was conducted for 1 hour to establish the pH gradient and gel was turned by 90 °C. Sample was deposited in a trench across the gel and electrophoresis done for 20 min. Staining with Coomassie Blue visualizes several protein populations separated by charge. Insets show IEF and PAGE patterns obtained in separate experiments.

The ε coefficients can be determined from ponderal, spectrophotometric, or refractive index measurements, as well as from colorimetric or dye binding assays (51).

In mixtures, protein concentrations are estimated using empirical formulas based on absorption differences at two wavelengths in order to eliminate the contribution of nucleic acids. For cells of *l* cm optical pathways (52, 53):

c (mg/ml) = $(1.55\ A_{280\ nm} - 0.76\ A_{260\ nm})/l$, or for a better estimate
c (mg/ml) = $0.3175\ (A_{228.5\ nm} - A_{234.5\ nm})/l$.

Approximate concentrations of RNA are obtained assuming that 1 A_{260nm} unit (for 1 cm path-length) corresponds to 0.040 mg/ml. Values are more accurate when extinction coefficients are known (54). They are obtained from absorbance and phosphorus content measurements (55).

Proteins can also be quantitated by colorimetric methods such as the Lowry or Bradford assays (51, 56). But results are skewed when macromolecular compositions deviate from that of the standard protein or when contaminants give interference. Activity assays are often more specific. For enzymes, active-site titration monitors the activity of individual molecules (57), but this method cannot detect microheterogeneous molecules whose activity is not affected. Immunological properties may be used to assess the ability of antibodies to recognize conformational states or parts of molecules.

4.2 Handling of macromolecules

Pure macromolecules require special care to ensure they are not damaged or lost before or during the crystallization assays. Diluted solutions are concentrated by ultrafiltration in devices using pressure or centrifugal force (in cylindrical cells or parallel flow plate systems), or by dialysis against hygroscopic compounds (e.g. dry high-M_r, PEG or gel-filtration matrixes). In stirred pressure cells, loss of material by adsorption can be prevented by choosing appropriate membranes with low binding capacity. Adsorption on to membrane surfaces and damage by shearing are prevented by optimizing stir-rate. Foaming and air bubbles should be avoided because they create an oxidizing environment where disulphide bonds in proteins can break. Aggregates forming as a consequence of a decrease in pH, of oxidation, or of an increase in salt or protein concentration must be removed by centrifugation. Concentration is also carried out by precipitation, e.g. by adding ammonium sulphate, followed by re-dissolution of the precipitate in a small volume. Adsorption on a chromatographic matrix followed by elution at high ionic strength is an alternative approach. Unfortunately some techniques may also concentrate contaminants, including hydrolytic enzymes. Other techniques, however, may be used as additional purification steps; among them concentration over membranes to eliminate small M_r compounds, provided appropriate cut-offs are chosen, and exchange of buffers in micro-scale set-ups. Finally, it must be mentioned that lyophilization leads to denaturation of many macromolecules.

Mild non-ionic detergents (e.g. octyl glucoside) help to solubilize proteins (see Chapter 8). High concentrations of denaturing agents, like guanidinium chloride, urea, or chaotropic detergents should be avoided because they inactivate or unfold macromolecules. (For potential applications of low concentrations of such reagents in crystallization assays see Chapter 9).

For better reproducibility, always adjust the pH of buffers after mixing of all ingredients since it may change after dilution or in the presence of other compounds. Also prepare buffers freshly with ultra-pure water and high-grade chemicals. Suspicious chemicals must be purified by crystallization, distillation, or chromatography (see also Section 4.2). Bactericidal agents (e.g. Na azide or Na ethyl-mercuri-thiosalicylate at 0.02% w/v) may be added in some solutions (some buffering compounds, like cacodylate ions, have bactericidal properties. Note that cacodylate is a *toxic* organo-arsenic compound which should be handled with care!).

Never repeatedly freeze and thaw macromolecules, to avoid denaturation, and do experimentation on aliquots to limit repeated handling of stock solutions. Remove undesired molecules that might hinder crystallization, by dialysis (e.g. glycerol), ultrafiltration, or size-exclusion chromatography. In practice, it may also be important to prepare macromolecules with or without their ligands (e.g. coenzyme, metal ions) or try additives (e.g. ions, reducing agents, chelators) because one or the other form may be more able to crystallize. To avoid contamination by nucleases, glass- and plastic-ware in contact with nucleic acids should be sterilized in the autoclave, and experimenters should wear gloves during all manipulations to prevent contact with fingers (see Section 4.5 for nuclease inhibitors; ref. 26 and Chapter 7 for additional recommendations).

5. The problem of purity and homogeneity

5.1 The concept of 'crystallography-grade' quality

The concept of purity takes a peculiar meaning in biological crystallogenesis (36). Not only must molecules be pure, i.e. deprived of unrelated macromolecules or of undesired small molecules, but they must be 'pure' in terms of structure and conformation. In other words crystallization trials should be done with homogeneous populations of single conformer species. This concept is based on the fact that good monocrystals can only be grown from solutions containing well-defined entities with identical conformations and physicochemical properties. Over the years it has been refined with the improvement of analytical biochemical micromethods. Today, for biological macromolecules, the term 'purity' means absence of contaminants, whereas 'homogeneity' implies absolute identity of all macromolecules in a sample.

The importance of purity may appear exaggerated and contradictory with earlier views on crystallization because crystallization can be used as a purification method in chemistry and biochemistry (58). For structural studies, however, the aim is to prepare large monocrystals, diffracting at high resolution, with a good mosaicity, and a prolonged stability in the X-ray beam. It is thus understandable that contaminants may compete for sites on the growing crystals and generate lattice errors leading to internal disorder, dislocations, irregular faces and secondary nucleation, twinning, poor diffraction, or early cessation of growth. Because of the high number of molecules in a single crystal (about 10^{20} in a cubic crystal measuring 1 mm on the edge), p.p.m. amounts of contaminant correspond to large numbers of molecules that can interfere with crystal growth.

5.2 Purity of macromolecular samples

The level of confidence for the purity of a macromolecule depends upon the resolution, specificity, and sensitivity of the methods used to identify contaminants.

5.2.1 Unrelated macromolecular contaminants

Protein or nucleic acid samples are seldom analysed for contamination by other classes of macromolecules. Although most of these contaminants are eliminated along the purification steps through which the proteins or nucleic acids have gone, traces of polysaccharides, lipids, proteases, or nucleases may be sufficient to hinder crystallization.

5.2.2 Small size contaminants

Small molecules like peptides, oligonucleotides, amino acids, carbohydrates, or nucleotides, as well as uncontrolled ions, should be considered as contaminants. Buffer molecules remaining from a former purification step can be responsible for irreproducibility of crystallization assays (e.g. phosphate ions are relatively difficult to remove and may crystallize in the presence of other salts). Ions act as counter-ions and play a critical role in the packing of macromolecules. Often macromolecules do not crystallize or yield the same crystal habits in the presence of various buffers adjusted at the same pH. For these reasons, 'purity' means also that all reagents used with pure macromolecules (e.g. precipitants, buffers, or detergents) should be of the highest grade. Commercial reagents contaminated by trace impurities should be purified. This is especially true for precipitants since they are used at relatively high concentrations. For instance, a contamination by Fe ions at a level of 0.001% w/w in a 2 molar ammonium sulphate solution corresponds to 0.1 mM impurities, which equals protein concentration of 0.1 mM (e.g. 5 mg/ml for a protein of M_r of 50 000). Purification techniques for common precipitants are listed in *Table 2*. Salts must be recrystallized. Several techniques have been developed for the purification of PEG (59, 60), but also

Table 2. Techniques for the purification of major crystallization agents

Chemicals	Major contaminants in commercial batches	Purification techniques and references
Ammonium sulphate	Ca^{2+}, Fe ions, Mg^{2+}, $PbSO_4$[a], CN^-, NO^{3-}	Recrystallization
PEG	Cl^-, F^-, NO^{3-}, PO_4^{2-}, SO_4^{2-}; peroxides, aldehydes	Column chromatography[b] (59, 60) Recrystallization[c]
MPD	Cl^-, K^+, Na^+, SO_4^{2-}	Distillation[d] Column chromatography (63)

[a] Non-soluble species; [b] See *Protocol 1* in Chapter 4; [c] Technique for the purification of PEG-6000: Dissolve 30 g PEG-6000 in 240 ml acetone (ACS grade) at 40 °C in the waterbath. Chill on ice to obtain a white precipitate. Add 240 ml diethylether and filter on sintered glass filter (n° 2). Wash the precipitate with 240–300 ml diethylether and let it dry in the air. Dry under vacuum. Store dry. [d] MPD is distilled under vacuum.

for detergents (61, 62, and Chapter 8). MPD is purified either by distillation or column chromatography (63). High-grade products (e.g. HPLC-grade ammonium sulphate) in which contaminants do not exceed a few p.p.m. are commercially available, but the label 'ultra-pure' is sometimes exaggerated or misleading. Molecules released from chromatographic matrices are another type of contaminants. They originate from non-inert matrices susceptible to enzymatic digestion (e.g. Sephadex, celluloses) or from desorption of organic compounds (e.g. organic phases bound to silica matrices).

5.3 Microheterogeneity of macromolecular samples

Heterogeneity in pure macromolecules is called 'microheterogeneity' because generally it is only revealed by high-resolution methods (see Section 4.4). As *Table 3* shows, its causes are multiple, the most common being uncontrolled fragmentation and post-synthetic modifications.

Table 3. Possible sources of microheterogeneity in proteins and ribonucleic acids

Variation in primary structure (genetic variations, degradations)
Variation in secondary structure (errors in folding or partial unfolding)
Variation in tertiary structure (conformers)
Variation in quaternary structure (oligomerization)
Molecular dynamics (flexible domains)
Variation in post-transcriptional or post-translational modifications
Partial binding of ligands or of foreign molecules
Various aggregation states
Fragmentation (i.e. degradations by hydrolysis)
Partial oxidation (e.g. sulfhydryl groups in proteins)
Ageing (e.g. deamidation in proteins)
Others

5.3.1 Fragmentation

Proteolysis is often the major difficulty to overcome in protein isolation (64). In cells, proteolysis is a controlled mechanism involved in major physiological processes (e.g. maturation, regulation of enzymic activity, and catabolism). For general reviews, see references 65–68. Proteases (also termed peptidases) are enzymes with M_r in the range 20 000–800 000, localized in various cellular compartments or secreted in the extracellular medium. Four classes have been characterized according to the structure of their catalytic site: the serine-, aspartic acid-, cysteine-, and metallo-proteases. They can be inhibited by a number of commercially-available compounds (see *Table 4*). Others, however, cannot be inhibited by the usual inhibitors. During isolation of intracellular proteins, control over proteolysis is lost after all disruption due to the mixing of the contents of cellular compartments and to contact with extracellular proteases. Decrease of protein size or stability, modification of charge or hydrophobicity, decrease of stability, partial or total loss of activity or of immunological properties, may result from proteolysis.

Hydrolysis by endo- or exonucleases is a frequent source of heterogeneity in nucleic acids. Contaminant proteases or nucleases may not be detectable even when overloading electrophoresis gels but can cause damage during concentration or storage of samples. Chemical hydrolysis is another problem encountered when working with RNAs. It occurs mainly in the presence of metal ions or at alkaline pH (69) (see Chapter 7).

Table 4. Some commercially available protease or nuclease inhibitors

Proteases or nucleases	Inhibitors[a]
All protease classes	possibly α_2-macroglobulin or DEPC
Serine proteases	DIFP, PMSF, Pefabloc® SC[b] (from Pentapharm Ltd), aminobenzamidine, 3,4-dichloro isocoumarin, antipain, chymostatin, elastinal, leupeptin, boronic acids, cyclic peptides, trypsin inhibitors (e.g. aprotinin, peptidyl chloromethyl ketone)
Aspartic acid proteases	Pepstatins and statine-derived inhibitors
Cysteine proteases	Peptidyldiazomethanes, epoxysuccinyl peptides (e.g. E-64) cystatins, all thiol binding reagents, peptidyl chloromethanes
Metallo-proteases	Chelators (e.g. EDTA, EGTA), phosphoramidon and phosphorus containing inhibitors, bestatin, amastatin and structurally related inhibitors, thiol derivatives, hydroxamic acid
Ribonucleases	RNAsin® (Promega), ribonucleoside-vanadyl complexes, DEPC
Deoxyribonucleases	DEPC, chelatants (e.g. EDTA, EGTA)

[a] These compounds are *dangerous* for human health and must be manipulated with caution. See also ref. 32. [b] According to the manufacturer, Pefabloc® SC (4-(2-aminoethyl)-benzen-sulphonyl fluoride) is a non-toxic alternative to PMSF and DIFP.

5.3.2 Post-synthetic modifications

Modifications occurring co- or post-translationally are other common sources of microheterogeneity in proteins. Either all modification sites are occupied but the added groups are heterogeneous (e.g. glycosylation introducing oligosaccharide chains or various sequences) or the modifications are complete but unevenly distributed over the polypeptide chains (e.g. when all sites are not substituted). Over a hundred modifications are known some of which are listed in *Table 5*. Most of them require special methods for their analysis (70–2 and refs therein). Some modifications are reversible (e.g. phosphorylation) whereas others are not (e.g. glycosylation, methylation). Heterogeneity in carbohydrate chains, either N-linked at asparagine or O-linked at serine or threonine residues, is frequent in eukaryotic proteins. Partial post-translational modifications may be minimized by removing added heterogeneous groups, as was done for carbohydrates using mild hydrolysis by oligosaccharide-cleaving enzymes (73). Microheterogeneities may appear during storage, e.g. by deamidation of asparagine or glutamine residues (74).

Numerous nucleotides are modified or hypermodified post-transcriptionally in tRNAs (75). Some of them modulate physicochemical properties, in particular charge and hydrophobicity, of individual tRNA species. Also, their amplitude varies with the physiological state of cells (16, 75), so that it is understandable that microheterogeneities in crude tRNA batches affect purification and crystallization of individual species.

5.3.3 Conformational heterogeneity

Conformational heterogeneity may have several origins: binding of ligands (76), intrinsic flexibility of macromolecular backbones, oxidation of cysteine

Table 5. Some co- or post-translational modifications of proteins

Amino acid residues or chemical groups	Modifications or chemical groups added
Amino terminal $-NH^{3+}$	formyl-, acetyl-, glycosyl-, amino acyl-, cyclization of Gln
Carboxy terminal -COOH	amide, amino acyl-
Arg	ADP ribosyl-, methyl-, ornithine
Asp, Glu	carboxyl-, methyl-
Asn, Gln	glycosyl-, deamination
Cys	seleno-, heme, flavin-
His	flavin-, phospho-, methyl-
Lys	glycosyl-, pyridoxyl-, biotinyl-, phospho-, lipoyl-, acetyl-, methyl-
Met	seleno-
Phe	hydroxyl-
Pro	hydroxyl-
Ser	phospho-, glycosyl-, ADP ribosyl-
Thr	phospho-, glycosyl-, methyl-
Tyr	iodo-, hydroxy-, bromo-, chloro-

residues (77), or partial denaturation. In the first case macromolecules should be prepared in both forms, the one deprived of and the other saturated with its ligands. In the second case controlled fragmentation may be helpful (see Section 4.5). In the last case, oxidation of a single cysteine residue can lead to complex mixtures of molecular species for which the chances of growing good quality crystals may be low (77); such redox effects can often be reversed by reducing agents (see Section 5.5).

5.4 Probing purity and homogeneity

Although macromolecules may crystallize readily in an impure state (58), it is preferable to achieve a high level of purity before starting crystallization trials. Biochemical quality controls can be performed at relatively little expense when compared to the time-consuming crystallization experiments. A combination of several independent analytical methods is always required to assert the absence of contaminants and microheterogeneities in a macromolecular sample (78); they are essentially those described in Section 3.1 for the characterization of macromolecules.

Spectrophotometry and spectrofluorimetry are helpful to obtain information about the quality of macromolecules if they or their contaminants have special characteristics in their absorbance or emission spectra. Chromatographic techniques are generally not resolutive enough to be used as the only reference. As illustrated in *Figure 1a*, occurrence of a sharp peak in HPLC is not sufficient to state about the quality of the product. Analysis of a protein sample by SDS–PAGE gives an idea about the presence and the size of polypeptidic contaminants if they are detectable by staining, autoradiography, or immunodetection (43, 45), but non-proteic contaminants are usually not detectable. Gel IEF gives an estimate of the isoelectric point of proteic components of a mixture and electrophoretic titration shows the mobility of proteins as a function of pH (*Figure 1b*) (for an application see ref. 79). Electrophoretic titration can also suggest the type of chromatography (i.e. anion, cation, exchange) or chromatofocusing suitable for a further purification (48), or guide experiments toward chromatographies based on adsorption, size exclusion, hydrophobic interaction, or affinity. Capillary electrophoresis is also well adapted for sensitive control of purity, and when known the amino acid composition can be used as a control. If the primary structure of the polypeptide is known, sequencing of its N- and C-termini is used to control its integrity (49). Homogeneity of nucleic acids (e.g. tRNAs) is checked by electrophoresis under denaturing conditions in the presence of urea (45, 80). After radioactive end-labelling of the molecules, low levels of cleavage in ribose-phosphate chains that could hamper crystallization can be detected (69). Purity can be evaluated by specific activity measurements (usually expressed in units/mg), the latter also giving good estimates of the efficiency of the individual steps in a purification procedure. It is recalled that macromolecules in microheterogeneous preparations can be fully active. Useful methods for detecting conformational heterogeneity are shown in *Table 6*.

Table 6. Selected techniques to detect macromolecular conformational heterogeneity

Techniques	**Expected information**
Activity assay	Biological activity
Active-site titration	Ligand binding, affinity
Gel electrophoresis	Mobility, size, charge, shape
Gel filtration	Size, shape
IEF/Titration curve	Charge
Immunological titration	Antigenic determinants
Scattering methods (light, X-rays, neutrons)	Size, shape
Spectrophotometry, fluorimetry	Absorption or emission properties
Ultracentrifugation	Size

5.5 Improving purity and homogeneity: some practical advice

Problems encountered in crystallization assays (e.g. no crystals, poor diffraction) are often solved by reconsidering the purification protocol. It is recommended to start with fresh material, to change the sequence of events (by inverting chromatographic steps), or to change the steps themselves (by using other chromatographic matrices). Avoid cross-contamination by never mixing batches of pure macromolecules even when they look identical. A small shift in the elution from a chromatography column, or a preparation done on the same column but at another scale or temperature, may introduce other contaminants in the fractions. Such variability can sometimes be detected by IEF (*Figure 2*). All solutions in contact with pure macromolecules should be cleaned and sterilized by filtration (e.g. over 0.22 μm porosity filters). To avoid release of molecules from chromatography columns, the best remedy is to use chemically inert and autoclavable matrices (e.g. Trisacryl supports from IBF Biotechnics or TSK gels from Merck).

A priori macromolecules can be rendered more homogeneous in various ways. Although there is no guaranteed method to prevent unwanted proteolysis occurring *in vivo* or *in vitro*, a convenient one is the addition of protease inhibitors. A variety of such inhibitors are commercially available. (*Table 4*). Their efficacy can be checked using assays based on the solubilization of clotted protein or employing labelled proteins (32, 81). A cocktail of inhibitors should contain at least one compound specific for each class of proteases; an example is given in *Protocol 4*.

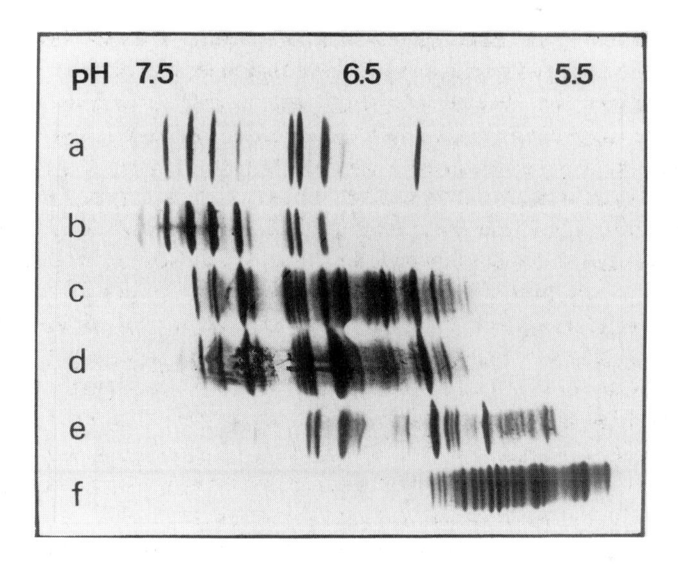

Figure 2. Batch dependent variation in the heterogeneity of a pure protein. Six batches of pure aspartyl-tRNA synthetase, purified according to a standard procedure and having the same specific activity, were compared in IEF under native conditions (samples of 5 μg protein were analyzed). The gel was prepared as in *Protocol 2* and proteins were stained with Coomassie Blue.

Protocol 4. Preparation of buffer solutions containing protease inhibitors and stabilizing agents

1. Prepare buffer solutions containing 10% glycerol and 10^{-3} M EDTA.
2. Prepare a stock solution of DIFP[a] by diluting a 1 ml commercial sample (~ 1 g) in 50 ml cold and anhydrous isopropanol to obtain a 0.1 M solution. This solution should be kept at −20 °C. The peptidic inhibitors[b] (pepstain, bestatin, and E-64) are prepared as 10^{-3} M stock solutions in ethanol/water (50/50).
3. Add DIFP, peptidic inhibitors, and 2-mercaptoethanol (also DTE or DTT[c]) in buffers just prior to use (final concentrations 5×10^{-4} M, 5×10^{-6} M, and 5×10^{-3} M, respectively).
4. Add inhibitors afresh at each step of the isolation procedure.

[a] **Important:** DIFP has to be handled with extreme caution since it is a powerful inhibitor of human acetylcholine esterase. [b] Experimenters should be aware of low solubility and *toxicity* as well as of limited stability, affinity, or reversibility of inhibitors. [c] Final concentration 10^{-4} M.

On a small scale, chromatography over a column of immobilized protease inhibitors (e.g. α_2-macroglobulin) or substrate-analogues (like arginine or benzamidin) may be designed to trap proteases. The major drawback of inhibitors lies in the possibility of their binding to or inactivating the proteins they should protect. An alternative solution to proteolysis or degradation is to express proteins in strains deprived of harmful protease activities (82).

Action of nucleases also may be minimized by addition of inhibitors (*Table 4*). Ribonucleoside–vanadyl complexes are efficient inhibitors (95). Diethylpyrocarbonate reacts with histidine residues at the surface of proteins and may lead to nuclease inactivation when blocking their catalytic site. RNasin® (trade name of Promega), a protein of M_r 51 000 isolated from human placenta, inactivates ribonucleases by stoichiometric non-covalent and non-competitive binding (84). It can be added in cell-free transcription media to protect *in vitro* synthesized RNAs but has to be removed prior to crystallization. An alternative way to prevent hydrolysis is to remove nucleases by affinity chromatography over blue dextran-agarose (85) or over 5′-(4-aminophenyl)-uridine-(2′,3′) phosphate-agarose (86). Metal ions involved in chemical hydrolysis of RNAs are removed from solutions by ion exchange or chelation (e.g. with EDTA, EGTA, or chelators immobilized on agarose beads). Magnesium at neutral pH is an exception to this rule and should be added to stabilize the structure of tRNAs. In all cases, the pH of the medium should never be alkaline, i.e. higher than 7.5 (for more details see Chapter 7).

Removal of undesirable parts of multidomain macromolecules by controlled fragmentation may help to obtain more compact and homogeneous structures. Indeed, some proteins crystallize or yield better crystals when proteolysed prior to crystallization (87). (In uncontrolled situations unsuspected contaminations by proteases may lead to a similar result.) After controlled treatment, the macromolecular core must be purified in order to remove proteases and small fragments. To be reproducible enzymic tools must be free of contaminant proteases that could generate undesirable cleavages. Unfortunately, preferential attack of accessible domains often introduces microheterogeneities that are detrimental to crystallization. Thus, purification of single species may be essential to grow large monocrystals, and indeed improved purifications by HPLC or IEF yielded better crystals in many instances (88–91).

6. Characterization of crystal content

Once crystals have been obtained, the major concern is to demonstrate that they are made of macromolecules and not of small molecules present in the crystallization medium. The X-ray diffraction pattern is of course the most convincing proof that crystals are of macromolecular nature. In this way crystallographers will immediately gain information about crystal quality in terms of resolution and about the space group to which they belong (see Chapter 12). Although obtaining well-diffracting crystals is stimulating and decisive for the start of a time-consuming crystallographic analysis, it should not be forgotten that simple analytical biochemical methods give additional information about the content of the crystals. Moreover, reproducibility of crystallization may depend upon such routine controls; insufficient knowledge may become a severe handicap for further work.

6.1 Handling of macromolecular crystals

Due to their nature macromolecular crystals are fragile edifices that need to be handled with special care so that they do not break or dissolve. Before any biochemical analysis can be undertaken the crystals should be stabilized to prevent dissolution, by slowly increasing the precipitant concentration as is usually done before crystal mounting in X-ray capillaries (see Chapter 12). Free soluble macromolecules or amorphous material (within the mother liquor or deposited on to the crystal faces) must be washed away from crystals by transferring them several times in large volumes of mother liquor. (Dilution factors should be estimated to minimize interference by contaminating uncrystallized macromolecules.)

6.2 Biochemical and biophysical micromethods

Besides X-ray diffraction methods used to characterize crystallographic parameters of macromolecular crystals, many biochemical and biophysical methods can be used to analyse crystals (see also Chapter 5). *Table 7* lists two classes of methods: those applicable when dealing with the crystalline material itself, and those requiring solubilized molecules. Since crystals contain rather small amounts of macromolecules (in the μg range) most methods must be scaled-down as was done for chromatography (92) and electrophoresis (93). Other methods require particular equipment, as for microspectrophotometry (94, 95). In all cases the aims are:

(a) to verify that crystals contain the desired macromolecules (done by PAGE or IEF);
(b) to check the presence of all components in the required stoichiometry in case of complexes (this needs usually a combination of methods like PAGE, IEF, HPLC, spectrophotometry, or activity assays);
(c) to measure the density of crystals (see Chapter 12);
(d) to check if macromolecules are in biologically active conformations within the crystals (for enzymes this can be asserted by *in situ* catalytical assays provided ligands can be diffused harmlessly within the crystal lattice and that active sites are accessible);
(e) to verify whether the crystallized conformations are identical to the ones in

Table 7. Selected methods for the biochemical and biophysical characterization of crystals

Methods[a]	**Expected information**
(a) *Analyses on crystalline material*	
Test for mechanical stability (with glass needle)	Macromolecular crystals are often soft
X-ray diffraction[b]	Space group, resolution[b]
Density measurements	Solvent content
EM	Crystallinity
Spectrophotometry (micromethods)	Absorption properties, activity
Raman spectroscopy	Conformational characteristics
Enzymatic digestion	Biochemical content
Binding of selective dyes, or ligands	Staining or affinity properties
In situ catalysis (for enzymes)	Activity
(b) *Analyses of dissolved crystals*	
Spectrophotometry/fluorimetry	Absorption properties, quantitation
Electrophoresis	Characterization of macromolecules, size
Isoelectric focusing (for proteins only)	Characterization of proteins, charge
Column chromatography (microscale)	Characterization of macromolecules
Various activity assays	Biological characterization

[a] Other methods may be employed in particular cases, e.g. dichroism, ultracentrifugation; [b] see Chapter 12.

solution (this can be done with appropriate spectroscopic methods like laser Raman spectroscopy).

A few examples of successful analyses taking advantage of microspectrophotometry (94, 95), catalytic activity assays (96), HPLC (92), dichroism (97), Raman spectroscopy (98), or of a combination of methods (93) may be useful to lucky crystal growers.

7. Concluding remarks

In conclusion, impurities and microheterogeneities are two important factors capable of influencing crystallization of biological macromolecules. Multiple origins of imperfections and the large size and complex structure of macromolecules mean that most parameters of their crystallization are not well controlled at present. Despite large efforts to prepare ultra-pure and homogeneous macromolecules, in some cases the actual reasons for the lack of success in crystallization are not fully understood, but in no case should the reader consider the situation hopeless. Stimulating examples come from studies on ribosomal subunits and the ribosome itself (99 and Chapter 7). These large nucleoproteic complexes, made of many protein and RNA molecules, have a high probability of being either heterogeneous or contaminated by impurities. Nevertheless, they are able to crystallize and this seems to occur with unexpected ease when compared to their individual protein or nucleic acid components or to other biological macromolecules of small size.

References

1. Wierenga, R. K., Lalk, K. H., and Hol, W. G. J. (1987). *J. Mol. Biol.*, **198**, 109.
2. Matthews, B. V. (1968). *J. Mol. Biol.*, **33**, 491.
3. McPherson, A. (1982). *Preparation and analysis of protein crystals*. John Wiley and Sons, New York.
4. Brennan, R. G., Wosniak, J., Faber, R., and Matthews, B. W. (1988). *J. Crystal Growth*, **90**, 160.
5. Zaccaï, G. and Eisenberg, H. (1990). *Trends Biochem. Sci.*, **15**, 333.
6. Kern, D., Giegé, R., Robbe-Saul, S., Boulanger, Y., and Ebel, J.-P. (1975). *Biochimie*, **57**, 1167.
7. Lorber, B. and DeLucas, L. J. (1990). *FEBS Lett.*, **241**, 14.
8. Oxender, D. L. and Fox, C. F. (ed) (1987). *Protein engineering.* Alan R. Liss, New York.
9. Sambrook, J., Fritsch, E. F., and Maniatis, T. (1989). *Molecular cloning: a laboratory manual*, 3 Vols. Cold Spring Harbor Laboratory Press.
10. Wosney, J. M. (1990). *Methods in Enzymology*, **182**, 738.
11. Schein, C. H. (1990). *Biotechnology*, **8**, 308.
12. Carter, C. W. (1988). *J. Crystal Growth*, **90**, 168.
13. Horwich, A. L., Neupert, W., and Hartl, F. U. (1990). *Trends in Biotechnology*, **8**, 126.
14. Marston, F. A. O. (1986). *Biochem. J.*, **240**, 1.
15. Spirin, A. S., Baranov, V. I., Ryabova, L. A., Ovodov, S. Y., and Alakhov, Y. B. (1988). *Science*, **242**, 1162.
16. Perona, J. J., Swanson, R., Steitz, T. A., and Söll, D. (1988). *J. Mol. Biol.*, **202**, 121.
17. Lowary, P., Sampson, J., Milligan, J, Groebe, D., and Uhlenbeck, O. C. (1986). In *Structure and dynamics of RNA* (ed. P. H. van Knippenberg and C. W. Hilbers), pp. 69–76. Plenum Press, New York.
18. Szewczak, A. A., White, S. A., Gewirth, D. T., and Moore, P. B. (1990). *Nucleic Acids Res.*, **18**, 4139.
19. Pace, C. N. (1990). *Trends in Biotechnology*, **8**, 93.
20. Sassenfeld, H. M. (1990). *Trends in Biotechnology*, **8**, 88.
21. Dao-Pin, S., Alber, T., Bell, J. F., Weaver, L. H., and Matthews, B. W. (1988). *Protein Eng.*, **1**, 115.
22. Deutscher, M. P. (ed.) (1990). *Methods in Enzymology*, **182**, 1.
23. Scopes, R. K. (1987). *Protein purification: principles and practice*, 2nd ed. Springer Verlag, Berlin, New York.
24. Harris, E. L. V. and Angal, S. (ed.) (1989). *Protein purification methods: a practical approach.* IRL Press, Oxford.
25. Gillam, I. C. and Tener, G. M. (1981). In *RNA and protein synthesis* (ed. K. Moldave), pp. 43–58. Academic Press, New York.

26. Dock, A.-C., Lorber, B., Moras, D., Pixa, G., Thierry, J.-C., and Giegé, R. (1984). *Biochimie*, **66**, 179.
27. Dudock, B. S. (1987). In *Molecular biology of RNA* (ed. M. Inouye and B. S. Dudock), pp. 321–329. Academic Press, Inc., New York.
28. Oliver, R. W. A. (ed.) (1989). *HPLC of macromolecules: a practical approach.* IRL Press, Oxford.
29. Wagner, H. (1989). *Nature*, **341**, 669.
30. Righetti, P. G., Barzaghi, B., and Faupel, M. (1988). *Trends in Biotechnology*, **8**, 121.
31. Sadana, A. (1989). *Biopharm*, **2**, (4) 14 and (5) 20.
32. Beynon, R. J. and Bond, S. (ed.) (1989). *Proteolytic enzymes: a practical approach.* IRL Press, Oxford.
33. Colpan, M. and Riesner, D. (1988). *Modern physical methods in biochemistry* Part B, (ed. A. Neuberger and L. L. M. Van Deenen) pp. 85–105. Elsevier, Amsterdam.
34. Kime, M. J. and Moore, P. B. (1983). *Biochemistry*, **22**, 2615.
35. Gulewicz, K., Adamiak, D., and Sprinzl, M. (1985). *FEBS Lett.*, **189**, 179.
36. Giegé, R., Dock, A. C., Kern, D., Lorber, B., Thierry, J.-C., and Moras, D. (1986). *J. Crystal Growth*, **76**, 554.
37. Blanchard, J. S. (1984). *Methods in Enzymology*, **104**, 404.
38. Stoll, V. S. and Blanchard, J. S. (1990). *Methods in Enzymology*, **182**, 24.
39. Creighton, T. E. (ed.) (1989). *Protein structure: a practical approach.* IRL Press, Oxford.
40. Creighton, T. E. (ed.) (1989). *Protein function: a practical approach.* IRL Press, Oxford.
41. Laemmli, U. K. (1971). *Nature*, **227**, 680.
42. Merril, C. R., Goldman, D., and Vankeuren, M. L. (1982). *Electrophoresis*, **3**, 17.
43. Weber, K, Pringle, J. R., and Osborn, M. (1972). *Methods in Enzymology*, **26**, 3.
44. Shapiro, A. L., Vinuela, E., and Maizel, J. V. (1967). *Biochem. Biophys. Res. Commun.*, **28**, 815.
45. Rickwood, D. and Hames, B. D. (1990). *Gel electrophoresis of nucleic acids: a practical approach.* IRL Press, Oxford.
46. Righetti, P. G., Gianazza, E., and Gelfi, C. (1988). *Trends Biochem. Sci.*, **13**, 333.
47. Hames, B. D. and Rickwood, D. (ed.). (1989). *Gel electrophoresis of proteins: a practical approach.* IRL Press, Oxford.
48. Haff, L. A., Fagerstam, L. G., and Barry, A. R. (1983). *J. Chromatogr.*, **266**, 409.
49. Findlay, J. B. C. and Geisow, M. J. (ed.) (1989). *Protein sequencing: a practical approach.* IRL Press, Oxford.
50. Edelhoch, H. (1967). *Biochemistry*, **6**, 1948.
51. Stoscheck, C. M. (1990). *Methods in Enzymology*, **182**, 50.
52. Warburg, O. and Christian, W. (1941). *Biochem. Z.*, **310**, 384.
53. Ehresmann, B., Imbault, P., and Weil, J. H. (1973). *Anal. Biochem.*, **54**, 454.
54. Guéron, M. and Leroy, J. L. (1978). *Anal. Biochem.*, **91**, 691.
55. Ames, B. N. and Dubin, D. T. (1960). *J. Biol. Chem.*, **235**, 769.
56. Neuhoff, V., Ewers, J. H., and Huether, G. (1981). *Hoppe Seyler's Z. Physiol. Chem.*, **362**, 1427.

57. Fersht, A. (1977). *Enzyme structure and mechanism.* Freeman and Company, Reading.
58. Jakoby, W. B. (1971). *Methods in Enzymology*, **22**, 248.
59. Ray, W. J. and Puvathingal, J. (1985). *Anal. Biochem.*, **146**, 307.
60. Jurnak, F. (1986). *J. Crystal Growth*, **76**, 577.
61. Kuhlbrandt, W. (1988). *Quarterly Reviews of Biophysics*, **21**, 429.
62. Lorber, B., Bishop, J. B., and DeLucas, L. J. (1990). *Biochim. Biophys. Acta*, **1023**, 254.
63. Bello, J. and Nowoswiat, E. F. (1965). *Biochim. Biophys. Acta*, **105**, 325.
64. Pringle, J. (1975). In *Methods in cell biology* (ed. D. M. Prescott), **12**, 149.
65. Reich, E., Rifkin, D. B., and Shaw, E. (1975). *Proteases and biological control*, Cold Spring Harbor Laboratory Press, New York.
66. Bond, J. S. and Butler, P. E. (1987). *Ann. Rev. Biochem.*, **56**, 333.
67. Neuberger, A. and Brocklehurst, K. (ed.) (1987). *Hydrolytic enzymes.* Elsevier, Amsterdam.
68. Neurath, H. (1989). *Trends Biochem. Sci.*, **14**, 268.
69. Moras, D., Dock, A.-C., Dumas, P., Westhof, E., Romby, P., Ebel, J.-P., and Giegé, R. (1985). *J. Biomol. Struct. Dyn.*, **3**, 479.
70. Wold, F. and Moldave, K. (ed.) (1984). *Methods in Enzymology*, **106**, 1; **107**, 1.
71. Seifter, S. and Englard, S. (1990). *Methods in Enzymology*, **182**, 626.
72. Kendall, R. L., Yamada, R., and Bradshaw, R. A. (1990). *Methods in Enzymology*, **185**, 398.
73. Kalisz, H. M., Hecht, H. J., Schomburg, D., and Schmid, R. D. (1990). *J. Mol. Biol.*, **213**, 207.
74. Rothstein, M. (1985). In *Modification of proteins during ageing*, (ed. R. C. Adelman and E. E. Dekker), pp. 53–67. Alan R. Liss, New York.
75. Björk, G. R., Ericson, J. U., Gustafsson, C. E. D., Hagervall, T. G., Jonsson, Y. H., and Wikstrom, P. M. (1987). *Ann. Rev. Biochem.*, **56**, 263.
76. Joachimiak, A. (1988). *J. Crystal Growth*, **90**, 201.
77. Van der Laan, J. M., Swarte, M. B. A., Groendijk, H., Hol, W. G. J., and Drenth, J. (1989). *Eur. J. Biochem.*, **179**, 715.
78. Rhodes, D. G. and Laue, T. M. (1990). *Methods in Enzymology*, **182**, 555.
79. Lorber, B., Kern, D., Mejdoub, H., Boulanger, Y., Reinbolt, J., and Giegé, R. (1987). *Eur. J. Biochem.*, **165**, 409.
80. Howe, C. J. and Ward, E. S. (ed.) (1989). *Nucleic acids sequencing: a practical approach*, pp. 1–250. IRL Press, Oxford.
81. Beynon, R. J. (1989). In *Protein purification methods: a practical approach*, (ed. E. L. V. Harris and S. Angal), pp. 40–51. IRL Press, Oxford.
82. Gottesman, S. (1990). *Methods in Enzymology*, **185**, 119.
83. Berger, S. L. and Birkenmeier, C. S. (1979). *Biochemistry*, **18**, 5143.
84. Scheele, G. and Blackburn, P. (1979). *Proc. Natl. Acad. Sci. USA*, **76**, 4898.
85. Thompson, S. T., Cass, K. H., and Stellwagen, E. (1975). *Proc. Natl. Acad. Sci. USA*, **72**, 669.
86. Wierenga, K., Huizinga, J. D., Gaastra, W., Welling, G. W., and Beintema, J. (1973). *FEBS Lett.*, **31**, 181.
87. Waller, J.-P., Risler, J.-L., Monteilhet, C., and Zelwer, C. (1971). *FEBS Lett.*, **16**, 186.
88. Bott, R. R., Navia, M. A., and Smith, J. L. (1982). *J. Biol. Chem.*, **257**, 9883.

89. Anderson, W. F., Boodhoo, A., and Mol, C. D. (1988). *J. Crystal Growth*, **90**, 153.
90. Boulot, G., Guillon, V., Mariuzza, R. A., Poljak, R. J., Riottot, M. M., Souchon, H., Spinelli, S., and Tello, D. (1988). *J. Crystal Growth*. **90**, 213.
91. Spangler, B. D. and Westbrook, E. M. (1989). *Biochemistry*, **28**, 1333.
92. Lorber, B. and Giegé, R. (1985). *Anal. Biochem.*, **146**, 402.
93. Lorber, B., Giegé, R., Ebel, J.-P., Berthet, C., Thierry, J.-C., and Moras, D. (1983). *J. Biol. Chem.*, **258**, 8429.
94. Michel, H. and Oesterhelt, D. (1980). *Proc. Natl. Acad. Sci. USA*, **77**, 1283.
95. Ottonello, S., Mozzarelli, A., Rossi, G. L., Carotti, D., and Riva, F. (1983). *Eur. J. Biochem.*, **133**, 47.
96. Kirsten, H. and Christen, P. (1983). *Biochem. J.*, **211**, 427.
97. Ford, R. C., Picot, D., and Garavito, R. M. (1987). *EMBO J.*, **6**, 1581.
98. Chen, M., Lord, R., Giegé, R., and Rich, A. (1975). *Biochemistry*, **14**, 4385.
99. Hansen, H. A. S., Volkmann, N., Piefke, J., Glotz, C., Weinstein, S., Makowski, I., Meyer, S., Wittmann, H. G., and Yonath, A. (1990). *Biochim. Biophys. Acta*, **1050**, 1.

3

Design of crystallization experiments and protocols

C. W. CARTER, Jr.

1. Introduction

Despite the fact that most pure proteins should be expected to crystallize readily under some set of conditions, that 'winning combination' of solution conditions may be rather hard to find *a priori*. Often, there is little apparent connection between the degree to which proper protocols are observed and the degree to which satisfactory results are obtained. For this reason one should assume the worst at the outset for any new macromolecule or intermolecular complex, and try to identify potentially important factors by setting up efficient screening trials. The best all-round experiments for this purpose are factorial designs used by several generations of industrial chemists (1–3). Factorial experiments have been applied explicitly to the problem of protein crystallization (4–11). They serve not only to screen conditions, but also to provide information simultaneously about the effects of a number of different possible 'factors'. Factorial experiments are the statistician's answer to the 'needle-in-the-haystack' problem in any of its many guises. The less one knows about the behaviour of a system, the harder it will be to find an optimum combination, and the more important it is to think carefully about the experiments to use in order to characterize it. This reasoning is especially relevant to the crystallization of a new macromolecule.

In this chapter the idea of taking averages of replicated experiments, and other underlying motivations for statistical experimental designs are reviewed, and different implementations of factorial experiments are compared. The most significant advantages of these designs are their efficiency, accuracy, and completeness. Of these advantages the efficiency of factorial experiments alone is sufficient to justify their use over other approaches to screening. The method has proven its efficiency in other applications, e.g. where variation of numerous reaction parameters was necessary to optimize the *in vitro* rate of a complex biochemical reaction (12). Another application of factorial designs in protein crystallography would be in screening for heavy-atom derivatives, where one might search simultaneously with different reagents, at different pH's, in different mother liquors, and so forth. Protocols for using factorial designs are basically the same for all these different applications.

2. The theory of factorial experimental designs

Why do particular processes, like crystallization of a protein, occur? Often, the 'causes' comprise a rather small number of specific factors, or combinations of factors, that are particular to the case in point. These specific factors constitute the heart of the 'explanation' or 'recipe' for the effect. They usually act somewhat independently of one another. Rarely do the contributing factors interact so intimately as to require simultaneous fine-tuning of each one. In the jargon of experimental design, the contributing factors are called *main effects*, while combinations of main effects are called *n-factor interactions*.

A *factorial design* is a set of experiments intended to identify important main effects and interactions. Factorial experiments are frequently used in industrial quality-control applications, where the goal is either to identify reasons why a particular product fails, or to optimize the yield or some other criterion associated with its manufacture. Crystal growth screening experiments differ from these situations in two interesting respects:

(a) industrial applications examine factors as sources of variation in existing processes;

(b) in industrial applications one has a standard way to quantify the outcome of each test.

At the outset of a crystal growth screening experiment, one does not know how to produce crystals. Nor is it obvious how to estimate quantitatively the success of different crystallization trials. The crystal grower is, relatively speaking, considerably more ignorant than the industrial chemist, for the latter can already measure the yield of a reaction when he begins to optimize it. Nevertheless, the parallels between screening for successful crystal growth conditions and the optimization of an industrial process are much more significant than these two differences. In both cases one is trying to select from a multiplicity of possible causes those that matter. The same characteristics that make factorial designs so powerful in optimization are therefore also very appropriate for screening:

(a) they are very efficient search vehicles;

(b) if one is willing to score the results, factorial experiments also offer a natural framework for using analysis of this quantitative information to optimize crystal growth.

The efficiency of factorial designs can be appreciated without understanding the bases for quantitative analysis of the results. The question of accuracy in statistical designs is more difficult to understand because factorial experiments are counter-intuitive: they involve simultaneous variation of all possible experimental variables systematically from experiment to experiment within a design. In the most efficient designs, this variation is achieved by wholesale randomization of the combinations to be tested. The surprising accuracy of randomized designs is hard to accept without first understanding the underlying theory of factorial experiments, their strengths, and their weaknesses. The following examples will therefore introduce useful aspects of data analysis for a full factorial experiment, and then show how many of the advantages of this design can be preserved in more efficient designs derived from it.

2.1 Averaging results from replicated experiments

Perhaps the most important underlying principle concerns the nature and power of averaging the outcomes from several experiments. The efficiency of properly designed screening experiments comes from the fact that one can extract useful information from averages of experiments which are not exact replicates of one another. The following hypothetical examples, adapted from a similar presentation in reference 2, illustrate the basic concepts.

There is always a practical limit to the total number of experiments one can do. How this number of experiments is deployed can greatly influence the quality and quantity of the information they will provide. The following discussion compares four different ways to use a total of eight experiments to probe the effects of different factors on crystal growth: one factor at a time, a full factorial, a fractional factorial, and an incomplete factorial experiment.

Suppose that one can evaluate quantitatively, on a scale from 1–10, results of a series of crystallization experiments carried out at two different pH's and, say, at 4 °C. Whatever the precision of the individual measurements, the confidence level of inferences based on the data increases the more times a given experiment is repeated. This means replicating the experiments at the two pH's and taking averages of all those for a given pH (*Figure 1*). This example illustrates averages of four experiments for each experimental point, together with the standard deviations of the mean values. The average of multiple determinations is a better estimate of the 'true' value than is any single result.

2.1.1 Main effects

If crystals reproducibly grow better at pH = 7.5 than they do at pH = 6.5, as indicated in *Figure 1*, then according to standard statistical terminology there is a *significant main effect* of pH. Based on these results, it would make sense to grow crystals at the higher pH. Moreover, knowing that it is a significant main effect, one can expect to find an optimum, not necessarily at pH = 7.5, by sampling the pH more finely and over a larger range.

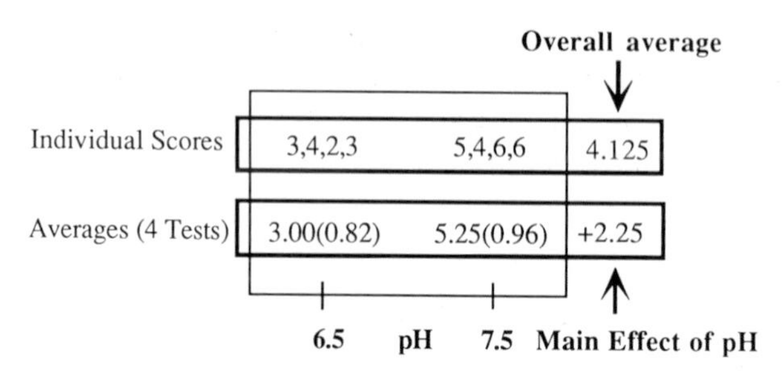

Figure 1. A 'main effect' of pH. Averages and their standard deviations are given in the second row. The standard deviation is the standard deviation of the mean:

$$\sigma = \sqrt{\frac{(y - y)^2}{(n - 1)}}.$$

Suppose now that it is also important to know whether or not temperature has a significant main effect. The 'one factor at a time' approach would require eight more experiments: four at 20 °C and another four at 4 °C. One immediately wonders, at which pH should these eight new experiments be done? Common sense might suggest doing them at the pH previously found to be 'better'. However, a better choice would be to do eight new experiments at 20 °C, four at pH = 6.5 and four at pH = 7.5 (*Figure 2*). In this way, both main effects can be evaluated from averages of the same number of experiments. Even better would be to do the 16 experiments in *Figure 2* in a random sequence, to avoid the possible effects of unknown factors that differed on two different days.

The results of these additional eight experiments appear in *Figure 2*,

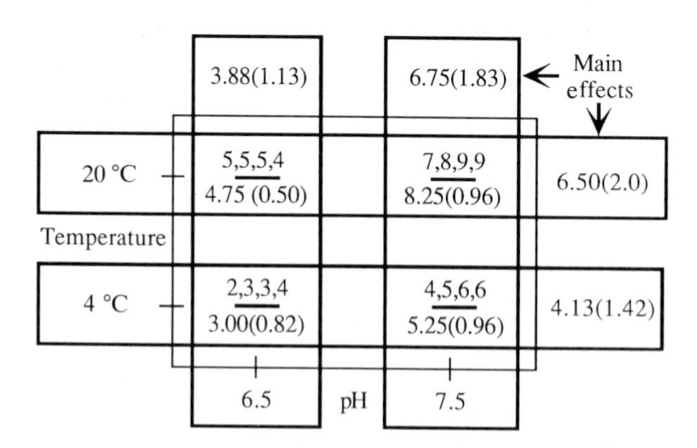

Figure 2. A two-level, complete factorial design with fourfold replication, showing significant main effects of temperature and pH. Sixteen experiments are distributed, four to each box. The eight experiments at 4 °C are identical to those in *Figure 1*. Averages and their standard deviations are as described for *Figure 1*.

together with those from *Figure 1*. They illustrate main effects of both pH and temperature. The pH main effect is estimated to be 2.87, from the difference between the average values for the eight experiments at pH = 6.5 and those at pH = 7.5. The temperature main effect is estimated to be 2.37 by a similar calculation.

2.1.2 Two-factor interactions

Careful examination of the averages in *Figure 2* reveals a more subtle effect: the temperature influences the main effect of pH, and vice versa. The pH main effect is 55% greater (3.5) at 20 °C than it is (2.25) at 4 °C. Similarly, the temperature main effect is 71% greater (3.0) at the higher pH than at the lower pH (1.75). This dependence of the magnitude of one main effect on the level of another factor is called a *two-factor interaction.* An estimate of the strength of this pH × temperature interaction is given by the difference between the average results for all experiments in which both factors are taken at the same level (i.e. both high or both low) and the average results for all experiments in which the factors are taken at different levels (i.e. one high and the other low). In this example, the estimated pH × temperature interaction is 1.25.

Figure 2 illustrates the layout of a factorial experiment. Tests are distributed evenly among each of the possible combinations of the two factors. Information obtained from a factorial experiment can be expressed succinctly by giving the important main effects and interactions. The main effects of pH and temperature are apparent in all one-dimensional rows and columns, as was the main effect of pH in *Figure 1*. Thus any two rows or any two columns will show approximately the same 'main effect' if there is no interaction, and these estimates of the main effects will differ when there is a significant interaction. The significance of any effect (main effect or interaction) can be evaluated approximately by comparing the size of the effect with the overall average effect, which in this case is 5.31. In this hypothetical example, the two main effects are about 50%, and their interaction is slightly more than 20% of this value.

2.1.3 Averaging experiments that are not exact replicates

The standard deviations of averages from different experiments in a balanced matrix of experiments such as that in *Figure 2* depend strongly only on the precision of individual measurements and only weakly on the number of different factors being screened. Consider the results in *Figure 3*, in which only half of the 16 experiments have been included. In this case, the power of averaging four results at each pH is achieved by using two experiments at 4 °C and two at 20 °C. The two main effects and their interaction all have approximately the same values (3.25, 2.25, and 0.75 versus 2.87, 2.37, 1.25) with only slightly less precision, yet they were estimated using only 8 experiments instead of 16.

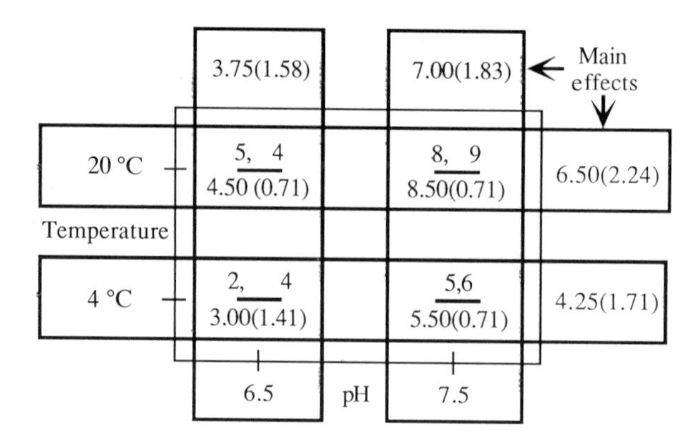

Figure 3. The design shown in *Figure 2* but using only eight of the sixteen experiments. The averages and their standard deviations have approximately the same values as do the corresponding ones in *Figure 2*. The same main effects can therefore be inferred with approximately the same confidence.

2.2 Factorial experiments

2.2.1 Full factorial designs

The design shown in *Figures 2* and *3* is called a two-level, two-factor factorial experiment. It is also called a full or complete factorial design to distinguish it from fractional and incomplete designs to be described below. The two factors, pH and temperature, are each tested at two different *levels*. The simplest example would involve only four experiments, because there are only 2^2, or 4, possible combinations. Since each combination is tested twice, this 2^2 factorial design is said to be carried out with two-fold replication. Factorial designs can involve more than two levels, but for reasons of simplicity and efficiency they are often limited to two levels for each factor.

It is useful to present the data from *Figure 3* in the form of a table (*Table 1*)

Table 1. Presentation of a complete factorial experiment: evaluation of contrasts for the main effects and the two factor interaction

Experiment	pH	Temperature	pH × temperature	Score
1	−	−	+	2
2	−	−	+	4
3	−	+	−	5
4	−	+	−	4
5	+	−	−	5
6	+	−	−	6
7	+	+	+	8
8	+	+	+	9
Contrast	3.25	2.25	0.75	5.375

in which the level treatments for each test are indicated in the appropriate columns, the scores in an additional column, and the effects themselves (also called *contrasts*) along the bottom row. There are several notable features of full factorial designs:

- They are easy to construct. Lay out a $2^N \times N$ matrix for the treatment of each factor in each of the 2^N experiments. All possible combinations of the N factors are tested in 2^N experiments. If there are only two levels for each factor, they can be represented conveniently by pluses for the high level (present or higher) and minuses for the low level (absent or lower).
- They are easy to evaluate. Add up the score for each experiment with its sign for each contrast column, as in *Table 1*, to estimate the useful information from all experiments about the factors and their interactions. Properly, contrasts should be estimated by analysis of variance, but often this is not done.
- Unlike traditional 'one-at-a-time' designs, factorial designs allow each experiment to do several jobs at the same time, contributing consistently to the estimation of all effects and interactions. In the example, the power of averaging four results at each pH is achieved by using two experiments at 4 °C and two at 20 °C. Eight experiments deployed as in *Figure 3* determine the main effects of pH and temperature, together with their interaction, with nearly the same confidence achieved with eight experiments at the same temperature shown in *Figure 1*. This is an important source of their high efficiency.

Full factorial experiments are especially useful in situations where a phenomenon is suspected to depend on the effects and interactions of a small number of factors. One example, described in references 2 and 8, showed how pH, temperature, and two low-molecular substrates influenced the crystal growth of an enzyme suspected previously to have both pH- and ligand-dependent conformational changes. Important and biochemically significant main effects (pH) and two-factor interactions (pH × substrate and substrate × substrate) were identified from a quantitative analysis of the crystal growth behaviour in a 2^4 factorial experiment similar to that shown in *Table 1*. A 2^4 experimental design provides estimates for the main effects of four factors as well as six two-factor interactions, four three-factor interactions, and one four-factor interaction.

2.3 Selecting subsets of experiments from a full factorial design

For many more than four separate factors a full factorial design becomes difficult to carry out because too many experiments are required. Various methods exist to increase the scope of factorial designs by judicious choices of a subset of the full design. In the fractional factorial design this is done by systematically deleting some of the individual experiments, leaving only those necessary to identify main effects (3). In an incomplete factorial design the experiments to be done are chosen randomly, subject to careful balancing to insure that two-factor interactions can also be identified.

2.3.1 Fractional factorial designs

To convert the design in *Table 1* into a fractional factorial design for three factors, one simply uses the pattern of pluses and minuses corresponding to the two-factor interaction to represent the testing pattern for the third factor (*Table 2*). The effect of this ruse is to intentionally identify the main effect of the third factor with the interaction of the other two factors.

The contrasts in *Table 2* reflect the possibility that the presence of substrate may also be an important factor influencing crystal growth, a situation which is not uncommon. However, what does the large third contrast really mean? Is the presence of substrate what is really important, or is it really the pH × temperature interaction? It is impossible to tell from this design because the two contrasts have been intentionally identified, one with the other. This procedure is called *aliasing*, and the resulting confusion regarding the true origin of the high contrast (is it substrate or the pH × temperature interaction?) is called *confounding* of the two contrasts. In a fractional factorial experiment confounding is done on purpose, by aliasing specific columns of experimental specifications with new factors. A consequence is that one may wind up confusing a hidden interaction between two factors for a strong effect of a third. The only way to resolve this ambiguity is to carry out additional experiments specifically designed to distinguish between the multiple possibilities (3).

In this case a better strategy would be to forego the twofold replication and arrange the eight experimental treatments as a proper full 2^3 factorial design, which requires a total of eight experiments. Treating the third column as

Table 2. Presentation of a fractional factorial experiment: evaluation of contrasts for three main effects

Experiment	pH	Temperature	Substrate	Score
1	−	−	+	5
2	−	−	+	6
3	−	+	−	5
4	−	+	−	4
5	+	−	−	5
6	+	−	−	6
7	+	+	+	9
8	+	+	+	10
Contrast	2.50	1.50	2.50	6.25

shown in *Table 3* avoids the situation encountered in *Table 2*, in which the arrangement of pluses and minuses in one of the main effect columns is the same as that of the product of two of the others. By sacrificing some of the precision obtained by replicating each experiment exactly, we recover a full description of the magnitudes of three main effects and their four interactions. Each contrast represents the difference between averages of four experiments, so the design makes efficient use of each experiment by using averages from experiments which are not exact replicates, as noted in Section 2.1.3.

Table 3. Presentation of a 2^3 full factorial experiment with eight experiments: evaluation of contrasts for three main effects, two two-factor interactions, and one three-factor interaction.

Experiment	pH	Temp	Subst	pH × T	T × S	pH × S	pH × S × T	Score
1	−	−	+	+	−	−	+	6
2	−	−	−	+	+	+	−	3
3	−	+	+	−	+	−	−	7
4	−	+	−	−	−	+	+	4
5	+	−	+	−	−	+	−	7
6	+	−	−	−	+	−	+	5
7	+	+	+	+	+	+	+	10
8	+	+	−	+	−	−	−	8
Contrast	2.50	2.00	2.50	1.00	0.00	−1.25	0.00	6.25

2.3.2 Incomplete factorial designs

In contrast to the fractional factorial design, which incorporates new factors by intentionally aliasing them to the interaction columns of a full factorial design, the incomplete factorial design approaches the problem from a completely different point of view. Confounding is expressly avoided by randomizing the patterns of pluses and minuses in the design, subject to the constraint that the design be balanced in two respects. Firstly, the design must provide equal (or nearly equal) numbers of tests at each level of all factors. Secondly, the first-order interaction matrices must have at least one experiment in all four positions. The example in *Table 4* was generated in this way. (The entries were random numbers, either 1 or 2, rather than − or +.) It represents an incomplete factorial design to test for the effects of seven factors, A–G. There are eight experiments (or 6.25%) of the 128 which would be required for a full 2^7 factorial design. Notice that owing to the balance of the design, the contrast sums for each main effect involve differences between averages of four experiments at each level. No column is the product of any other two columns, so none of the seven main effects is aliased to any two-factor interaction. The contrasts estimate the seven main effects without confounding.

Table 4. Presentation of an incomplete factorial experiment with eight experiments, testing for main effects of seven factors

Experiment	A	B	C	D	E	F	G	Score
1	+	−	+	+	+	−	+	4
2	−	−	−	−	−	−	−	5
3	+	+	−	+	−	+	−	7
4	−	+	+	−	+	+	+	9
5	−	+	+	−	−	+	−	10
6	+	−	−	+	+	+	+	6
7	+	+	−	−	+	−	−	8
8	−	−	+	+	−	−	+	6
Contrast	−1.25	3.25	0.75	−2.25	−0.25	2.25	−1.25	6.88

In fact, the design in *Table 4* can, under many circumstances, also be used to obtain a reasonable estimate of important two-factor interactions. As an example, consider the C × F interaction. The eight experiments in the incomplete factorial design constitute a balanced probe of this interaction, with two tests carried out for each combination of C and F. The experimental results themselves can therefore be used to calculate a contrast for the C × F interaction by the procedure illustrated in *Tables 5* and *6*. The treatments of factors C and F and the scores from *Table 4* are given for reference. The interaction column is prepared as usual from the products of the factor C and F treatments, and the contrast calculated in the usual way. The resulting pattern of signs does not appear anywhere in *Table 4*, so this is a distinct estimate, which matches the pattern of the C × F interaction. One can also compare the average effect of C at the two levels of F, or the average effect of F at the two levels of C. All three calculations give the same answer, namely, 2.25.

In *Table 4* 12 of the 21 interactions are balanced, with two experiments for each combination, and 9 are unbalanced. However, all of the two-factor interactions have at least one experiment for each combination, so that an

Table 5. Evaluation of the C × F interaction for the design in *Table 4*

Experiment	C	F	C × F	Score
1	+	−	−	4
2	−	−	+	5
3	−	+	−	7
4	+	+	+	9
5	+	+	+	10
6	−	+	−	6
7	−	−	+	8
8	+	−	−	6
Contrast	0.75	2.25	2.25	6.88

Table 6. Calculation of the C × F interaction contrast for the data from the incomplete factorial experiment in *Table 5*. Scores for individual experiments are given in bold face in parentheses beside each experiment number, and their averages are given in the second row of each entry

C \ F	−	+
−	2(**5**), 7(**8**) **6.5**	3(**7**), 6(**6**) **6.5**
+	1(**4**), 8(**6**) **5.0**	4(**9**), 5(**10**) **9.5**

estimate of the interaction can be obtained by comparing the average main effects of one factor at the two levels of the other. In the case of the nine unbalanced interactions, however, this will involve comparing 'averages' of only two experiments with 'averages' of six experiments.

2.4 Pros and cons of incomplete factorial designs

The difference between these two strategies is important. One way to think of the difference between incomplete factorials and fractional factorials is this: an incomplete factorial design permits unambiguous examination of as many potential main effects and interactions as possible, at the expense of lower precision. Since the experimental noise level is higher, it detects only the more important effects and interactions; less important ones are harder to distinguish. Nevertheless, the more important ones often stand out, main effects are not confounded with interactions, and interactions are rarely confounded with each other.

The incomplete factorial design was developed for applications in industrial quality control (13). It shares some properties with saturated fractional factorial designs (14), but has not been described extensively in the literature of statistical experimental design. For the purpose of screening for crystal growth conditions it has several important advantages:

(a) It provides a very efficient sampling of the experimental space to be explored. Randomly chosen experiments are as evenly distributed as possible.

(b) In crystal growth experiments, a single screening experiment can often turn up a broader selection of possible directions to pursue; including not only diffraction quality crystals, but a variety of crystal forms with different ligands, conformations, non-crystallographic symmetries, and/or diffraction qualities. (Too much of a choice may be a disadvantage unless one is willing to characterize all results.)

(c) The most important main effects and two-factor interactions can generally be estimated. None of the main effects or interactions can be estimated as accurately as they can with complete factorial designs, since there are in general fewer experiments per effect. However, since important effects usually stand out, this means that one can identify the important conditions to vary and/or combine in subsequent optimization experiments.

(d) Confounding of main effects and interactions is kept to a practical minimum. Examination of the experiment in *Table 4* reveals only two confounded interactions: A × E with D × G, and A × G with D × E. Thus, off 378 possible confoundings among the 7 main effects and 21 interactions, only 0.5% are confounded.

3. Practical implementation of factorial experimental designs

Using a designed experiment involves five stages.

(a) Design the experimental matrix.

(b) Carry out the individual tests as specified by the experimental matrix, scoring each test.

(c) Analyse the scores in order to estimate the most important main effects and interactions.

(d) Verify the information obtained from the experiment by additional experiments

(e) Use it to optimize crystal growth.

In many cases suitable crystals grow in one or more of the initial tests, and one only needs to complete the first two stages. These two stages will be described in this section; the subsequent analysis will be described in Section 4. Examples can be found in references 4–10.

3.1 Preparation of the experimental matrix

It should be obvious from Section 2 that the heart of a designed experiment is the experimental matrix. This matrix specifies how each individual experiment or 'test' is to be carried out. Once the tests have all been completed and scored, the matrix is augmented by a column with the score for each test. Construction of the experimental matrices for full and fractional factorial designs can be done by following the samples given in *Tables 1–3*. The experimental matrix for an incomplete factorial experiment is constructed by a more elaborate procedure, which is summarized in *Protocol 1*.

Protocol 1. Preparation of the experimental matrix for an incomplete factorial design

1. Gather information about the protein to be crystallized from the following sources.
   - Everyone involved with the crystallization project, including those involved with expression and purification of the sample.
   - The Biological Macromolecule Crystallization Database (15,16), which may have information about crystallization of related molecules.
   - Previous screening experiments for the same or related molecules.
2. Prepare a list of possible factors that might influence crystal growth, and from these decide on the factors to be screened in the current experiment and the levels to test.
3. Decide on the number of tests. In general, this number should be somewhat more than the number of factors to be screened. There is no hard and fast rule, but the number should be manageable in terms of carrying out and evaluating the different tests.
4. Compile the experimental matrix itself.
   (a) Choose the level of each factor at random, proceeding down each column, and across from column to column.
   (b) Balance each column by readjusting the levels, if necessary, so that there are equal or as nearly equal as possible numbers of tests at each level.
   (c) Balance each two-factor interaction (submatrices similar to *Table 6*) by compensating readjustments to two columns. In this case, it is adequate to insure that each combination is represented by at least one test.
   (d) Verify the balance of all columns and two-factor interactions.
5. Examine the experimental treatments for main effects and interactions, and identify any possible confounding. Confounding is indicated whenever identical patterns of level assignments are observed for two different effects. In the event that a confounded effect turns out to be large, this knowledge is useful in further experimentation to distinguish between the two possibilities.

It is worthwhile elaborating on the steps in *Protocol 1*. If a screening experiment is to produce crystals, the conditions tested must include some that will support crystal growth. Even the best design cannot produce crystals unless it is given a chance to do so. To give it this chance, one must examine carefully everything that is known about the macromolecule and distill from this database a list of possible factors to be tested. This is the essence of step 1.

3.1.1 Data gathering and decision-making

i. The Biological Macromolecular Crystallization Database and selection of crystallizing agents

The Biological Macromolecular Crystallization Database (15, 16) is an important resource for the data-gathering stage of the process. It is a compilation of data from published crystallization papers, providing for each crystal form information about the macromolecule itself, the methods used to crystallize it, and the crystal data (unit cell parameters and symmetry). A retrieval program has been recently implemented (17). A cursory analysis of crystallizing agents (15) shows it to be a rich resource for detailed analysis of crystallization conditions. *Figure 4* illustrates that ammonium sulphate and potassium phosphate (~4%) are disproportionately useful crystallizing agents; other popular ones include PEG 6000 (~10%), low ionic strength (~9%), MPD (~7%), and Ethanol (~6%). Overall, more than 60% of all crystals obtained used salting out, 16% used organic solvents, and 15% used some size range of PEG. These proportions provide a good indication of the types of crystallizing agents to include in any screening experiment. Jancarik and Kim have recently proposed a 'sparse matrix' set of crystallizing agents that attempts to match this distribution in greater detail using 48 different conditions (18). Although not a factorial experiment, success using this matrix (numerous unpublished results from several laboratories) suggests that some attempt to reflect the distribution in *Figure 4* might be worthwhile. The incomplete factorial design reported in (4) was a balanced treatment that approximately reflected this distribution.

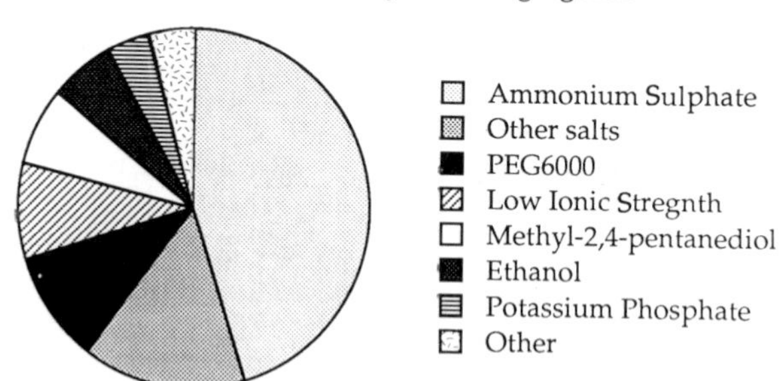

Figure 4. Fractional distribution of crystal forms obtained with different crystallizing agents. The agents most used in growing crystals tabulated in the Biological Macromolecule Crystallization Database (12). 'Other salts' represents 54 different salts. 'Other' represents 33 different crystallizing agents.

ii. Mutants and designed variants

Very often mutant or variant proteins will crystallize readily under the conditions used for the parent protein. Over 80% of the single site variants tests in one study could be crystallized under conditions closely related to those that were successful with the native protein (19).

iii. pH, ionic composition, and temperature

Macromolecular crystals are stabilized by interactions similar to those that stabilize protein quaternary structure, and these in turn resemble those that stabilize tertiary structure. With the progression from tertiary to quaternary structures to lattice packing, however, the degree of hydration in interfaces increases, and the relative importance of different types of interactions changes. In particular, complementary non-polar surfaces play a decreasing role [membrane proteins may be exceptions; (20) and Chapter 8], and complementary electrostatic forces (hydrogen bonding and ion-pairing) play an increasingly important role. In crystal lattices many of the latter forces are mediated by bound solvent molecules, either water or counterions (21). Specific ion pairs have rarely been observed in crystal packing contacts, except in the case of oligonucleotides, where magnesium has been observed to cross-link different molecules (22). However, this type of interaction may be more important that previously realized, even for proteins (23). These considerations underscore the paramount importance of such factors as pH that affect charged surface groups, and the nature of different ions in solution.

The net charge on a protein may play an important role in determining which ions are most influential in promoting crystal growth (see Chapter 9 and reference 23). Screen carefully among different anions for cationic and among different cations for anionic macromolecules.

Many interactions within and between macromolecules in water are temperature dependent because they involve significant entropy changes. For this reason it is worthwhile varying the temperature over the range in which the molecule is stable. For the tryptophanyl-tRNA synthetase from *Bacillus stearothermophilus* there seems to be less competition from precipitation if crystals are nucleated at 30 °C, rather than at either higher or lower temperatures.

iv. Sampling ranges and intervals

It is very unlikely that a protein will crystallize in any situation in which it denatures. This constraint limits the ranges of pH and temperature to be explored. As a rule of thumb, we sample the pH in a range that includes both the pH optimum and the isoelectric point of an enzyme, and in increments of 0.8–1.5 pH units. However, recent quantitative (24) evidence suggests that pH need not be sampled more finely than about 1.5 pH units because most crystals grow over a pH range of about 1.5–2.5 pH units. These experiments also suggest that crystallization will occur over a fairly broad range of crystallizing agent concentrations.

v. Ligands and additives

Potential ligands of all kinds (substrates, inhibitors, effectors) should be included in the initial search for crystallization conditions. Ligands may change either the stability or the conformation of the macromolecule. There is little extra cost in including these as additional two-level factors, and much to gain from the knowledge of how these changes impact on crystal growth (4,8,10). In contrast, additives [β-octyl glucoside (25), the various PEG (26)] which are useful in improving or changing crystal size or morphology need not be tested at this stage. MPD has been proposed as such an additive (27). Testing it as a crystallizing agent may simultaneously determine its effect as an additive. Otherwise there may be little gain from testing it only as an additive.

vi. Protein concentration

There are probably good reasons to vary protein concentration as a factor at two or three levels. Variation of protein concentration will change supersaturation conditions and hence nucleation and growth kinetics. This variation may, in turn, cause conditions which give microcrystalline precipitates at one concentration to produce macroscopic crystals at another (28).

3.1.2 Constructing the experimental matrix

The experimental matrix of an incomplete factorial design is determined in accordance with three principles. The list of factors to be screened should be comprehensive; the level at which each factor is tested is chosen at random for each experiment; and the distribution of experiments among the factor levels (for main effects) and among the first-order interactions should be balanced. Imagine experiments as being located in a multi-dimensional space whose basis vectors are factors and whose coordinates are different levels. Choosing the coordinates for each experiment by chance will produce an even distribution ensuring efficient sampling.

i. Attributes or interacting factors?

There are two different ways to treat such factors as crystallizing agents and different types of ions. They can be treated as factors whose levels are the different components (called attributes), with each level corresponding to a different component; or treated separately, as different factors whose levels indicate the presence or absence of a particular component. To see this distinction explicitly, consider two different ways to use the design in *Table 4*. Column A can represent the choice between two different crystallizing agents (e.g. 1 = ammonium sulphate, 2 = potassium phosphate). In this case, the two levels of the factor 'crystallizing agent' are called *attributes*. Alternatively, we could use column A to indicate the presence or absence of ammonium sulphate and column B to represent the presence or absence of potassium phosphate. This difference is important. If all crystallizing agents are attributes of the same factor, they cannot interact with each other. In general, using factors with attributes allows screening of more factors with the same number of columns, but it degrades the data analysis.

Which treatment of factors like 'crystallizing agent' is better? This question is hard to answer. Treating different components as attributes of a single factor leads to simpler experiments, both conceptually and in practice. Simple solutions with fewer components generally work better than complex mixtures, and approach more closely the ideal protocol for crystallizing small molecules (29). However, treating such factors separately is significantly more in line with the precepts of statistical experimental design, and permits more straightforward data analysis (30). For designs in which crystallizing agents or ions are treated as attributes one must recast the experimental matrix in a different form (10) that sacrifices balanced design characteristics. Treatment of all components as separate factors preserves these characteristics throughout the analysis. Both approaches therefore have significant merits and disadvantages. More experiments should be done to elucidate this point.

ii. Randomization and balance: the INFAC program

Although the procedures in step 4 of *Protocol 1* can be done by hand (4) it becomes a rather tedious job when the number of experiments exceeds about 10. In practice, they are readily carried out by a computer program, INFAC (10). This program continues to generate designs according to input variables; including the total number of experiments, the number of factors, and the number of levels for each factor; until it finds one in which there are no unbalanced two-factor interactions and in which no two experiments are identical. The INFAC program is available as Fortran source code on IBM PC or Macintosh diskettes (31).

INFAC does not check for such unfortunate combinations as Ca^{2+} and phosphate because it has no way of knowing how the experimental matrix it outputs will be used. Thus, in order to carry out experiments we usually examine the matrix output by INFAC and choose level assignments to minimize such combinations. It should be noted that such combinations are mandatory if one is screening simultaneously for Ca^{2+} and phosphate as factors. In such cases, we simply use very dilute solutions of Ca^{2+} and tolerate the resulting inorganic crystals.

3.2 Experimental set-ups

The tests should be set up according to the design and carried out by systematically approaching supersaturation, manipulating either the precipitant (in dialysis experiments) or total (in vapour diffusion experiments) concentration to ensure that each experiment is taken to completion. This should be done repeatedly for conditions producing precipitates, to verify so far as possible that precipitation was not due simply to excessive precipitant concentrations. Microdialysis buttons have obvious advantages from this point of view, but vapour diffusion can be used if desired. Vapour diffusion is essential for experimental conditions, such as organic crystallizing agents and high molecular weight polymers, for which dialysis is inappropriate. In these cases, it is useful either to titrate the concentration of crystallizing agent in preliminary trials, or to set up multiple experiments over a range of concentrations for each test.

Factorial experiments preclude using the same buffer for any two experiments, so some thought should be given to rational preparation of stock solutions, and so on. We find that there is no good alternative to simply making up each buffer separately, in order to keep the ionic compositions consistent with design requirements.

3.3 Quantitative evaluation of crystallization results

The scores, or Q values for experiments in a design form the basis for all subsequent data analysis. Unfortunately, the problem of evaluating crystallization experiments quantitatively is still without a general solution. Despite its problems, evaluation of all experiments in a design using an arbitary scale such as that in *Table 7* often gives significant and useful information when subjected to regression analysis. Precipitates differ somewhat in their appearance under the microscope. Fluffy or filamentous precipitates have little likelihood of being crystalline, but uniform, granular, and/or particulate precipitates often are microcystalline. Similarly, among different crystalline samples, one can readily distinguish spherulites (radially symmetric aggregates of microscopic needles) from needles, plates, and prisms, with obvious preference for the latter over the former. The major weakness of this scale is the tendency to confuse microcrystalline and amorphous precipitates (6, 10). One can achieve more accurate scoring of experimental results using the following methods:

- Examine precipitates for birefringence. Amorphous and microcrystalline precipitates can often be distinguished on the basis of their birefringence. If examined between crossed polarizing lenses, a microcrystalline precipitate may 'glow' as a result of the rotation of the plane of polarization that occurs when light passes through the crystal. True amorphous precipitates do not affect the plane of polarization in a coherent fashion, and so will not glow.

Table 7. Scale of crystal quality

Result	Score, Q
Cloudy precipitates	1.0
Gelatinous or particulate precipitates	2.0
Spherullites	3.0
Needles	4.0
Plates	5.0
Prisms	6.0

Unfortunately, use of this method to assay precipitates for microcrystallinity requires that the optical path consist entirely of glass as plastic/air interfaces induce birefringence that masks the sample birefringence. So it is difficult to use this method either with plexiglas dialysis cells or tissue culture plates. Instead, transfer a sample of the precipitate to a glass slide for analysis. Work rapidly, and use a multiple spot depression slide with reservoir solutions in the depressions surrounding the sample with a large glass slide as a coverslip over the entire dish to avoid crystallization of the crystallizing agent.

- An alternative and complementary approach is to streak seed from precipitates to assay for microcrystallinity (see Chapter 5). This is an excellent way to confirm a subjective assessment that one set of precipitates 'looked promising'.
- The most promising quantitative measurement is the method of dilution curves (see Chapter 10 and references 6 and 32–35). The co-operative three-dimensional interactions that stabilize crystals (21) imply that very small crystallites should have a much more protein concentration-dependent stability than do small precipitates. This protein concentration dependence of aggregate size in solution provides the basis for the dilution curve assay for pre-crystalline nuclei. Any convenient method can be used for determining the aggregate size, although quasi-elastic laser light scattering appears to be a method of choice (32–35). Although these methods have not been combined systematically with statistically designed experiments, the degree of co-operativity of the dilution curve can be estimated quantitatively (32), nevertheless, and used to score crystal growth experiments.

4. Practical analysis of factorial experiments

The main justification for statistical analysis of Q values from incomplete factorial experiments is that even semi-quantitative evaluations more often than not prove to be informative and useful. This means that, given improvements in quantification methods, it should work even better. The following guidelines are offered in this spirit. Two different kinds of analysis can be done:

(a) Estimation of the main effect and interaction contrasts.

(b) Construction and testing of linear models for Q.

A close connection exists between these two approaches as the first involves the analysis of variance while the second involves multiple regression (36). Full descriptions of how these two procedures should be applied are beyond the scope of this article. However, most of the procedures necessary for rudimentary analyses are available in many of the standard statistical packages for personal computers (37). An engineering-friendly program suite explicitly for the analysis of such experiments (13), including setting up incomplete factorial designs, is also available from The Product Integrity Company, Enfield, CT 06082.

4.1 Estimation of contrasts

The idea of contrasts, introduced in Section 2, is that averages of specific, balanced sets of experiments, treated at different levels of a particular factor, can be analysed for information about whether or not that factor is a *significant source of the variation* observed in the results. The null hypothesis is that these averages, the treatment means, are equal within the error of the measurements. As the phrase in italics suggests, the simple contrast sum calculations presented in Section 2 can be misleading, and should be verified by a full analysis of variance whenever possible. Generally, this is easiest for complete factorial experiments. Incomplete factorial experiments do not lend themselves readily to the analysis of variance.

4.2 Multiple regression analysis

The most robust tool for analysis of Q-values from incomplete factorial designs is multiple regression analysis (37–40), in which a linear model for Q is first developed, of the form:

$$Q = \text{constant} + \Sigma\beta_i \times F_i. \quad [1]$$

The factors, F_i, are assumed to contribute in a linear fashion to the experimental outcome, Q. The β_i coefficients represent magnitudes of the individual dependencies of Q on F_i. As with most mathematical models, this has the virtue of providing a simple means of comparing the values calculated on the basis of the model with those actually observed.

In order to utilize this model decide first which subset of the factors are to be included in the model. This can be done in one of several ways, including examination of the correlation matrix and stepwise multiple regression. In a separate step, estimate the coefficients by minimizing the sum of the squared differences between observed and predicted values of Q. A good statistical package both for the preliminary analysis used in generating the models and for the estimation of the β_i is called SYSTAT (37). In the following discussion data from *Table 4* are used as an example of how to proceed.

4.2.1 Prepare the data

If the design includes factors whose levels are attributes, prepare the data for analysis by translating the original design parameters into a form describing each experiment by a row of numbers, one for each component in the design. Let us suppose that column A in *Table 4* represented the choice between ammonium sulphate and potassium phosphate, as suggested in Section 3.1.2.

In this case column A was replaced in *Table 8* by two new columns, A1 and A2, representing the presence or absence of the two crystallizing agents. Additional columns can also be introduced, if desired, to represent other aspects of the design, such as the presence of divalent cations or ligands of any kind. *Table 8* suggests the analysis to be performed. Numerical values of 0 or 1 for each level allow prediction of Q values in terms of the entries in *Table 8* according to the linear model (*Equation 1*). Multiple regression provides β_i values that minimize the sum of squared differences

$$\sum_{i=1}^{8} (Q_{obs,i} - Q_{pred,i})^2.$$

Table 8. Reformulation of columns representing factors with attributes (column A1 and A2 derive from column A in *Table 4*) for multiple regression

Experiment	A1	A2	B	C	D	E	F	G	Score
1	0	1	0	1	1	1	0	1	4
2	1	0	0	0	0	0	0	0	5
3	0	1	1	0	1	0	1	0	7
4	1	0	1	1	0	1	1	1	9
5	1	0	1	1	0	0	1	0	10
6	0	1	0	0	1	1	1	1	6
7	0	1	1	0	0	1	0	0	8
8	1	0	0	1	1	0	0	1	6

4.2.2 Choose a regression model

In general, not all factors contribute to Q. The significant contributors can be selected initially by analysis of the matrix of correlation coefficients for all variables including Q; or, independently, by stepwise multiple regression. A good discussion of how regression models are chosen can be found in Chapter 11 of reference 40. The correlation matrix is simply the matrix of correlation coefficients for each pair of columns. In this case the correlations between Q and each of the independent variables are the most useful to examine, although it is also important to look for any two columns that are highly correlated with each other. In cases where factor–factor correlations are high it may be useful to choose a more general factor to represent together all those which are highly correlated.

Stepwise multiple regression is a process of including the most important factors one-by-one from the total list of factors for a sequence of models. It is not a reliable procedure for statistical estimation (37), but is useful for sorting the factors into those which should and those which should not enter a regression model. There is a stringency parameter in the stepwise process that relaxes or tightens the criteria for accepting new factors in the model, and by varying this value one can generate a larger or smaller number of factors in a descending order of their importance to the overall prediction of Q.

Although the choices and the criteria for validity are subjective, useful regression models have resulted from including factors whose correlation coefficients with the Q column are > 0.15–0.20, and for stepwise multiple regression models that produced 3–6 factors. For the data in *Table 8* stepwise multiple regression indicated that the strongest predictors of Q were factors A2, B, and F.

4.2.3 Estimate the β_i-coefficients for main effects.

A linear model Q = constant + (β1 × A2) + (β2 × B) + (β3 × F) formed the basis for a multiple regression estimation of the regression coefficients of the independent variables of the model in standardized form. This means that they are all transformed by a scaling operation to a state where their mean value is equal to 0 and their standard deviation are both equal to 1. The β-coefficients were then evaluated by conventional multiple regression analysis. Significant main effects are indicated by the largest coefficients, positive coefficients indicating that the factor promotes crystal growth and negative coefficients that the factor impairs crystal growth. They are conveniently represented as a histogram in *Figure 5*. Note that the indications from this 'factor profile' are only moderately consistent with the contrast sums in *Table 4*. Both indicate that factor B is the strongest positive contributor, and they agree qualitatively with respect to the nature of factors A and F. However, the strength of the A main effect is not evident from the contrast sums, nor is it obvious from these sums that the factor D is less significant a negative factor than factor F is a positive factor. Analysis of variance for the linear model involving A2, B, and F shows that the F-ratio for the regression is 9.4, which in this case is significant at the 0.025 level.

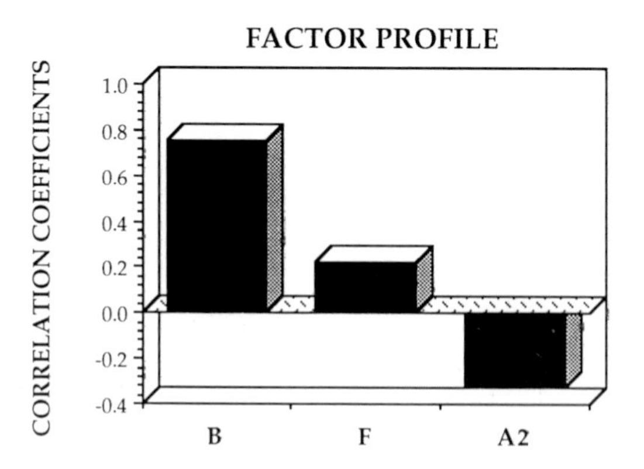

Figure 5. Factor profile derived from multiple regression analysis of the scores in *Table 8*, as modified from *Table 4*. The regression model, Q = constant + (β1 × A2) + (β2 × B) + (β3 × F), was obtained from a stepwise multiple regression using a model Q = constant + (β1 × A1) + (β2 × A2) + (β3 × B) + (β4 × C) + (β5 × D) + (β6 × E) + (β6 × F) + (β7 × G) with an α value of 0.5 for entry or removal of factors (reference 37).

5. Iterative use of designed experiments

With an experimental design covering so broad a range of conditions as those suggested here, it is surprising how many times one or more experiments produces crystals that can be used immediately for diffraction work. This most frequent outcome underscores the efficiency of the designs. Nevertheless, the points sampled by a design are not optimized, so supplementary experiments should be done to find out where the best crystals can be grown; what causes polymorphisms, what influences crystal habit, size, and morphology, and so on. Re-investigation of the quantitative optima for factors and/or combinations of factors identified in an initial screening experiment should be done using the same kinds of strategies used in the initial screening experiments. Full factorial design can characterize more fully how different factors interact (10). One-factor-at-a-time experiments may also have a role, if it is evident that one factor exerts a dominant influence on crystal growth. This iteration certainly benefits from quantitative analysis of the initial screening experiments as outlined in Section 4.

The first round of screening tests for factors, such as pH and ionic composition, that influence the thermodynamics of interactions between molecules. If a solution is headed toward precipitation there is little one can do to change the outcome by varying the level of supersaturation during nucleation and subsequent growth. On the other hand, if one already knows that a set of solution conditions will promote three-dimensional interactions, then it makes sense to investigate that set of conditions carefully, with particular emphasis on controlling nucleation and growth phases. An important example is titration of supersaturation by changing protein concentration, which will help to optimize crystal size (5, 8, 41).

References

1. Fisher, R. A., (1949). *The design of experiments*, (3rd edn). Olivier and Boyd, London.
2. Hendrix, C. D. (1979). *Chemtech*, **9**, 167.
3. Box, G. E., Hunter, W. G., and Hunter, J. S. (1978). *Statistics for experimenters*. Wiley Interscience, New York.
4. Carter, C. W., Jr. and Carter, C. W. (1979). *J. Biol. Chem.*, **254**, 12219.
5. Betts, L., Frick, L., Wolfenden, R., and Carter, C. W., Jr. (1989). *J. Biol. Chem.*, **264**, 6737.
6. Carter, C. W., Jr., Baldwin, E. T., and Frick, L. (1988). *J. Crystal Growth*, **90**, 60.
7. Carter, C. W., Jr., Green, D. C., Toomim, C. S., and Betts, L. (1985). *Anal. Biochem.*, **151**, 515.

8. Bell, J. B., Jones, M. E., and Carter, C. W., Jr. (1991). *Proteins, Structure, Function and Genetics*, **9**, 143.
9. Lewit-Bentley, A., Doublié, S., Fourme, R., and Bodo, G. (1989). *J. Mol. Biol.*, **210**, 875.
10. Carter, C. W., Jr. (1990). *Methods, A Companion to Methods in Enzymology*, **1**, 12.
11. Abergel, C., Loret, E., Rochat, H., and Fontecilla-Camps, J. C. (1990). *J. Mol. Biol.*, **214**, 637.
12. Gilly, M. and Pellegrini, M. (1985). *Biochemistry*, **24**, 5781.
13. Carter, C. W. and Brockway, G. W. (1990). *Quality*, **29**, 42.
14. Plackett, R. L. and Burman, J. P. (1946). *Biometrika*, **33**, 305.
15. Gilliland, G. (1988). *J. Crystal Growth*, **90**, 51.
16. Gilliland, G. and Bickham, D. M. (1990). *Methods, A Companion to Methods in Enzymology*, **1**, 6.
17. Roussell, A., Serre, L., Frey, M., and Fontecilla-Camps, J. C. (1990). *J. Crystal Growth*, **106**, 405.
18. Jancarik, J. and Kim, S. H. (1991). *J. Appl. Cryst.*, **24**, 409.
19. Wozniak, J. A., Faber, H. R., Dao-pin, S., Zhang, X-J., and Matthews, B. W. (1990). *Methods, A Companion to Methods in Enzymology*, **1**, 100.
20. Roth, M., Lewit-Bentley, A., Michel, H., Deisenhofer, J., Huber, R., and Oesterhelt, D. (1989). *Nature*, **340**, 659.
21. Salemme, F. R., Genieser, L., Finzel, B. C., Hilmer, R. M., and Wendoloski, J. J. (1988). *J. Crystal Growth*, **90**, 273.
22. Wang, A. H.-J. and Teng, M.-K. (1988). *J. Crystal Growth*, **90**, 295.
23. Ducruix, A. and Ries-Kautt, M. (1990). *Methods, A Companion to Methods in Enzymology*, **1**, 25.
24. Weber, P. C., (1990). *Methods, A Companion to Methods in Enzymology*, **1**, 31.
25. McPherson, A., (1982). In *Preparation and analysis of protein crystals*, p. 82. John Wiley and Sons, New York.
26. McPherson, A., (1985). *Methods in Enzymology*, **114**, 120.
27. Garber, M. B., Yaremchuck, A. D., Tukalo, M. A., Egorova, S. P., Berthet-Colominas, C., and Leberman, R. (1990). *J. Mol. Biol.*, **213**, 631.
28. Boistelle, R. and Astier, J. P. (1988). *J. Crystal Growth*, **90**, 14.
29. Rosenberger, F. (1986). *J. Crystal Growth*, **76**, 618.
30. Baldwin, E. T. (1990). Crystallization and X-ray diffraction analysis of *Lactobacillus plantarum* Manganese Catalase. Ph. D. Thesis Dissertation, University of North Carolina at Chapel Hill.
31. INFAC. correspondence to C. W. Carter, Jr., Department of Biochemistry and Biophysics, CB # 7260, University of North Carolina at Chapel Hill, Chapel Hill, NC 27599-7260.
32. Kam, Z., Shore, H. B., and Feher, G. (1978). *J. Mol. Biol.*, **123**, 539.
33. Wilson, W. W. (1990). *Methods, A Companion to Methods in Enzymology*, **1**, 110.
34. Kadima, W., McPherson, A., Dunn, M. F., and Jurnak, F. A. (1990). *Biophysical J.* **57**, 125.
35. Mikol, V., Hirsch, E., and Giegé, R. (1990). *J. Mol. Biol.*, **213**, 187.

36. Edwards, A. L. (1979). *Multiple regression and the analysis of variance*. W. H. Freeman and Co., San Francisco.
37. Wilkinson, L. (1987). SYSTAT, The system for statistics SYSTAT, Inc., 2902 Central ST., Evanston, IL 60601.
38. Morrison, D. F., (1990). *Multivariate statistical methods*, (3rd edn). McGraw Hill, New York.
39. Mardia, K. V., Kent, J. T., and Bibby, J. M. (1979). *Multivariate analysis*. Academic Press, London.
40. Neter, J. and Wasserman, W. (1974). *Linear statistical models: regression analysis of variance and experimental designs*, Richard D. Irwin, Inc. Homewood, IL.
41. Ataka, M. and Tanaka, S. (1986). *Biopolymers*, **25**, 337.

4

Methods of crystallization

A. DUCRUIX and R. GIEGÉ

1. Introduction

There are many methods to crystallize biological macromolecules (for recent reviews see references 1–3), all of which aim at bringing the solution of biological macromolecules to a supersaturation state (see Chapters 9 and 10). Although vapour phase equilibrium and dialysis techniques are the two methods most favoured by crystallographers and biochemists, batch and interface diffusion methods will also be described.

Many parameters influence nucleation and crystal growth of biological macromolecules (see *Table 1* of Chapter 1). Crystal growth and nucleation will in addition be affected by the method used. Thus it may be wise to try different methods, keeping in mind that protocols should be adapted (see Chapter 3). As solubility is dependent on temperature (it could increase or decrease depending on the protein), it is strongly recommended to work at constant temperature (unless temperature variation is part of the experiment), using commercially thermoregulated incubators. Refrigerators can be used, but if the door is often open, temperature will vary, impeding reproducibility. Also, vibrations due to the refrigerating compressor can interfere with crystal growth. This drawback can be overcome by dissociating the refrigerator from the compressor. In this chapter, crystallization will be described and correlated with solubility diagrams as described in Chapter 9.

Observation is an important step during a crystallization experiment. If you have a large number of samples to examine then this will be time consuming, and a zoom lens would be an asset. The use of a binocular generally means the presence of a lamp; use of a cold lamp avoids warming the crystals (which could dissolve them). If crystals are made at 4 °C and observation is made at room temperature, observation time should be minimized.

2. Sample preparation

2.1 Solutions of chemicals

2.1.1 Common rules

Preparation of the solutions of all chemicals used for the crystallization of biological macromolecules should follow some common rules:

- when possible, use a hood (such as laminar flux hood) to avoid dust;
- all chemicals must be of purest chemical grade (ACS grade);
- stock solutions are prepared as concentrated as possible with double distilled water. Solubility of most chemicals are given in Merck Index. Filter solutions with 0.22 μm minifilter. If you use a syringe, do not press too hard as it will enlarge the pores of the filter. Filters of 0.4 μm will retain large particles whereas 0.22 μm filters are supposed to sterilize the solution. Label all solutions (concentration, date of preparation, initials) and store at 4 °C. Characterize them by refractive index from standard calibrated solutions. Use molar units (mole per litre) in preference to percentage. This avoids confusion between weight to weight (w/w), weight to volume (w/v), and volume to volume (v/v). Quite often crystallization articles refer to percentage without any information, making the results difficult to reproduce. As an example, a 20% (w/v) stock solution twice diluted will give a 10% solution whereas this would not be the case if starting from a 20% (w/w) solution.

2.1.2 Buffer

The chemical nature of the buffer is an important parameter for protein crystal growth. It must be kept in mind that the pH of buffers is often temperature dependent (see Section 3.3 in Chapter 2); this is particularly significant for Tris buffers. Buffers, which must be used within one unit from their pK value, are well described in standard text books (4).

2.1.3 Purification of PEG

PEG is available in a variety of polymeric ranges; the most used are compounds with mean molecular weights of 2000, 4000, and 6000. These are polydisperse mixtures and their composition around the mean may vary from one producer to the other; it is better to always use the same brand. Molecular weights higher than 10 000 are rarely used because of excessive viscosity of their solutions. Reproducibility and quality of crystals may depend on PEG molecular weight.

The optimal range of PEG concentration for crystallization of a given protein depends on PEG molecular weight and it may be very narrow (about 1.5% w/v). As the viscosity of PEG solutions may lead to pipetting errors, it is better to routinely verify the concentration of PEG in reservoirs for dialysis or vapour diffusion by measuring the refractive index with an Abbe refractometer. A reference curve is established on known amounts of PEG dissolved in the same buffer. From a practical point of view, commercial PEG does contain contaminants, either ionic (5) or derived from peroxidation. Repurification as shown in *Protocol 1* is strongly recommended before use (6).

Protocol 1. Purification of PEG[a]

1. Pour a column (2.5 × 10 cm^2) with a mixed bed strong ion-exchange resin in the H^+–OH^- form (e.g. Biorad AG501X8). Wash with 300 ml methanol/water 3:7 (v/v) then with 500 ml water.
2. Dissolve 200 g PEG in water (500 ml final volume). Measure the refractive index of the solution. Degase 30 min under vacuum (water aspirator) with gentle magnetic stirring. Add 1.24 g $Na_2S_2O_4,5H_2O$ and let stand for 1 hour.
3. Pass the solution through the column at a flow rate of 1 ml/min. Discard the first 30 ml and collect the following eluate.
4. Check the concentration of PEG by refractometry and store frozen by small aliquots at −20 °C.
5. Before use, an antioxidant can be added (para-hydroxyanisole, stock solution in isopropanol, 1.3 mg/ml; add 1 µl per ml PEG stock solution).

[a] See *Table 2* in Chapter 2 for another method.

2.1.4 Mother liquor

Throughout this paragraph, mother liquor is defined as the solution containing all crystallization chemicals (buffer, salt, crystallizing agent, and so on) except protein or nucleic acid at the final concentration of crystallization.

2.2 Preparing samples of biological macromolecules

2.2.1 Removing salts

Proteins and nucleic acids often contain large amount of salts of unknown composition when first obtained. Thus it is wise to dialyse a new batch of a biological macromolecule against a large volume of well-characterized buffer of given pH, to remove unwanted salts and to adjust the pH. Starting from known conditions helps to ensure reproducibility.

Commercially available dialysis tubings are generally composed of cellulose or polyacetate. They should be prepared as described in *Protocol 2*. Molecular cut-off (i.e. the pore size limit) is given by manufacturers for spherical particles. As most of proteins and nucleic acids are better described as ellipsoids or cylinders, a cut-off far enough from the molecular weight should be chosen. As an example, a 12 000 cut-off is not appropriate for lysozyme (M_r 14 600) and if used will allow the protein to leak slowly through the membrane, thus diluting it in the external chamber. You can check the impermeability of the membrane towards the protein by placing a dialysis tube containing the biological macromolecule at a given concentration in a beaker (outer reservoir) containing a small volume (a few ml). After 24 hours, check the biological macromolecule concentration in the outer reservoir. In practice, it should not contain more than 1% of the amount of the biological macromolecule.

Depending on the choice of the commercial membrane, it should be either prepared and demetallized (following *Protocol 2*) or washed with distilled water. The membrane must be kept cold. If kept for a long period (even in the form of dry tubes) contamination problems may arise leading to leaks of protein. Dialysis membranes are fragile and it is quite easy to puncture them with nails, so do wear plastic gloves, remembering to rinse them because they are often treated with talc.

Protocol 2. Preparation of dialysis tubing

You will need a solution containing $NaHCO_3$ 5% (w/v), 50 g/l and 50 mM EDTA (18.6 g/l) (solution A) and a bunsen burner.

1. Boil the tubing for 30 min in solution A. Avoid puncture at this stage when mixing with glass rods or magnetic stirrers.
2. Rinse several times with distilled water.
3. Store in 50% (v/v) ethanol solution.
4. Check each tubing integrity for possible puncture.
5. Prior to crystallization, rinse membranes several times with distilled water and then with buffer.

2.2.2 Concentration

Whatever the crystallization method used, it requires high concentrations of biological macromolecules as compared to normal biochemistry conditions. Before starting a crystallization experiment, a concentration step is generally needed. Keep pH and ionic strength at desired values, since pH may vary when the concentration of the biological macromolecule increases. Also, low ionic strength could lead to early precipitation (see Section 4.2 in Chapter 2). It could be very frustrating when the biological macromolecule precipitates irreversibly or adsorbs on to the concentration apparatus membrane and/or support. Many commercial devices are available; they are based on different principles and operate:

(a) under nitrogen pressure

(b) by centrifugation

(c) by lyophilization (because it may denature some proteins, test first on a small amount). Non volatile salts are also lyophylized and will accumulate.

Choice of the method of concentration depends on the quantity of biological macromolecule available. Dialysis against high molecular PEG proved to be successful in our hands. We use a dialysis chamber (volume 50–500 μl), the top of which is covered by a glass cover slip which is sealed to the plastic chamber with grease. *Figure 1* describes the apparatus. This allows for an easy access from the top of the dialysis chamber.

2.2.3 Removing solid particles

Before a crystallization experiment, solid particles such as dust, denatured proteins, and solids coming from purification columns (beads) or lyophylization should be removed. This could be achieved by centrifugation or filtration, depending on the available quantity.

2.2.4 Measuring protein concentration

The most common method to measure biological macromolecule concentration is to sample an aliquot, dilute it with buffer, and measure absorbance at 280 nm or 260 nm (for proteins or nucleic acids, respectively) within the linear range of a spectrophotometer. Proper substraction with the reference cell should be made especially when working with additives absorbing in the 260–300 nm wavelength range. When working with enzymes, an alternative method to measure the concentration of protein is to perform activity tests, otherwise colourimetric methods can be used. This can be done either by a modification (7) of the assay described by Winterbourne (8) or by the modification (9) of the reagent assay developed by Bradford (10). See also Section 4.1 in Chapter 2 and Section 2.3.4 in Chapter 7.

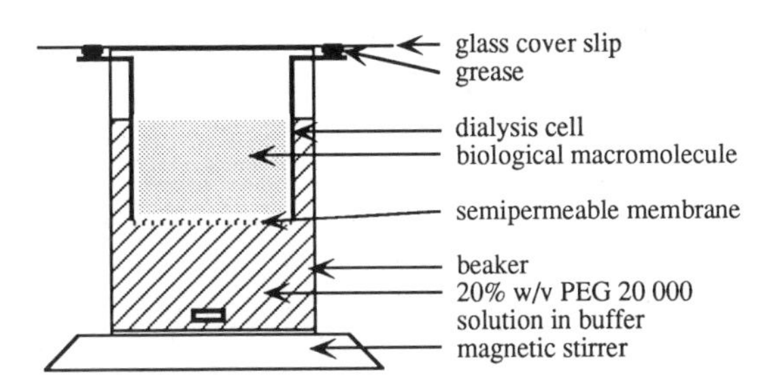

Figure 1. Dialysis apparatus used for the concentration of biological macromolecule. Prepare a solution of 20% (w/v) of PEG 20 000 in an appropriate buffer and against it dialyse your biological macromolecule. Check the biological macromolecule concentration using optical absorbance, or colorimetric or enzymatic assay on a small aliquot.

3. Crystallization by dialysis methods

3.1 Principle

This method allows to an easy variation of the different parameters which influence the crystallization of biological macromolecules. Different types of dialysis cells are used but all follow the same principle. The biological macromolecule is separated from a large volume of solvent by a semi-permeable membrane which gives small molecules (ions, additives, buffer, and so on) free passage but prevent biological macromolecules to circulate. The kinetics of equilibrium will depend on the membrane cut-off, the ratio of the concentration of crystallizing agent inside and outside the protein chamber, the temperature, and the geometry of the cell.

3.2 Examples of dialysis cells

3.2.1 Macrodialysis

The most simple technique is to use a dialysis bag. Large crystals are occasionally obtained. It is very convenient for successive recrystallization. Commercially available dialysis tube such as Spectrapor of inner diameter 2 mm can be used to limit the amount of protein. However, each assay requires about 100 μl at least per sample.

3.2.2 Microdialysis

i. Zeppenzauer cells

Crystallization by dialysis was first adapted to microvolumes by Zeppenzauer (11). The microdialysis cells are made from capillary tubes closed either by dialysis membranes or polyacrylamide gel plugs. Those cells require only 10 μl or less of macromolecule solution per assay. A modified version of the Zeppenzauer cell was described by Weber and Goodkin (12).

ii. Dialysis buttons

Commercially available (such as from Cambridge Repetition Engineers Ltd), microdialysis cells are made of transparent perspex (*Figure 2*). They can be accommodated at your local workshop. The protein chamber should be filled so that it forms a dome. The membrane is then placed over the button and hold by an O-ring of appropriate diameter. Installing the membrane has a reputation of difficulty, and beginners often trap air bubbles between the protein solution and the membrane. To avoid the problem, one can use either a piece of plastic (*Figure 3a*) having the same diameter as the button, or when working with flat buttons use a plastic cork. A more sophisticated apparatus is described in *Figure 3b* (courtesy of M. Pontillon). The cell is then immersed in a vial, and an inexpensive way is to use transparent scintillation counting vials.

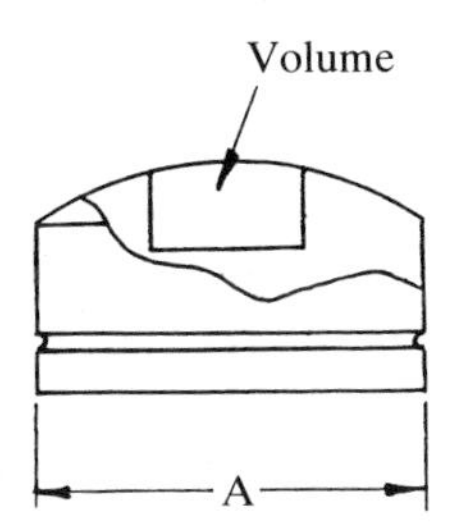

Figure 2. Dialysis button. Diameter (A) generally varies between 10–20 mm and the volume of the biological macromolecule chamber is 5–350 μl.

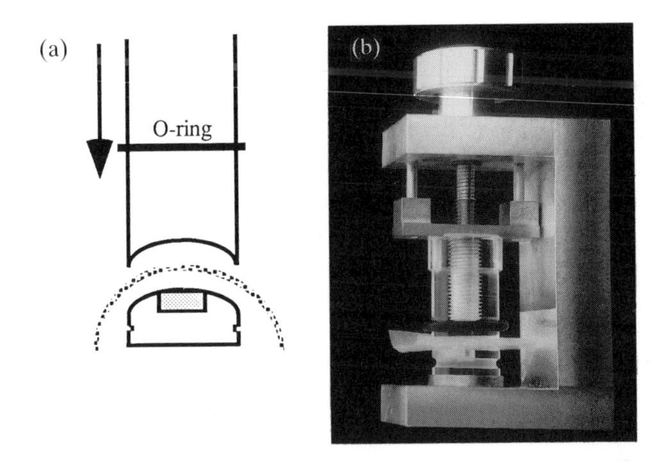

Figure 3. How to install a dialysis membrane. (a) Use a cylinder with a convex extremity adapted to the concave shape of the dialysis button. (b) For flat buttons, use a cork or a press as shown.

Observation with a binocular or microscope through the membrane is easy. However, if you use cross-polarizer, the membrane will depolarize light. For crystal mounting O-ring and membrane should be removed gently. Problems occur when crystals stick to the wall of the chamber. In this case a whisker can be used to gently free the crystal.

iii. Microcaps dialysis

The technique, described in *Figure 4* and adapted from (13), was useful for membrane proteins (see Chapter 8). Although it is more difficult to observe crystal growth with this method it is very convenient for storing, and the microcaps are disposable. The method is quite easy to use (*Protocol 3*) and you can play with the ratio of the diameter versus height of the microcap to

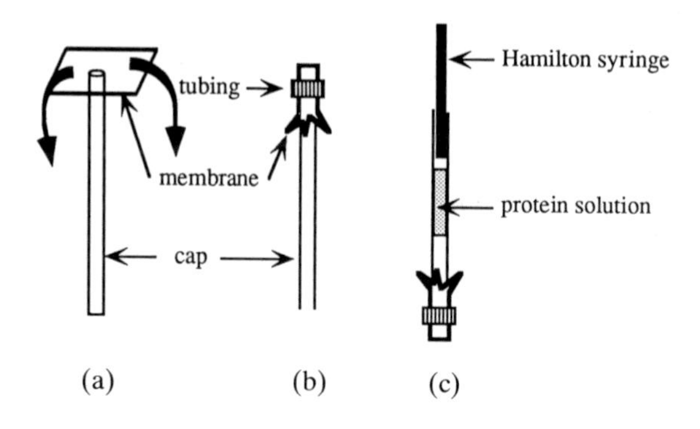

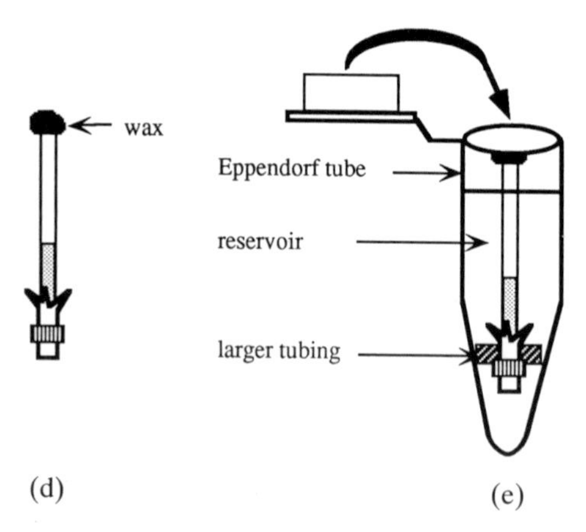

Figure 4. Crystallization by microcap dialysis. (a) Place the dialysis membrane on the microcap; (b) secure with tygon ring; (c) load the protein; (d) close the extremity with wax; (e) fill up a 1.5 ml Eppendorf tube with crystallizing agent and insert microcap.

influence the kinetics of crystallization. It should be noted that when the biological macromolecule is not entirely filling the microcap chamber, the presence of air (which is compressible) allows osmotic pressure to develop, and thus modifies the biological macromolecule concentration.

Protocol 3. Crystallization by microcaps dialysis

You will need the following parts: microcaps, Hamilton syringe, tygon tubing of 1.3 and 3 mm diameter, 1.5 ml Eppendorf tubes, low melting wax, dialysis membrane cut in small squares, and a low temperature soldering bit.

1. Commercial microcaps are cut with a glass saw (for instance a 50 μl cap is cut in three parts to fit in an Eppendorf tube of 1.5 ml).

Protocol 3. *Continued*

2. Wrap a piece of dialysis membrane around one end (the one which is smooth) and secure with a piece of tubing of diameter 1.3 mm.
3. Load biological macromolecule with a Hamilton syringe.
4. Shake the assembly to bring the biological macromolecule solution in contact with the membrane.
5. Seal the free microcap end with wax molten by soldering bit.
6. Split a second ring of tubing of diameter 3 mm and superpose it to the first one. The aim is to prevent the membrane from touching the bottom of the Eppendorf tube which would limit the exchange with the reservoir.
7. Insert in an Eppendorf tube (volume 1.5 ml) containing 1 ml of the crystallizing solution.
8. Close cap and wrap top of Eppendorf tube with parafilm (American Can Company).

3.2.3 Double dialysis

The purpose of double dialysis (14) is to reduce the rate of equilibration and therefore to provide a better control of crystal growth. Apparatus is shown in *Figure 5*. Large crystals of delta toxin of *Staphylococcus aureus* were obtained (14) this way. Equilibration time is rather long (could be several weeks) as the gradient concentration of crystallizing agent is low. It is therefore more geared toward production of large crystals than screening. *Protocol 4* describes the methodology. One can manipulate the different parameters (membrane cut-off, distance between dialysis membranes, relative volumes, and so on) to optimize crystallization.

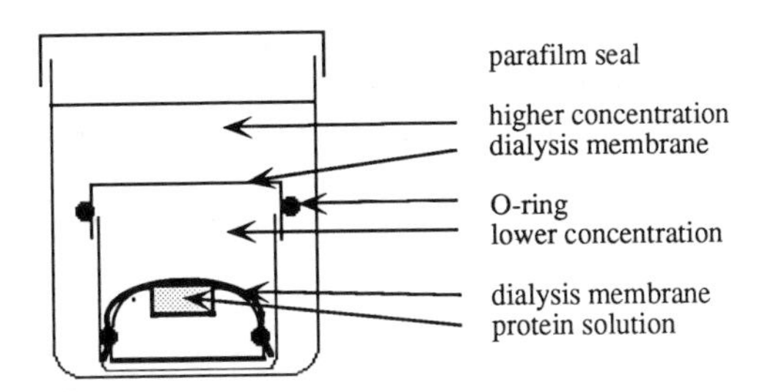

Figure 5. Double dialysis set-up (adapted from reference 14). Macromolecule is contained in a conventional dialysis button placed in a second dialysis set-up. The equilibration rate depends upon the volumes of buffers in the different compartments.

Protocol 4. Crystallization by double dialysis

1. Prepare dialysis button as in Section 3.2.2 with a solution of crystallizing agent at a concentration in which the biological macromolecule is undersaturated. This is called 'inner compartment'.
2. Insert the conventional dialysis button in a vial (about 10 ml) called 'middle compartment' containing solution of crystallizing agent at a concentration in which the biological macromolecule is supersaturated.
3. Cover with a dialysis membrane maintained by an O-ring.
4. Place in a larger vial (for instance a beaker of 50 ml) containing solution of crystallizing agent at a concentration in which the biological macromolecule will precipitate completely. This is the 'outer compartment'.
5. Cover with parafilm or a stopper.

4. Crystallization by vapour diffusion methods

Among the crystallization micromethods, vapour diffusion techniques are probably the most widely used throughout the world. They were first used for the crystallization of tRNA (15) and are probably the most widely used crystallization methods throughout the world.

4.1 Principle

The principle of vapour diffusion crystallization is indicated in *Figure 6*. It is very well suited for small volumes (down to 2 µl or less). A droplet containing the biological macromolecule to crystallize with buffer, crystallizing agent, and additives, is equilibrated against a reservoir containing a solution of crystallizing agent at a higher concentration than the droplet. Equilibration proceeds by diffusion of the volatile species (water or organic solvent) until

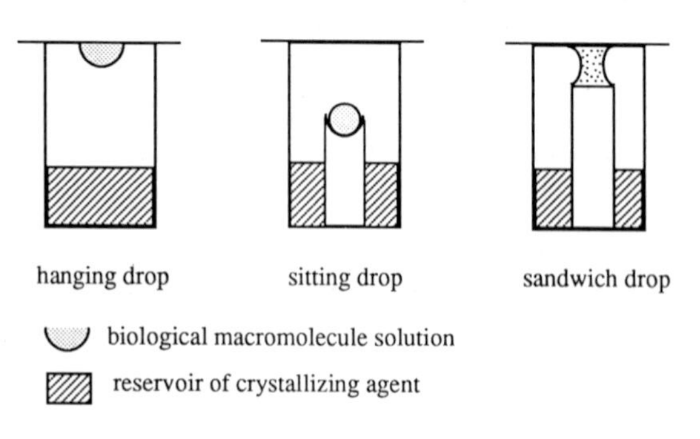

Figure 6. Schematic representation of hanging drop, sitting drop, and sandwich drop.

vapour pressure in the droplet equals the one of the reservoir. If equilibration occurs by water exchange (from the drop to the reservoir), it leads to a droplet volume change. Consequently, the concentration of all constituents in the drop will change. For species with a vapour pressure higher than water, the exchange occurs from reservoir to drop. The same principle applies for hanging drops, sitting drops, and sandwich drops.

Glass vessels in contact with biological macromolecule solution should be treated in a way to obtain an hydrophobic surface. It is done following *Protocol 5*.

Protocol 5. Preparation of glass cover slips[a]

A. **Silanization:** 10 min in a bath of toluene containing 1% of dimethyl-dichlorosilane at 60 °C for 10 min. Cover slips are then washed with soap solution and rinsed with distilled water and ethanol. The same procedure is used for pyrex plate. Dry overnight at 120 °C to sterilize vessels.

B. **Siliconization:** can be achieved with commercially available reagent solutions (e.g. Sigmacoat). Cover slips are washed in the solution, and dried overnight at 120 °C.

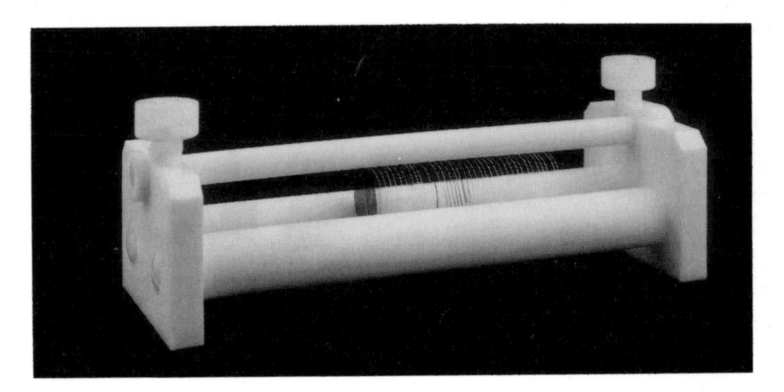

Figure 7. Device for treating cover slips (perspex or teflon made). The set-up displayed (in Teflon) can be manufactured in the laboratory workshop. Cover slips are held in the threading of the two bottom axis; a smaller unthreaded axis secures the coverslips. The set-up displayed is about 20 cm long and permits handling of about 60 coverslips.

A device shown in *Figure 7* helps to treat cover slips. If made in perspex, the device is only used for drying cover slips; if Teflon made, it can be used for the all silanization process.

[a] All operations can be performed under vacuum when dealing with narrow vessels like capillaries.

4.2 Experimental set-ups

4.2.1 Hanging drops in Linbro boxes

Commercially available Linbro boxes are plastic boxes (*Figure 8*) normally used for tissue culture. Plastic boxes will depolarize light; so unless for a particular orientation of a crystal, no birefringence will be observed. Boxes contain 24 wells labelled A, B, C, D, vertically, and 1–6 horizontally. It is convenient, to avoid confusion, to use them always in the same orientation. Each box must be carefully labelled (date, experiment number, operator, and so on). Each well has a volume of approximately 2 ml and an inner diameter of 16 mm. There is a small rim which will be used for sealing the system. Each well will be covered by a glass cover slip of 22 mm diameter treated as described in *Protocol 5*.

Drops are set-up following *Protocol 6*. Most of the people use a 'magic' ratio of 2 between the concentration of the crystallizing agent in the reservoir (well) and in the droplet. This is conveniently achieved by mixing a droplet of protein at twice the desired final concentration with an equal volume of the reservoir at the proper concentration. To avoid local overconcentration, it can be achieved by placing the two drops (protein and reservoir) on each side of an Eppendorf tube and vortex it quickly.

Figure 8. Linbro box for vapour diffusion crystallization in hanging drops. The photograph shows the box with its cover which is held by Plasticine in the corners. For a better display, two drops were prepared with dyes (at the left).

Protocol 6. Crystallization in Linbro boxes

1. Grease rims with silicon grease[a].
2. Fill-up reservoir with one ml of filtered (0.22 μm) crystallizing agent.
3. Spray glass cover slip with antidust.
4. Mix 2–10 μl drop of filtered (0.22 μm) biological macromolecule solution with an equivalent volume of reservoir.

Protocol 6. *Continued*

5. Layer the drop on the 22 mm diameter cover slip (do not touch the coverslip with the extremity of the tip of the pipetor or it will spread) so that a nearby hemispherical drop is formed[b].
6. Return cover slip with a pair of brussel (or fingers). First train yourself with water!
7. Set on the grease rim and gently press to seal the well with the grease. Do not press too firmly, otherwise the coverslip will break.
8. Check the sealing by inspecting the rim in an azimuthal way. If sealing is not properly done, the drop will concentrate as well as the reservoir. Crystallization will occur eventually, but will be very difficult to reproduce.
9. Adjust glass cover slips tangent to each other otherwise they overlap.
10. Put plasticene in the corners to avoid the contact between cover and grease, otherwise cover slips get stuck at the cover.

[a] To dispend grease, fill-up a syringe and replace needle by an Eppendorf yellow tip.
[a] You may layer several microdrops on a cover slip.

When no crystal or precipitate is observed, either supersaturation is not reached or one has reached the metastable region (see Chapter 9 for definition). In the latter case, changing the temperature by a few degrees is generally sufficient to initiate nucleation. For the former, the concentration of crystallizing agent in the reservoir must be increased. In the former case, gently rotate the cover slip in the plane of the rim to ease the grease (which becomes 'stiff' with time), then lift it, suck the reservoir entirely, and replace it by a more concentrated solution. More grease is added and the coverslip sealed. The volume of the drop will decrease again and all constituants concentrations in the drop increase.

Problems:
For membrane proteins (see Chapter 8), the presence of detergent tends to spread out the drops and lower surface tension. In all cases gravity will tend to sink the drops containing the biological macromolecule in the reservoir for volumes exceeding 25 μl. Shaking Linbro boxes will give the same results. Boxes must be transported horizontally and carefully. It is always painful for beginners to ruin an experiment when transporting a box. If you prepare boxes at room temperature and transfer them to 4 °C, condensation will occur on the surface of the cover slip. Water droplet will surround the biological macromolecule drop. If it mixes, protein will dilute and probably stay in undersaturated state. To avoid this problem, set up boxes at final experiment temperature and cover boxes with polystyrene sheets.

It is possible to recrystallize macromolecules using hanging drops as described in *Protocol 7*.

Protocol 7. Recrystallization of macromolecules

A. For purification:

1. Remove mother liquor.

2. Wash crystals with fresh buffer.

3. Redissolve in fresh crystallizing agent solution.

4. Recrystallize.

B. For crystal growth:

1. Redissolve crystals by replacing crystallizing agent in the reservoir by buffer. Depending on the biological macromolecule, it may take a few hours or a few days. Drop volume will increase so control the process, otherwise the drop will sink.

2. Filter drop on a minifilter (e.g. Costar, Millipore) by centrifugation. Warning: dead volume is at least 5 μl.

3. Place drop on a clean glass cover slip.

4. Add some grease to the rim.

5. Fill reservoir with crystallizing agent and recrystallize.

4.2.2 Crystallization with ACA CrystalPlates®

The American Crystallographic Association (ACA) sponsored a new vapour diffusion plate called CrystalPlate® (manufactured by ICN Flow). Although dedicated for use with automated systems (see Chapter 13), it is equally useful for manual crystallization. As shown by *Figure 9*, it may be used to setup crystallization by hanging drops, sitting drops, or sandwich drops, depending on the thickness of the lower glass slip and the volume of the drop. Each box contains 15 wells. The glass cover slips should be prepared as described in *Protocol 5*. If desired, the breakaway plug in the bottom of a reservoir—see tank in *Figure 9*—may be removed with pliers and a rubber septum inserted so that the reservoir solution may be changed with an hypodermic syringe.

Advantages:

Because of the glass windows, when looking at drops under a polarized binocular, birefringence of crystals can be observed. Drops containing biological macromolecule are no longer above the reservoir thus eliminating sinking. Large drops can be prepared with the sandwich method.

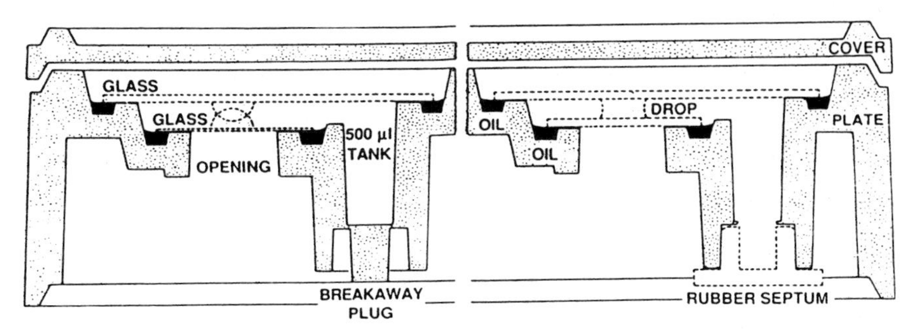

Figure 9. ACA CrystalPlate® (courtesy of ICN Flow). This is a new versatile system for vapour diffusion crystallization allowing individual experiments on sitting, hanging, or sandwiched drops under classical or automated conditions.

Warning:
If you use rubber septum, plates stick when you translate them during observation, eventually provoking disasters.

Protocol 8. Crystallization with ACA CrystalPlates®

1. Prepare the plate by filling up the upper and lower troughs of each well with ordinary hydrocarbon vacuum pump oil or grease.
2. To dispense oil or grease, fill-up a syringe and replace needle by an Eppendorf yellow tip.
3. Put 0.5 ml of crystallizing agent into each reservoir.
4. Position one of the 14×14 mm^2 glass slips over the hole in each well. The slips should seal quickly if there is enough oil or grease.
   - For hanging or sitting drops use $14 \times 14 \times 0.2$ mm^3 glass cover slips.
   - For sandwich drops (5–25 µl) use $14 \times 14 \times 1$ mm^3 glass cover slips.
   - For sandwich drops (15–75 µl) use $14 \times 14 \times 1.5$ mm^3 glass cover slips.
5. Put a drop of the biological macromolecule solution in the centre of the lower (for sitting or sandwich drops) or upper (for hanging drops) glass slip and then set one of the $24 \times 30 \times 1$ mm^3 glass cover slips in position on the upper trough.

4.2.3 'Sandwich' versus 'couturière' boxes

In *Figure 10*, a pyrex plate (Corning Glass 7220) with three or nine depressions is sitting on a plastic petri dish in a sandwich box (16). Drops are carefully formed in the depressions and reservoir is poured around the petri dish. An example of variation of the MIT system (16) was developed by

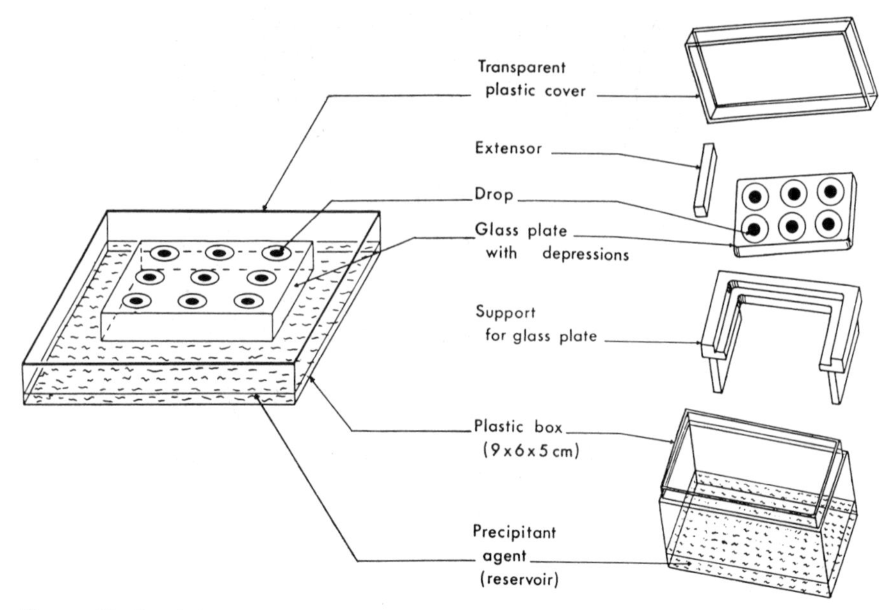

Figure 10. Sandwich box (left) and couturière box (right). These set-ups are used to conduct simultaneously several crystallization trials by vapour diffusion on sitting drops. *Caution*: If crystals are desired in each drop of the box, only faint modifications will be tolerated in the buffer content of the drops. Usually drops are duplicated and variations are brought by additives, the concentration of the precipitating agent remaining constant.

the Strasbourg group (17) as shown in *Figure 10* with sandwich boxes replaced by 'couturière' boxes.

4.2.4 Other systems

Cryschem MVD24 plates are commercially available (manufactured by Cryschem Inc. or C. Supper Company). Each well contains a plastic post in the centre to hold the protein sitting drop. The cup has been designed (*Figure 11*) to provide maximum surface area for free diffusion during equilibration. The reservoir solution is held within the narrow moat surrounding the support port. The plates are sealed with a special tape which is supplied by the manufacturer.

A variety of other set-ups have been designed in many laboratories, allowing, for instance, the use of Linbro boxes for sitting drop experiments (with the depressions on a small plastic bridge or on a glass rod, as on the Oxford or Perpetual Systems Corporation set-ups, respectively), or doing vapour phase equilibration in capillaries (18, 19), or even directly in X-ray capillaries as was described for ribosome crystallizations (20). For extremely fragile crystals, transferring from crystallization cells to X-ray capillaries (see Chapter 12) can lead to internal damage and mechanical cracks of crystals, so this last method may be best adapted.

Figure 11. Crystal growth multi-chamber plate for vapour diffusion crystallization on individual sitting drops (courtesy of C. Supper Company).

4.3 Varying parameters

Although not unique in this respect, vapour diffusion methods permit easy variations of physical parameters during crystallization, and many successes were obtained by modifying supersaturation by temperature or pH changes (21) (see also Section 5). With ammonium sulphate as the crystallizing agent, it has been shown that the pH in the droplets is imposed by that of the reservoir (22). Consequently, varying the pH of the reservoir permits gentle adjustment of that in the droplets (see also Chapter 10). From another point of view, sitting drops are well suited for attempting epitaxial growth of macromolecule crystals on appropriated mineral matrices (23).

In vapour diffusion crystallizations, the contamination by microorganisms can be prevented by placing a small grain of thymol, a volatile organic compound, in the reservoir. Thymol, however, can also have specific effects on crystallization, as shown with glucose isomerase (24), and may thus represent a useful additive to assay in crystallization screenings.

4.4 Kinetics of evaporation

4.4.1 Final concentrations

Calculating the final concentration of constituants in an equilibrated drop is often a source of misunderstanding in protocols. Sometimes it refers to final concentration in the drop before vapour diffusion process is initiated, sometimes it describes the final concentrations in the drop at equilibrium at the end of the process. So, except for species of vapour pressure higher than that of water, at equilibrium, if the ratio of crystallizing agent concentrations between reservoir and drop is 2, final concentrations and volumes are as follows:

- final drop volume = 1/2 initial volume
- final concentration of all constituants of the drop (protein, additive, and so on) equals twice the initial concentration.

Many other ratios can be used. Varying the volume of the droplet will influence the kinetics of crystallization and so the protein crystal size.

4.4.2 Equilibration kinetics

Using Linbro boxes, Mikol *et al.* (25) investigated water evaporation rates from hanging drops using ammonium sulphate, PEG, and MPD (see also Chapter 10). The kinetics of water evaporation will determine the kinetics of supersaturation and accordingly affect the nucleation rate. The main parameters which determine the rate of water equilibration are temperature, initial drop volume (and initial surface to volume ratio of the drop and its dilution with respect to the reservoir), water pressure of the reservoir, and chemical nature of the crystallizing agent. The drop to reservoir distance and the presence of biological macromolecule do not seem to affect the water evaporation rate.

From the practical point of view, the time for water equilibration to reach 90% completion can vary from about 25 hours to more than 25 days, the fastest equilibration occurring in the presence of ammonium sulphate, that in the presence of MPD being slower, while the equilibrations in the presence of PEG being by far the slowest (see *Figure 12*). Estimations of the minimal duration of equilibration under several standard experimental conditions can be obtained from an empirical model (25).

The particular behaviour of PEG may explain crystallization successes using this precipitating agent (26). Indeed crystal growth may be favoured when supersaturation is attained very slowly. This fact is corroborated by independent experiments in which the terminal crystal size was significantly increased by reducing the vapour pressure of the reservoir as a function of time (27, 28).

5. Crystallization by batch methods

The biological macromolecule to be crystallized is mixed with crystallizing agent at a biological macromolecule concentration such that supersaturation is instantaneously reached. This could be achieved with all methods previously described. For hanging drop or sitting drops, the reservoir no longer acts to concentrate the drop but is only present to maintain constant vapour

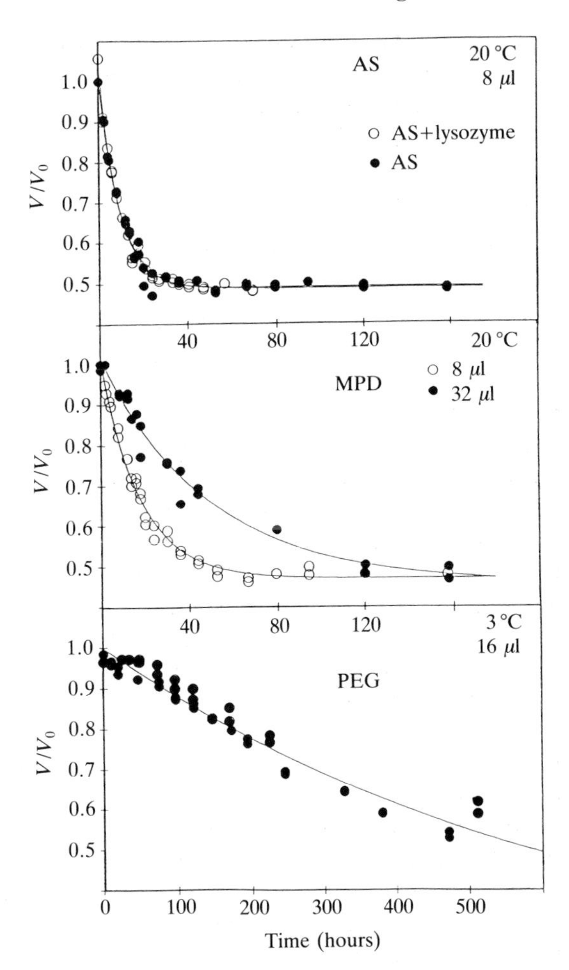

Figure 12. Water evaporation kinetics in the presence of ammonium sulphate (AS), MPD, and PEG. The experiments show also the influence of protein (in AS experiment), and of initial drop volume (in MPD experiment) on final drop volume. V_o is the initial volume of the drop; experiments were conducted with a concentration of crystallizing agent in the reservoir twice that in the drop time at zero. Adaped from reference 22.

pressure. Because one starts from supersaturation, nucleation tends to be too large. However, in some cases fairly large crystals can be obtained when working close to the metastable region. This is illustrated in Chapter 9. An automated system for micro-batch macromolecule crystallizations and screening has recently been described (29) allowing the set up of samples of less than 2 μl. Reproducibility of experiments is guaranteed because samples are dispensed and incubated under oil, thus preventing evaporation and uncontrolled concentration changes of the components in the micro-droplets.

An interesting variation of classical batch crystallization is the sequential extraction procedure of Jakobi (30), based on the property that many proteins (not all) are more soluble in concentrated salt (e.g. ammonium sulphate) when lowering temperature. The method can be adapted for microassays and was successfully applied for the crystallization of a proteolytic fragment of methionyl-tRNA synthetase from *Escherichia coli* (31).

6. Crystallization by the interface diffusion

This method was developed by Salemme (32) and used to crystallize several proteins. As shown by *Figure 13*, in the liquid/liquid diffusion method, equilibration occurs by diffusion of the crystallizing agent into the biological macromolecule volume. To avoid rapid mixing, the less dense solution is poured very gently on the most dense (salt in general) solution. Sometimes, crystallizing agent is frozen and the protein layered above to avoid rapid mixing.

One generally uses tubes of small inner diameter in which convection is reduced. This could be achieved more easily by using gels as described in Chapter 6. It follows the same diffusion method without the inconvenience of the metastability of two liquids sitting on the top of each other. This method gained new attention because of microgravity experiments (see Chapter 6).

7. Correlations with solubility diagrams

Even if it is not possible to determine the solubility (or phase) diagram for each biological macromolecules, it is important to understand the correlation between solubility diagrams and the method used to reach supersaturation and crystallization, using schematic diagrams (for more details see Chapter 9).

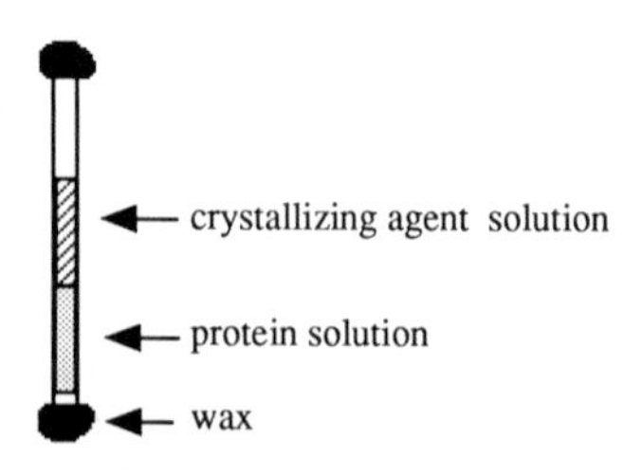

Figure 13. Liquid/liquid crystallization set-up. The equipment usually consists of capillaries of small diameter in which convection is reduced. The most dense solution should be in the bottom and capillaries should be maintained vertical during equilibration; they are sealed with low-melting wax plugs.

7.1 Dialysis

In the case of dialysis buttons, if one consider that stretching of the membrane is negligible, the biological macromolecule concentration will remain constant. However, if the biological macromolecule solution does not fill the chamber entirely, leaving room for air, it is no longer exactly true since the biological macromolecule concentration may vary (increase or decrease depending on the situation). The initial concentration of the crystallizing agent in the reservoir (this could be buffer) leaves the biological macromolecule in an undersaturated state. As shown in *Figure 14*, when increasing crystallizing agent concentration, supersaturation state is reached after passing through point S which is the equilibrium point on the solubility curve. Then, depending on final crystallizing agent concentration, it will crystallize or precipitate.

7.2 Vapour diffusion

In a classical case where concentration of crystallizing agent in the reservoir is twice the one in the drop, the protein will start to concentrate from an undersaturated state A (at concentration C_i) to reach a supersaturated one B (at concentration C_f) with both protein and crystallizing concentrations increasing by a factor of two. Two hypothetical cases are represented in *Figure 15a* and *15b* corresponding to experiments not leading (*15a*) or leading (*15*b) to crystals. Since no crystals are obtained in *Figure 15a*, the equilibrium

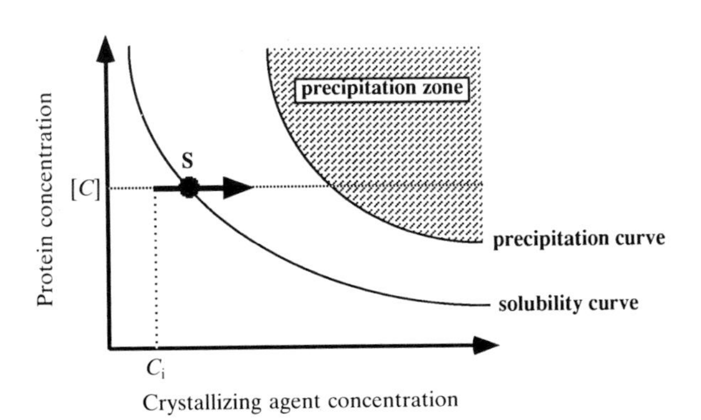

Figure 14. Schematic solubility diagram and correlation between macromolecule and precipitating agent concentrations in a crystallization experiment using a dialysis set-up. C_i is the initial concentration of crystallizing agent and C the constant protein concentration. The area between precipitation and solubility curves is the supersaturated region where crystallization can occur. Precipitation and solubility curves can be determined experimentally, although for the latter one crystals should be obtained first. For more details see Chapter 9.

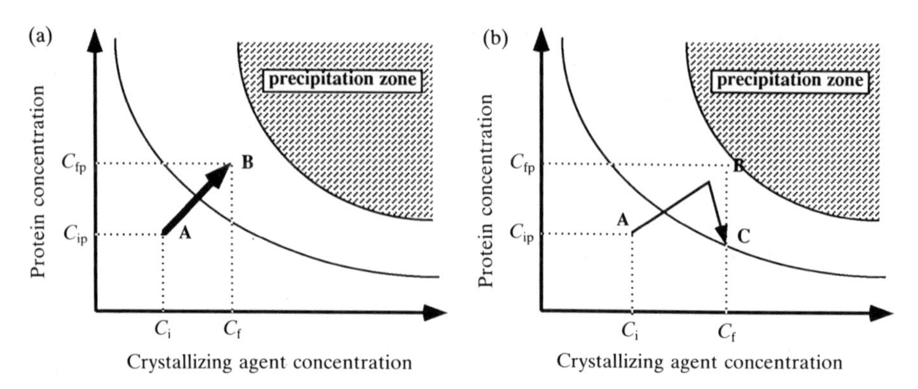

Figure 15. Schematic solubility diagram and correlation between macromolecule and precipitating agent concentrations in crystallizing experiments using vapour diffusion set-ups. Situation (**a**) without and (**b**) with crystallization. See legend to *Figure 14* and text for further explantions.

at point B will be located in the metastable region; when the first crystals appear (at the break of the arrow) the trajectory of equilibration is more complex. In that case the remaining concentration of protein in solution will converge towards point C located on the solubility curve.

7.3 Batch crystallization

In batch crystallization using a closed vessel, three cases can be considered as shown in *Figure 16*. If the protein concentration is chosen in such a way that the solution is undersaturated (point A), crystallization will never occur (unless an other parameter such as temperature is varied). The protein

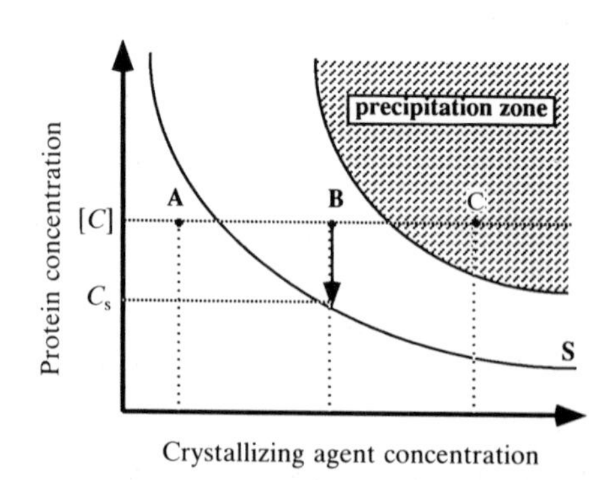

Figure 16. Schematic solubility diagram and correlation between macromolecule and precipitating agent concentrations in crystallization experiments using a batch method (in closed vessels). See legend to *Figure 14* and text for further explanations.

concentration may belong to the supersaturated region between solubility and precipitation curves (point B). In that case the arrow describes the variation of the remaining concentration of protein in solution. In the last case (point C) the protein will precipitate immediately because supersaturation is too high. In some cases, however, crystals may grow from the precipitates.

8. Exercises: crystallization of lysozyme

It is a good exercise to train oneself (*Protocols 9* and *10*) with a cheap, easily accessible protein. Lysozyme is a good candidate which is commercially available from various manufacturers, and it crystallizes readily. An example of crystallization using the vapour phase method in hanging drops is given in *Protocol 9*, and using dialysis buttons in *Protocol 10*. The experimental conditions described are adapted from reference 33. Some crystal habits frequently obtained are displayed in *Figure 1* of Chapter 1.

Protocol 9. Testing crystallization using hanging drops

1. Prepare stock solutions of 3 M NaCl and 50 mg/ml (3.43 mM) lysozyme in 50 mM acetate at pH 4.5 and buffer stock solution (50 mM sodium acetate at pH 4.5). Filter all solutions with a microfilter of 0.22 μm.
2. Prepare a Linbro box as described in Section 4.2.1.
3. Fill up reservoirs of row A with solutions of NaCl ranging from 0.6–1.1 M by steps of 0.1 M.
4. On a coverslip, mix 4 μl of protein stock solution with 4 μl of reservoir. Flip it and set it on the greased rim.
5. Fill up reservoir of row B with solutions of NaCl ranging from 0.8–1.3 M by steps of 0.1 M. Repeat the experiment on row B after diluting protein of stock solution by a factor 2 to obtain a new one of 20 mg/ml (1.37 mM).
6. Fill up reservoir of row C with solutions of NaCl ranging from 1.0–2.0 M by steps of 0.2 M. Repeat the experiment after diluting protein stock solution by a factor 2.
7. Use row D for duplicata or testing particular parameters (e.g. volume of drops to use the influence of kinetic effects on growth).
8. Store the experiments at 18 °C.
9. Observe the experiments once a day for a week.
10. Train yourself mounting crystals (see Chapter 12).

Protocol 10. Testing crystallization using dialysis button

1. Prepare stock solutions as in *Protocol 9*.
2. Fill the protein chamber of a dialysis button with protein stock solution diluted at 10 mg/ml. The solution must form a dome above the entry. Install the dialysis membrane (cut-off 5000) as described in Section 3.2.2.
3. Fill up a 5 ml beaker with the buffer solution and drop the button in it with the aperture of the protein chamber on the top.
4. Store the experiments at 18 °C.
5. Increase the concentration in the beaker by 200 mM of NaCl every day.
6. Observe the next day.
7. Repeat steps 5 and 6 until crystallization occurs.

9. Concluding remarks

In this chapter we have described the most common crystallization methods; all of them have advantages and drawbacks. Most crystallographers favour vapour phase diffusion which provides an easy way to practise crystallization. It is also the method of choice for robotics. Dialysis has the advantage that biological macromolecule concentration remains constant, so that only one parameter varies at a time and the nature of the buffer or crystallizing agent can be changed easily. It differs from a classical vapour phase equilibrium crystallization experiment where all constituants in the drop are concentrated.

In conclusion, it must be stressed that less standard crystallization methods may be useful in particular cases. These can be methods based on old ideas, not yet well explored, as in pulse-diffusion of precipitant, combining dialysis and free diffusion in capillaries (34), or combinations of dialysis and electrophoresis (35). Also, crystallization in a particular environment should be considered, such as at high pressure (36), under levitation (37) or microgravity (38), or in centrifuges (39). Particular attention should be given to crystallization in gels which may in time become a popular method for macromolecule crystallization (see Chapter 6). Methods based on temperature diffusion, which are widely used in material sciences (40), may be adapted under certain conditions for macromolecule crystallization. Finally, the use of novel types of crystallization cells may represent an interesting alternative for growing better crystals (see refs 38 and 41) for principles of cells developed for microgravity experiments). It is our hope that the methods and ideas discussed in this chapter will help readers, not only to solve their crystallization problems, but also to improve existing methods and even to develop new crystallization methodologies.

References

1. McPherson, A. (1990). *Eur. J. Biochem.*, **189**, 1.
2. Giegé, R. (1987). In *Crystallography in molecular biology* (ed. D. Moras, J. Drenth, B. Strandberg, D. Suck and K. Wilson), NATO ASI Series, Vol. **126**, pp. 15–26. Plenum Press, New York and London.
3. McPherson, A. (1985). *Methods in Enzymology*, **114**, 112.
4. Perrin, D. D. and Dempsey, B. (1974). *Buffer for pH and metal ion control.* Chapman and Hall Ltd, London and New York.
5. Jurnak, F. (1986). *J. Crystal Growth*, **76**, 577.
6. Ray, W. J. and Puvathingal, J. (1985). *Anal. Biochem.*, **146**, 307.
7. Chayen, N., Akins, J., Campbell-Smith, S., and Blow, D. M. (1988). *J. Crystal Growth*, **90**, 112.
8. Winterbourne, D. J. (1986). *Biochem. Soc. Trans.*, **14**, 1179.
9. Mikol, V. and Giegé, R. (1989). *J. Crystal Growth*, **97**, 324.
10. Bradford, M. M. (1976). *Anal. Biochem.*, **72**, 248.
11. Zeppenzauer, M. (1971). *Methods in Enzymology*, **22**, 253.
12. Weber, B. H. and Goodkin, P. E. (1970). *Arch. Biochem. Biophys.*, **141**, 489.
13. Pronk, S. E. Hofstra, H., Groendijk, H., Kingma, J., Swarte, M. B. A., Dorner, F., Drenth, J., Hol, W. G. J., and Witholt, B. (1985). *J. Biol. Chem.*, **260**, 13580.
14. Thomas, D. H., Rob, A., and Rice, D. W. (1989). *Prot. Engineering*, **2**, 489.
15. Hampel, A., Labananskas, M., Conners, P. G., Kirkegard, L., RajBhandary, U. L., Sigler, P. B., and Bock, R. M. (1968). *Science*, **162**, 1384.
16. Kim, S.-H. and Quigley, G. (1979). *Methods in Enzymology*, **59**, 3.
17. Dock, A.-C., Lorber, B., Moras, D., Pixa, G., Thierry, J.-C., and Giegé, R. (1984). *Biochimie*, **66**, 179.
18. Phillips, Jr., G. N. (1985). *Methods in Enzymology*, **104**, 128.
19. Luft, J. and Cody, V. (1989). *J. Appl. Cryst.*, **22**, 386.
20. Yonath, A., Müssig, J., and Wittmann, H. G. (1982). *J. Cell. Biochem.*, **19**, 145.
21. McPherson, A. (1985). *Methods in Enzymology*, **114**, 125.
22. Mikol, V., Rodeau, J.-L., and Giegé, R. (1989). *J. Appl. Crystallogr.*, **22**, 155.
23. McPherson, A. and Shlichta, P. (1988). *Science*, **239**, 385.
24. Chayen, N. E., Lloyd, L. F., Collyer, C. A., and Blow, D. M. (1989). *J. Crystal Growth*, **97**, 367.
25. Mikol, V., Rodeau, J.-L., and Giegé, R. (1990). *Anal. Biochim.*, **186**, 332.
26. McPherson, A. (1976). *J. Biol. Chem.*, **251**, 6300.
27. Gernert, K. M., Smith, R., and Carter, D. C. (1988). *Anal. Biochem.*, **168**, 141.
28. Przybylska, M. (1989). *J. Appl. Crystallogr.*, **22**, 115.
29. Chayen, N. E., Shaw Stewart, P. D., Maeder, D. L., and Blow, D. M. (1990). *J. Appl. Crystallogr.*, **23**, 297.
30. Jakoby, W. B. (1971). *Methods in Enzymology*, **22**, 248.
31. Waller, J. P., Risler, J.-L., Monteillhet, C., and Zelwer, C. (1971). *FEBS Lett.*, **16**, 186.
32. Salemne, F. R. (1972). *Arch. Biochem. Biophys.*, **151**, 533.
33. Ries-Kautt, M. and Ducruix, A. (1989). *J. Biol.Chem.*, **264**, 745.
34. Koeppe, R. E., Stroud, R. M., Pena, V. A., and Santi, D. V. (1975). *J. Mol. Biol.*, **98**, 155.

35. Chin, C.-C., Dence, J. B., and Warren, J. C. (1976). *J. Biol. Chem.*, **251**, 3700.
36. Visuri, K., Kaipainen, E., Kivimäki, J., Niemi, H., Leissla, M., and Palosaari, S. (1990). *Biotechnology*, 547.
37. Rhim, W.-K. and Chung, S. K. (1990). *Methods, A Companion to Methods in Enzymology*, **1**, 118.
38. DeLucas, L. J. and Bugg, C. E. (1990). *Methods, A companion to Methods in Enzymology*, **1**, 109.
39. Barynin, V. V. and Melik-Adamyan, V. R. (1982). *Sov. Phys. Crystallogr.*, **27**, 588.
40. Feigelson, R. S. (1988). *J. Crystal Growth*, **90**, 1.
41. Sieker, L. C. (1988). *J. Crystal Growth*, **90**, 349.

5

Seeding techniques

E. A. STURA and I. A. WILSON

1. Introduction

A seed provides a template for the assembly of molecules to form a crystal with the same characteristics as the crystal from which the seed originated. Seeding has often been used as a method of last resort rather than a standard practice. It has been used in situations when, after the first crystallization event, further attempts at crystallizing the same or subsequent batches of the same protein, under apparently identical crystallization conditions, either failed to yield any crystals at all, or provided crystalline precipitates or very small crystals. The use of seeding in crystallization can simplify the task of the crystallographer even when crystals can be obtained by other means. We will explore the various seeding techniques, and their applications in the growth of large single crystals.

1.1 Seeding

Crystallogenesis can be divided into two distinct phases. The first is the screening of crystallization conditions to obtain the first crystals; the second consists of the optimization of these conditions to improve crystal size and quality. Seeding can be used advantageously in both these situations. The first stage in crystallogenesis consists of the discovery of initial crystals, crystalline aggregates, or microcrystalline precipitate. This may result from a standardized screening method (1), an incomplete factorial search (see Chapter 3 and references 2 and 3), or by extensive screening of many conditions. This can be a very laborious search, which in some cases can be bypassed by starting with seeds from crystals of a related molecule that has been previously crystallized. Molecules that have been obtained by genetic or molecular engineering of a previously crystallized macromolecule fall into this category. This method is termed cross-seeding. It has been used to obtain crystals of pig aspartate aminotransferase, starting with a crystal from the chicken enzyme (4), and between native and complexed Fab molecules (1).

Whatever the method used to obtain the initial crystals, seeding may provide a fast and effective way to facilitate the optimization of growth conditions without the uncertainty which is intrinsic in the process of spontaneous nucleation. The streak seeding technique can be used to carry out a search quickly and efficiently over a wide range of growth conditions. Later, the use of macroseeding and microseeding methods can be used to grow large crystals with a good degree of reproducibility.

1.2 Supersaturation and nucleation

To understand some of the processes involved in crystallogenesis we must consider each step separately. In order to achieve nucleation, a state of supersaturation must be induced. Supersaturation is a metastable condition where the solvent holds more protein in solution than it would at the minimum free energy. This state is created by a variety of methods, such as the salting-out effect of high ionic strength salts like ammonium sulphate, the entropic segregation caused by polymers such as PEG, the solvent effect of organic molecules such as MPD, by de-salting macromolecules, and by varying other conditions such as pH and temperature. Details on the use of these precipitants, together with general and theoretical considerations and practical methods for macromolecule crystallization, are to be found in the preceding chapters and in other publications (5–9). Here we will consider only the aspects of crystallization that directly involve the application of seeding techniques.

It is useful to separate the events leading to the spontaneous formation of a crystal nucleus and those conditions that allow a crystal or nucleus to grow. While both events depend on the degree of supersaturation of the protein, the physical processes involved are very different and the degree of supersaturation required for nucleation is generally higher than that required for growth on to an existing crystal plane.

In the case of spontaneous nucleation, a new seed must be generated whilst other events are taking place which are driven by the requirement to lower the free energy of the supersaturated state. These other events involve the aggregation of molecules into various phases. During aggregation, reversible and irreversible processes are at work simultaneously. The formation of ordered nuclei may be competing for protein with irreversible processes that produce amorphous aggregates (such as precipitates and protein skins) and hence constantly lower the degree of supersaturation of the macromolecule. As supersaturation is decreased the chance of forming a stable nucleus is reduced. Since the occurrence of spontaneous nucleation depends on the relative rates at which these various competing events take place, crystals might never form even under conditions which might otherwise support crystal growth.

The principle that there is a lower energy requirement in adding to an existing crystal surface than in creating a new nucleus has important consequences. If the inverse were true, the protein would partition into a very large number of small nuclei and large crystals would never grow. From this principle we may also understand why in many cases it is possible to grow large crystals in the absence of seeding. Since spontaneous nucleation is a statistical phenomenon, whose probability increases with increasing degree of supersaturation, the nucleation and growth of crystals is a process with negative feedback. As a nucleus is formed its growth reduces the degree of supersaturation of the solution, and hence decreases the probability that other nuclei will form. To take advantage of this, supersaturation must be achieved slowly. The degree of supersaturation should be just sufficient to obtain a small number of nucleation centres. With the proper choice of crystallization conditions and good control over the environmental factors, it is often possible to fulfil all of the above conditions. Determining the appropriate conditions can require many crystallization trials and is consequently time consuming. In addition, the determination may involve the use of many milligrams of macromolecule and other materials. Seeding can be used efficiently and effectively during the crystallization trials to minimize the quantity of protein required for this analysis.

1.3 Crystal growth

When crystal seeds are added to an equilibrated protein solution, the protein partitioned between the soluble phase and irregular aggregate phases will redistribute. The final equilibrium will be achieved only when crystal growth has ceased. From this we may understand why seeding can be used not only in situations where the protein remains in the soluble phase, but also in situations where most of the protein has precipitated. In fact, as crystals grow the degree of supersaturation is reduced and protein may be transferred from the various (reversible) amorphous phases to the soluble phase, and from this phase it can accrete on to crystal surfaces. Several factors affect the quality of the crystals obtained, including the rate of growth, the internal order of the initial nucleus, and the purity of the sample. For some macromolecules the initial growth following nucleation may be too fast, because of the high degree of supersaturation, resulting in the incorporation of crystal defects which may eventually lead to the premature termination of crystal growth and to poorly-formed crystals. In certain cases, the nuclei which are generated spontaneously may be only partially ordered. Growth from such nuclei may result in the formation of crystal clusters rather than individual single crystals. Seeding may provide a method to avoid such difficulties. By uncoupling crystal growth from nucleation seeding provides a means by which growth conditions may be tailored for crystal growth rather than nucleation and lead towards the production of large, regular crystals. Seeding allows the experimenter to control not only the number of seeds but also reduce the supersaturation level of the protein and to decrease the incorporation of defects detrimental to crystal quality. Seeding provides a preformed, regular crystal surface on to which further molecules may aggregate in an orderly fashion. The seeds to be used in the seeding experiments should be carefully selected from the best crystals previously grown in order to obtain the best results.

1.4 Seeding techniques

Seeding consists of three stages; a pre-seeding stage, an analytical stage to refine crystal growth conditions, and the final production stage using the refined conditions to produce large single crystals. The pre-seeding stage is essential for seeding to work in consistent and reproducible manner. It can be separated into three important aspects:

(a) the environment, and the associated precautions necessary for seeding;

(b) pre-equilibration of the protein solution to be seeded;

(c) the determination of the appropriate supersaturation level for seeding.

The design of the experiment is important since during seeding there may be a slight alteration in the pre-equilibrated state of the protein. An increase in supersaturation at this stage may result in spontaneous nucleation giving rise to a shower of small crystals.

Seeding can be either homogeneous or heterogeneous. The two most common methods of homogeneous seeding are:

(a) Microseeding, which involves the transfer of microscopic crystals from a seed source to a non-nucleated protein solution.

(b) Macroseeding, in which pre-grown crystals are washed and introduced individually into a pre-equilibrated protein solution. This method has been widely applied to tackle the problem of enlarging small crystals into crystals of a suitable size for X-ray diffraction studies (10, 11).

Heterogeneous seeding can be divided into:

(a) Cross-seeding, a form of seeding in which seeds come from a different protein or protein complex than that being crystallized.

(b) Epitaxial nucleation, where a regular surface rather than a three-dimensional lattice induces the growth of new crystals. The nucleation of protein crystals on cellulose fibre impurities, which often end up in protein solutions, is a commonly observed case of epitaxial nucleation.

In both these methods the lattice dimensions of the crystals obtained may be different from those of the seeds from which they grew. Heterogeneous seeding should be followed by homogeneous seeding.

2. Crystallization procedures

Crystallization by vapour diffusion is a relatively simple technique. A crystallizing solution is mixed and placed in the reservoir. A protein sample is then mixed in a definite proportion with some reservoir solution, placed in close proximity to the reservoir within a sealed unit, and allowed to equilibrate with the reservoir solution at constant temperature over a period of a few days, or until crystals appear. However, it is important to have some degree of control over environmental parameters. This section deals with the establishment of crystallization procedures which are suited to the application of seeding.

2.1 Pre-seeding

The crystallization procedures used in conjunction with seeding techniques may be different from those that would be used otherwise. The design of the crystallization experiment must allow for the introduction of seeds at some stage in the equilibration phase. This must be done with the minimum disruption to the crystallization environment. A compact variation of the sitting drop vapour diffusion method has been used successfully in the seeding of many proteins (12, 13, 15, 19, 30, 32, 37).

2.1.1 Sitting drop vapour diffusion

Vapour diffusion provides a controlled and relatively slow method of equilibration by the transfer of vapour between the protein and precipitant solutions.

i. Temperature

To achieve success in crystallization and seeding it is important to control the overall environment of the set-up. This must include temperature. Temperature regulation can be achieved by the combined use of a sitting drop vapour diffusion set-up and a constant temperature incubator. The sitting drop environment provides better temperature control than its hanging drop counterpart because of heat conduction between the reservoir solution and the protein solution in the inverted glass pot (see *Figure 1*). In addition, differential distillation caused by temperature gradients and convection currents in the sealed well is greatly reduced since the protein drop is situated close to the surface of the reservoir. This contrasts with the hanging-drop environment where the drop is effectively in thermal contact with the outside air. The thin coverglass does not provide adequate insulation from temperature fluctuations. Short-lived changes in temperature (such as opening the door of the constant-temperature incubator containing the experiment) will rapidly vary the temperature of the hanging drop but not that of the reservoir, which because of its higher heat capacity, will remain at a relatively constant temperature. During a rise in temperature, vapour will distill away from the drop, increasing the degree of supersaturation, and consequently this may result in a shower of microcrystals. This is a common observation with hanging drop vapour diffusion experiments, where crystals are grown at high salt concentrations. When the temperature decreases more vapour condenses

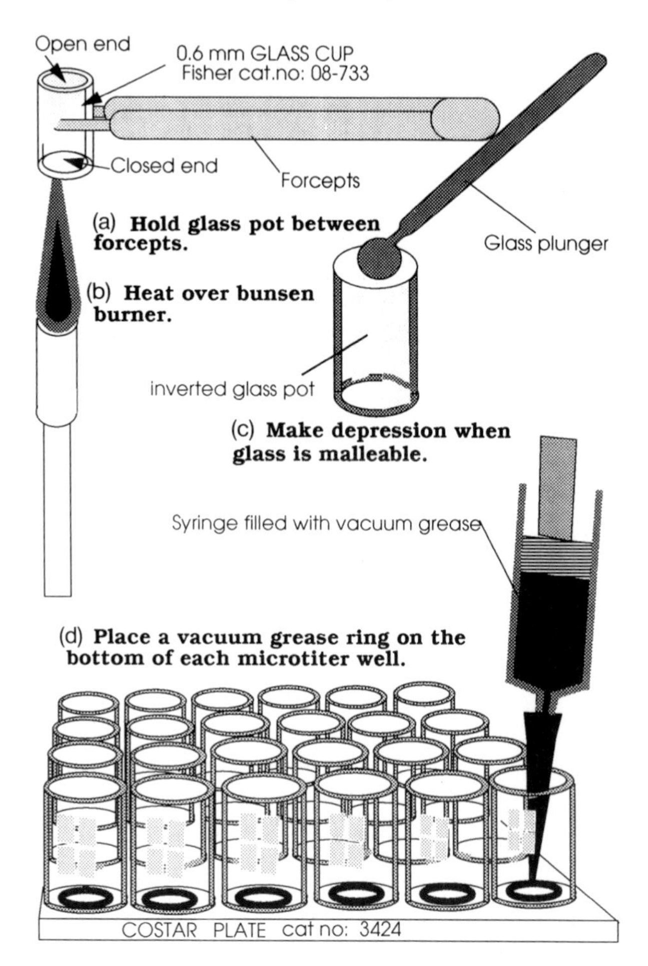

Figure 1. Schematic illustration of the sitting drop vapour diffusion plates. (a) A glass cup is held between long-sized forceps over a bunsen burner with the closed end down towards the flame. (b) The cup is heated until the glass becomes soft. (c) When the bottom is malleable the pot is inverted and a depression is made in it with a glass plunger. (d) Trays are prepared for inserting the pots by placing a ring of silicone vacuum grease in the bottom of each of the wells of the microtiter plate. (e) Each depression is siliconized, washed repeatedly with distilled water, and the coat baked in an oven. The siliconized pots can now be placed in each of the wells in the multiwell cluster on top of the silicone grease ring which holds them in position. (g) The rim of the individual well is smeared with petroleum jelly to seal the well once the coverglass is placed on top. (h) The precipitant solution is placed around the inverted pot, the protein in the depression, and the desired amount of precipitant mixed with it before sealing the experiment.

on to the drop diluting the protein solution. It is not uncommon in low-salt crystallization trials, using hanging drop vapour diffusion and using polyethylene glycol as a precipitant, to achieve an increase rather than a decrease

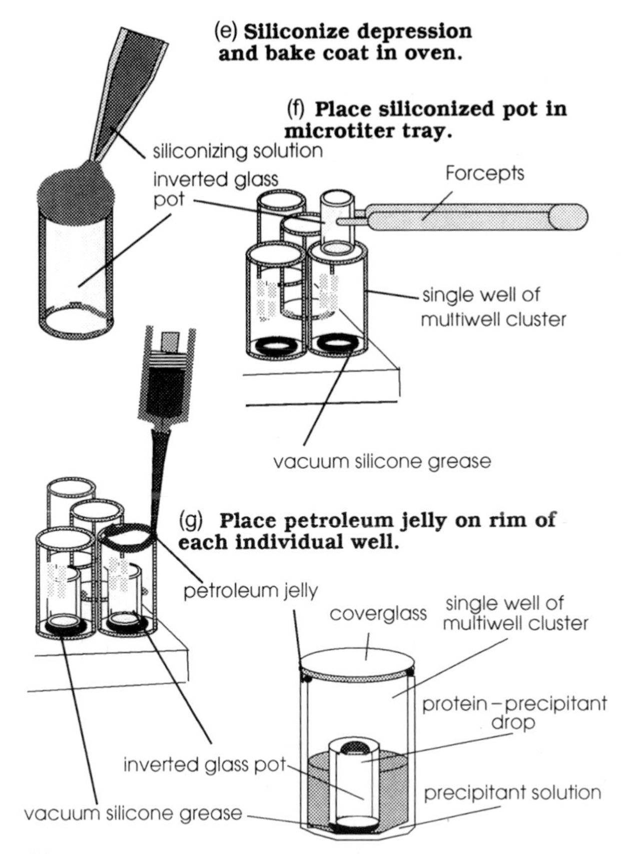

(h) Place precipitant solution in the annulus around the pot. Place protein inside the depression in the pot and mix with the desired amount from well. The well is finally sealed with a coverglass.

in the volume of the protein–precipitant drop. The use of the sitting drop method reduces these problems.

ii. Multiwell sitting drop vapour diffusion plate

This has been designed and fabricated in our laboratory to shield against short-lived temperature fluctuations (*Figure 1*). Such a plate can be used from the initial trials to the final refinement of the crystallization conditions. It offers several advantages over conventional sitting drop, as well as hanging drop set-ups. Its compact size makes more efficient use of incubator space. It enables for easy access to the protein drop and the reservoir. This allows the precipitant concentration to be varied during the course of the experiment as required. By lifting the coverglass the protein–precipitant drop can be seeded, and the plate design enables minimum disturbance to the seeding environment. These features can greatly enhance growth of quality single crystals. The tray consists of a 24-well tissue culture plate from Costar combined with 0.6 ml glass cups from Fisher. The wells are significantly smaller than those in the more commonly used Linbro plates, which with 22 mm circular coverglass slips instead of the 18 mm required for the Costar plates. The plates are made as described in *Protocol 1* and illustrated in *Figure 1*.

Protocol 1. Making of multiwell sitting drop plates

1. Make a depression in the bottom of the 0.6 ml cups by heating over a Bunsen flame and pressing down on the cylindrical base of each of the inverted cups with a rounded-end glass plunger.
2. Hold the resulting cup in place at the centre of each well of the tissue culture plate with Corning silicon vacuum grease. The open end of the cup sits on the bottom and the depression in the cup is at the volumetric centre of each individual well (ready made sitting-drop rods are now available from Perpetual System Corporation).
3. Siliconize the cavity created by the depression. Up to 100 μl of protein–crystallizing mixture can be used with this set-up. The reservoir solution, typically 1 ml, occupies an annulus around the inverted cup.
4. Smear the edges of each well in the multiwell plate with petroleum jelly. Silicone grease can be used instead of petroleum jelly to give a better seal, although this will make covers harder to remove for seeding.
5. Place a 18 mm round microscope coverglass on top of each well ensuring that an air tight seal is achieved.

Since this method is similar in many ways to the hanging-drop method, crystallization conditions determined for hanging drop experiments require little modification for implementation with sitting drop vapour diffusion. Another added advantage is that larger volumes can be used. Other sitting drop methods may be easily adapted to provide the temperature stability and controlled humidity in order for the seeding methods presented here to work reliably.

2.2 Analytical seeding

It is important to first determine under what conditions seeding will be effective. This is done by the use of an analytical seeding method such as streak seeding.

2.2.1 Streak seeding technique

i. Making the probe

A probe for analytical seeding is easily made with an animal whisker mounted with wax to the end of a small thick-walled capillary. Human hair has been tried with inconsistent results. Since the cross-section of the whisker varies along its length, it is possible to obtain several probes from the same whisker by repeatedly cutting the whisker to lengths of 5–20 mm from the end of the wax. The probes so obtained will be of different strength and thickness.

ii. Cleaning the probe

To clean the probes prior to their use the fibres are degreased using ethanol or methanol, and then washed in distilled water and wiped dry. Probes can be used several times by cleaning them with distilled water and tissue paper, in between experiments. After a period of time the whiskers lose the property that makes them superior to human hair. A good probe should be able to transfer seeds to 6–12 drops consecutively without being dipped in seed solution. A good probe should always be used when microseeding for production crystal growth. Old whiskers deposit many seeds in the first two drops and virtually none in further drops. This test can be used to determine which hair or whiskers are suitable for streak seeding. To date we have been successful with whiskers from rabbits, cats, and chinchillas.

iii. Seeding

The end of the fibre is then used to touch an existing crystal and dislodge seeds from it (*Figure 2a*) (gentle friction against the crystal is normally sufficient). Some of the seeds that are dislodged will remain attached to the fibre as it is drawn out from the drop containing the crystal. The probe is now used to introduce seeds into a pre-equilibrated drop by rapidly running the fibre in a straight line across the middle of the protein–precipitant drop (*Figure 2b*).

Sitting drop set-ups are preferable since hanging drops tend to dry out when exposed to the ambient air, even in the short time interval between collecting the seeds on the fibre and streaking the new drop, typically 1–30 seconds. The precipitant collected from the first drop increases in concentration as it travels to the next drop through the air, and this can affect the conditions in the seeded drop. The distance the whisker has to travel should be kept to a minimum, therefore, in most circumstances less than 10 cm. Both the source well and the receiving well are re-sealed immediately after the transfer. Seed nucleated crystals grow along the streak line; any self-nucleated seeds will occur elsewhere in the drop. The pre-incubation time and the range of supersaturation which allows for sufficient crystal growth without self nucleation is determined experimentally by observing the growth (or lack of growth) of seeded crystals along the streak line (*Figure 2c*).

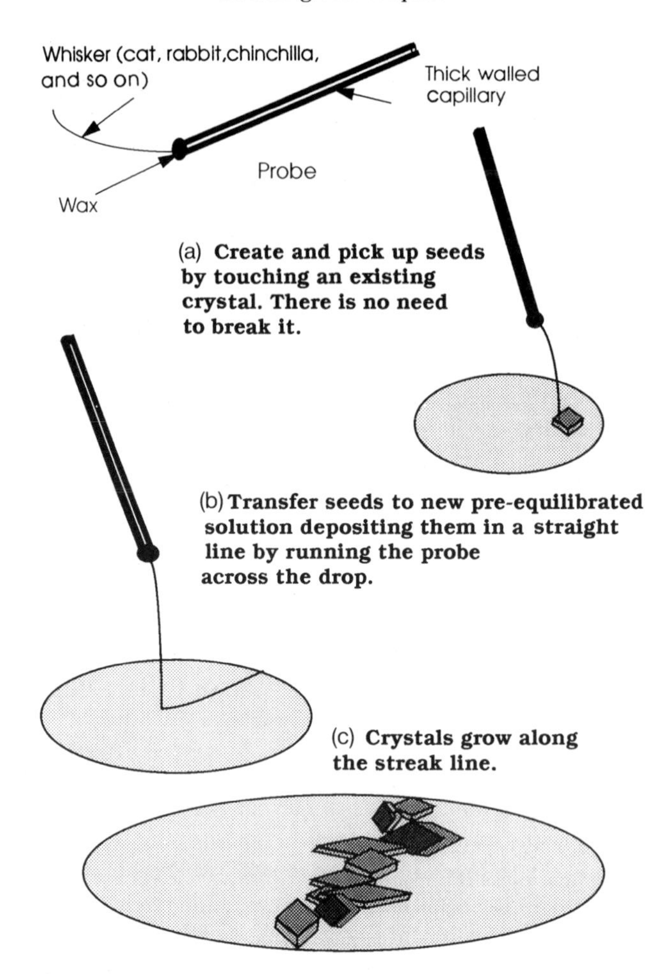

Figure 2. Schematic drawing of the stages in the application of the streak-seeding technique for analytical seeding. (a) A probe consisting of a thick walled capillary, on the end of which a short segment of an animal whisker is attached with molten wax, is used to pick up seeds from an existing crystal, precipitate, or other ordered aggregate, by simply touching it and displacing seeds from it. (b) The seeds remain attached to the whisker and can be transferred to a pre-equilibrated drop by running the end of the probe across the drop. Some seed are deposited along the path, where they will either grow into large crystals or dissolve into the solution. (c) Growth of crystals along the streak line indicates that the conditions may be suitable for the application of other techniques such as micro or macroseeding. Self-nucleated crystals will appear away from the streak line.

2.2.2 Protein pre-equilibration

In batch crystallization the precipitant concentration is slowly increased until the protein solution turns cloudy; further solvent is then added until it is clear again. The protein is continuously stirred until seeds are added. When seeding small volumes, it is best to avoid producing spontaneously nucleated seeds, as could be produced by exceeding the solubility threshold, since it is difficult to control their number, or ensure that they will later dissolve. By introducing the seeds well before sufficient supersaturation is reached, additional nuclei which might form spontaneously are prevented. However, if the protein is not suitably pre-equilibrated, the seeds will dissolve. Streak seeding should be repeated at different pre-equilibration times for different drops to determine suitable sets of conditions.

2.2.2 Determining the degree of supersaturation for seeding

Initially, during the screening phase of a crystallization, we want to obtain results quickly, in order to determine the many parameters that control the growth and morphology of the crystals. In the production phase, when conditions are optimized, we want to slow down the growth rate. Drops are generally allowed to equilibrate fully before seeding. This is a good procedure as long as protein is not being lost to amorphous phases. Optimizing the conditions for this type of seeding can be done by setting up drops under conditions which vary only slightly from those previously determined in the fast-growth experiments. Drop size, which was kept to a minimum to save material, should be increased at this stage, while simultaneously reducing the precipitant/protein ratio, and the precipitant concentration in the reservoir. This will have the effect of slowing down the rate of equilibration and the desired state of supersaturation will be approached more slowly. The reduced precipitant concentration is best determined experimentally by allowing the drops to equilibrate, usually three to five days, and then streak seeding to determine the suitable precipitant concentration for the reservoir. The streak will not appear in drops where the concentration of the precipitant in the reservoir is too low, and where the precipitant concentration is too high crystals will initially appear along the streak line, followed later by the formation of others away from the streak line. Caution should be used when making this determination, as some seeds may drift away from the line along which they were deposited and sink down to the bottom of the drop. The experimental conditions where crystals grow only along the streak line determine the precipitant concentration range for production seeding. This method is also well suited for testing minor changes to growth conditions, such as adding a co-precipitant, testing new additives, or simply finely analysing the pH range. Small changes at this stage can result in significant improvements in the quality of the crystals obtained and can be essential for growing suitable crystals for high-resolution X-ray structure analysis.

2.2.4 Assaying for micro-crystallinity

Micro-crystalline precipitates are often indistinguishable from their amorphous counterparts. Streak seeding can be used to distinguish between these two possibilities, by using particles from an uncharacterized precipitate as a source of seeds. For example, the initial crystallization trials of a complex between the Fab′ fragment from an anti-peptide mouse antibody B13I2 (14) with its 19 amino-acid peptide antigen (myohemerythrin residues 69–87) gave a precipitate composed of round or oval particles of roughly equal size (*Figure 5a;* (15)). No indication of micro-crystallinity could be deduced from microscopic observations of the precipitate since it was not appreciably birefringent. Three adjacent drops in which protein had not precipitated were streaked with this precipitate. Hexagonal shaped crystals appeared in one of the drops as a result of the streak seeding experiment (*Figure 5b*). These crystals were then used for macroseeding on to other drops. After an adjustment in the crystallization conditions (from 1.5 M Na citrate, pH 6.0, to 1.6–2.0 M phosphate, pH 5.0–6.0) it was possible to grow crystals (*Figure 5c*) that diffracted 2.6 Å resolution (15). The X-ray structure of this Fab′–peptide complex has been solved to 2.8 Å resolution (16). It is interesting to note that the presence of the peptide in the solution is essential to obtain these crystals and cross-seeding from the Fab′–peptide complex on to native Fab′ solutions does not produce crystals or precipitate.

3. Production seeding methods

3.1 Microseeding

In microseeding microscopic crystal fragments are introduced into a prepared protein solution. By using an analytical seeding technique (Section 2.2) the supersaturation threshold can be accurately determined by scanning a range of precipitant concentrations. The streak seeding technique was initially developed for this purpose. Microseeding is composed of three stages:

(a) preparation of the seed stock;

(b) repeated dilution of the seeds;

(c) the seeding itself.

3.1.1 Preparation of the seed stock

Seed stock is produced by washing three or four small crystals in a slightly dissolving solution to remove defects or amorphous precipitate from the crystal surfaces. After stabilizing the washed crystals in an appropriate precipitant solution they are transferred to a glass tissue homogenizer and crushed (*Figure 3*). The seeds from the crushed crystals are washed from the sides of the homogenizer into the bottom by adding further solution. The solution is now transferred from the homogenizer to a test tube. This is the seed stock, which for most proteins can be stored for future use.

3.1.2 Repeated dilution of the seeds

The seed stock is normally diluted with further solution as it contains too

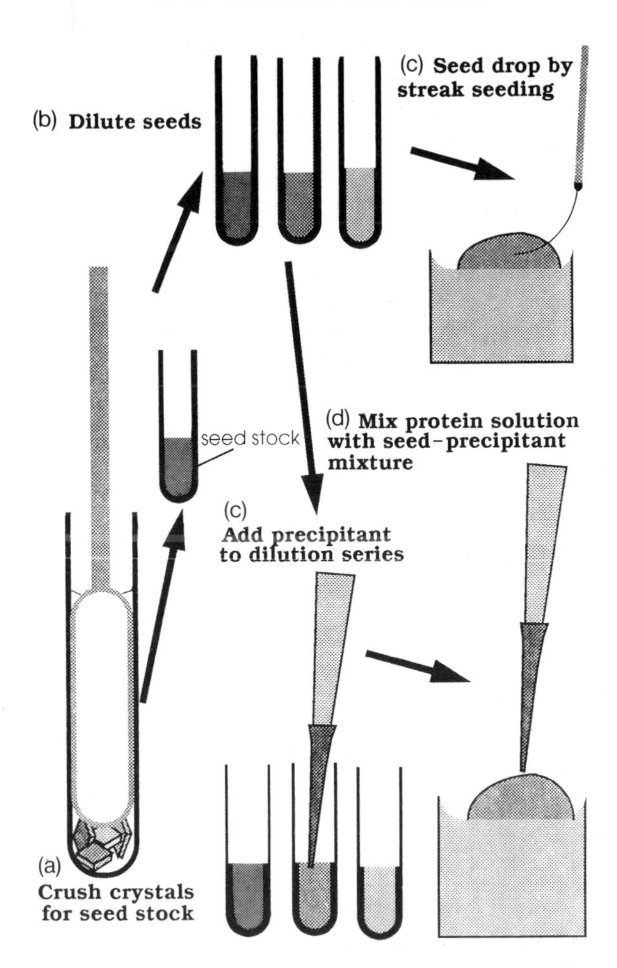

Figure 3. Diagrammatic illustration of the steps involved in microseeding. (a) Crystals of good morphology are crushed in a glass tissue homogenizer. The resulting seeds are washed into the bottom of the tube and stored in a test tube. (b) The seed stock is diluted to produce a dilution series. (c) Seeds can be picked up from the diluted solutions by using a probe, or precipitant can be added to these, so that they may be mixed with protein solution. (d) The wells are sealed by replacing the coverglass and the seeds are allowed to grow for several days.

many nuclei to be useful in the nucleation of only a small number of crystals. Dilutions, typically in the range 10^{-3}–10^{-7} are done sequentially and tested experimentally. Between dilutions the tube containing the microseeds should be vortexed to evenly distribute the seeds. One of the diluted seed solutions should be suitable to supply a small number of seeds into each drop that is seeded. Before the diluted seed solution is used for seeding, precipitant is added in such an amount that the resultant solution remains at a precipitant concentration above that in which the seeds will dissolve. To determine this, several drops are streaked and the streak is allowed to develop. Solvent (generally buffer) is then added to the reservoir solution to reduce the precipitant concentration. Some streaks will dissolve, others will remain clearly visible. The threshold of precipitant required to preserve the streak can be determined from these results.

3.1.3 Streak seeding as a microseeding technique

Since the seeds that are transferred from crystal to drop in the use of the streak seeding technique are microscopic, the technique is technically a microseeding method. But while in the analytical streak seeding technique the deposition of many seeds along the path of the whisker is essential for the subsequent visualization, when growing large crystals for X-ray structure analysis only a few seeds should be deposited, as seeds compete with each other for the available protein. When changing the use of streak seeding from an analytical to a production-seeding mode, we must find a way of diluting the seeds quantitatively in a reproducible manner. The same probe is thus used in each repeat experiment to ensure that the volume of liquid and the number and size of the seeds, which are transferred from one drop to another, remains constant. Thicker whiskers transfer more liquid containing microseeds and can potentially carry larger size seeds.

Protocol 2. Streak seeding

1. Note the angle at which the whisker is drawn out of the solution. It is important as this will affect the size and number of seeds loaded on the fibre. This should be kept constant in future experiments. It is also suggested that the whisker be lifted vertically upwards, maintaining it perpendicular to the drop's surface.
2. Pre-equilibrate the drops before seeding under conditions previously determined analytically by streak seeding.
3. Streak subsequent drops without loading the probe with new seeds to achieve seed dilution. To obtain greater dilutions the probe can be dipped in and out of the reservoir in-between streaks, to allow some seeds to drop into the precipitant solution.
4. Reduce the time the probe spends in the air by opening all the chambers to be seeded just before picking up the seeds. Drops are streaked sequentially, as speed is most important to prevent drying of the solution on the fibre.
5. Cover all the chambers without delay to reduce evaporation from the seeded drops.

The seed stock and seed dilutions, prepared as previously described, can be used in streak seeding reliably by dipping the whisker into each of the diluted solutions including the seed stock and applying the seeds to new drops (*Figure 3*). It is common to start by dipping the probe in the most dilute solution first, streaking one drop, then progressing up the dilution series to the seed stock. Typically, the results are analysed two days to one week after streaking.

3.2 Macroseeding

In macroseeding a single crystal is introduced into a suitably pre-equilibrated solution. Single prismatic crystals, which are free from twinning or any other crystallites, are most suitable for this technique. As in other seeding protocols it is important to take steps to maintain constant conditions, as even a slight dehydration of the drop being seeded could temporarily change the state of supersaturation and induce unwanted nucleation. Performing the experiment in a very humid environment and using large drops can reduce dehydration. A beaker with a filter paper cylinder soaked in distilled water is such an environment. However, it is more practical to use sitting drop multiwell trays, which have been used with high success in our laboratory. Macroseeding is done under a dissecting microscope where the small amount of heat generated from the microscope stage light bulb may actually be slightly beneficial in increasing the humidity level around the drop. The heat raises the temperature of the reservoir faster than that of the drop, increasing the rate of evaporation from the reservoir and counteracting evaporation into the room. Seeds are washed in a slightly dissolving solution to remove the top layer of protein, which contains possible defects, from the surface of the seed, without causing excessive etching or cracking. They are then transferred to a stabilizing solution to re-equilibrate the crystals (*Figure 4*). Older seeds benefit the most from this treatment, whereas freshly-grown crystals may be put directly through a series of washes in stabilizing solution. From the final wash solution each seed is then transferred to the protein–precipitant drop to be seeded (*Figure 4*).

3.2.1 Details of crystals handling in macroseeding

The handling of crystals, described in *Protocol 3*, is especially important in avoiding the generation of microseeds or unwanted nuclei in the transfer to the drop being seeded.

Protocol 3. Handling of crystals

1. Connect a glass or quartz capillary to a 1 ml glass syringe with a short piece of rubber tubing such as c-flex (Fisher #14–169–5c) which gives an excellent seal.

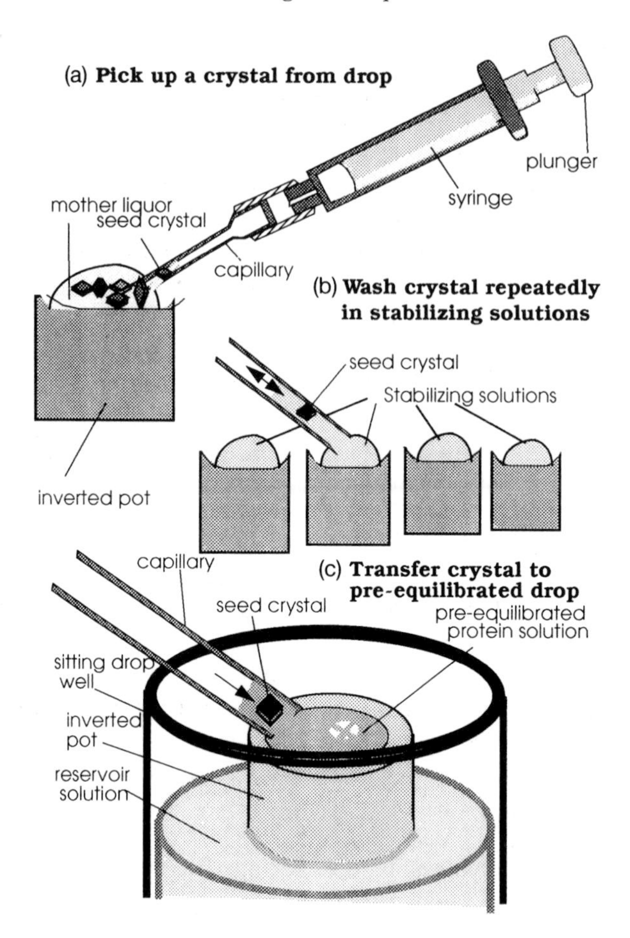

Figure 4. Illustration of the steps involved in macroseeding. (a) A single crystal is picked up from a drop. Crystals should be of good morphology and free from defects. (b) A series of washes is performed by repeatedly transferring the crystal from one depression to another, taking care not to damage the seed. (c) The seed is finally transferred to a pre-equilibrated drop for further enlargement.

Protocol 3. *Continued*

2. Snap open the end of the capillary with tweezers or scissors. Capillary may be siliconized if the experimental situation can benefit from a diminished adhesion of the solution to the glass capillary. After siliconizing it should be extensively washed.
3. Pick up crystals into the capillary under a dissecting microscope, using a magnification of 10–100 ×.

   Crystals from hanging drops should first be washed into a secondary

Protocol 3. *Continued*

vessel; in a sitting drop vapour diffusion set-up it is possible to pick up crystals directly from the depression in which they have been growing. After the coverglass sealing the vapour diffusion chamber is lifted off, the tip of the capillary is inserted into the drop, and a crystal is drawn into the capillary.

If the crystals are adhering to the well, by withdrawing liquid from the drop and gently ejecting it on to a chosen crystal, it is often possible to dislodge the crystal. Unfortunately some crystals have severe adhesion problems and cannot be dislodged without breaking them. For such problematic crystallizations, the depressions in glass pots should be coated with a thin film of Corning vacuum silicone grease before the protein–precipitant drop is added. This should be done throughout all experiments involving that protein, so that seeds will remain suspended on top of the grease, and the final crystals are mounted for X-ray diffraction use without the recurrence of this problem.

4. Once in the capillary the crystal is brought to the middle, and then allowed to sink and adhere sufficiently to the inside wall of the capillary so that the liquid can be moved over the crystal.
   (a) For crystals that fail to sink to the bottom and adhere to the capillary wall, a long hair or whisker may be wedged against the crystal to stop movement whilst liquid is drawn out.
   (b) For soft crystals there is the danger that during this procedure microseeds may be dislodged from them by the hair with obvious consequences. A series of washes will minimize the number of microseeds that will be transferred but not eliminate the risk that one or more will still be present in the solution which is transferred together with the crystal.
5. Once the crystal has been separated from the bulk of the mother liquor, the hair withdrawn, and the mother liquor ejected from the capillary and returned to the original drop, the crystal should remain in a small pool of liquid inside the capillary. Removing more of the remaining solution from around the crystal may help diminish the number of microseeds and aggregated material, and since the solution may now be at a higher precipitant concentration due to evaporation during the handling, transferring less of this solution may avoid creating conditions which are unsuited to the seeding.

The crystal can be repeatedly washed in a stabilizing solutions (typically the reservoir solution is used) prior to transferring it to the new drop as described in *Protocol 4*.

Protocol 4. Washing crystals

1. Fill four depressions of a multiwell sitting drop plate with about 100 μl of solution from the reservoir where the seeds originated (or a solution identical to this is prepared).
2. Fill the reservoirs around these drops with 1.5 ml of distilled water to maintain a high degree of moisture around the solution.
3. The crystal is repeatedly transferred and picked up from each of these stabilizing solution drops until finally it is picked up into the capillary. Because the addition of stabilizing solution to the new drop would unnecessarily modify the equilibrium, or dilute the equilibrated protein–precipitant solution, it is best to minimize the amount of liquid that remains around the crystal.
4. Remove excess liquid with a small thin strip of filter paper or a very thin capillary.
5. The crystal can now be resuspended in the new mother liquor drawn in from the drop to be seeded and returned to the well for equilibration and further growth. Alternatively, after the series of washes, the crystal may be allowed to sink down towards the open end of the capillary, so that when the capillary touches the solution of the drop being seeded, the crystal falls directly into this solution with little transfer of wash solution.

3.2.2 Macroseeding of needles

Macroseeding using needles as seeds is more complicated since needles have a tendency to bend while being transferred from the original growth solution to the new solution. The stress created in the crystals during this process may result in defects at each stress point, and each of these points may act as a nucleation site for growth of new needles. By breaking the needles with a sharp object (glass or metal) into smaller segments, the resulting fragments can be used for macroseeding. The sharper the instrument the less damage will be done to the seeds. A small number of these needle fragments are then transferred to a stabilizing solution. From this point the procedure is essentially the same as for prismatic crystals, as each is then transferred from this solution to another container with the same stabilizing solution, and then to a third and a fourth, to wash away any microseeds that may be transferred to the pre-equilibrated growth medium.

4. Heterogeneous seeding

The principle that there is a lower energy requirement in adding to an existing surface than in creating a new nucleus (Section 1.2) may hold for many surfaces. Such aggregation on to surfaces may be considered more of a problem than an advantage. However, regular surfaces, offering a charge distribution pattern which is complementary to a possible protein layer, could provide a suitable starting point for the nucleation of new crystals. The work of Alex McPherson (17) with various inorganic minerals provides strong support for the idea that regular planes are able to catalyse the nucleation of crystals of macromolecules, even if the lattice dimensions of the crystalline minerals differ from those of the resulting protein crystals. In these studies, nucleation occurred preferentially on the mineral substrate at a lower degree of supersaturation than was required for the same crystals to nucleate in the absence of the minerals. Crystals of related macromolecules can also be used to induce nucleation of proteins; the resulting crystals may maintain some, but not all, of the lattice dimensions or symmetry axes of the initial seeds. In such cases, where the protein in the crystals from which the seeds are obtained is related to the protein in the solution being seeded, the operation is termed cross-seeding.

4.1 Cross seeding

4.1.1 Cross-seeding between Fab–peptide complexes

In λ-type light-chain dimers, the dimers pack so as to form an *infinite β-sheet* maintaining one cell dimension in common, 72.4 (± 0.2) Å, along one of the 2_1 axes (18). Such packing in preferred planes for certain classes of protein molecules may indeed provide suitable surfaces for nucleation for other members of that class. A similar observation was made in the course of our work with different anti-hemagglutinin monoclonal Fab–peptide complex crystals, where it was noticed that these have a common crystal lattice plane with cell dimensions 73.0 (± 1.0) Å along the 2_1 axis and 66.4 (± 2.5) Å along one of the other axes (12, 13, 19, 20). This group of antibodies come from a panel of monoclonals raised against residues 75–110 of influenza virus hemagglutinin (HA1) (23, 24), and they recognize a short linear sequence in the 36-residue immunizing peptide as the major immunodominant site (HA1 101–106, DVPDYA) (23–25). Anti-peptide antibodies 26/9 and 21/8 also belong to this panel. Cross-seeding experiments with Fabs from these monoclonals as complexes with peptides of different lengths were performed by streak seeding. Fab 26/9 with a nine residue peptide (HA1 100–108) crystallizes by spontaneous nucleation from 15–18% PEG 10 000 in 0.2 M imidazole malate, pH 6.5–8.5. The quality of these crystals required improvement, and this was achieved by using the streak seeding technique as a microseeding method. Seeds from these crystals were used to search for growth conditions of complex crystals of Fab 26/9 with longer peptides, for which no conditions for spontaneous nucleation had been found. Both the 13-mer (HA1 98–110) and the 23-mer (HA1 88–110) peptide–Fab mixtures responded positively to the seeding. Seeds obtained from this cross-seeding

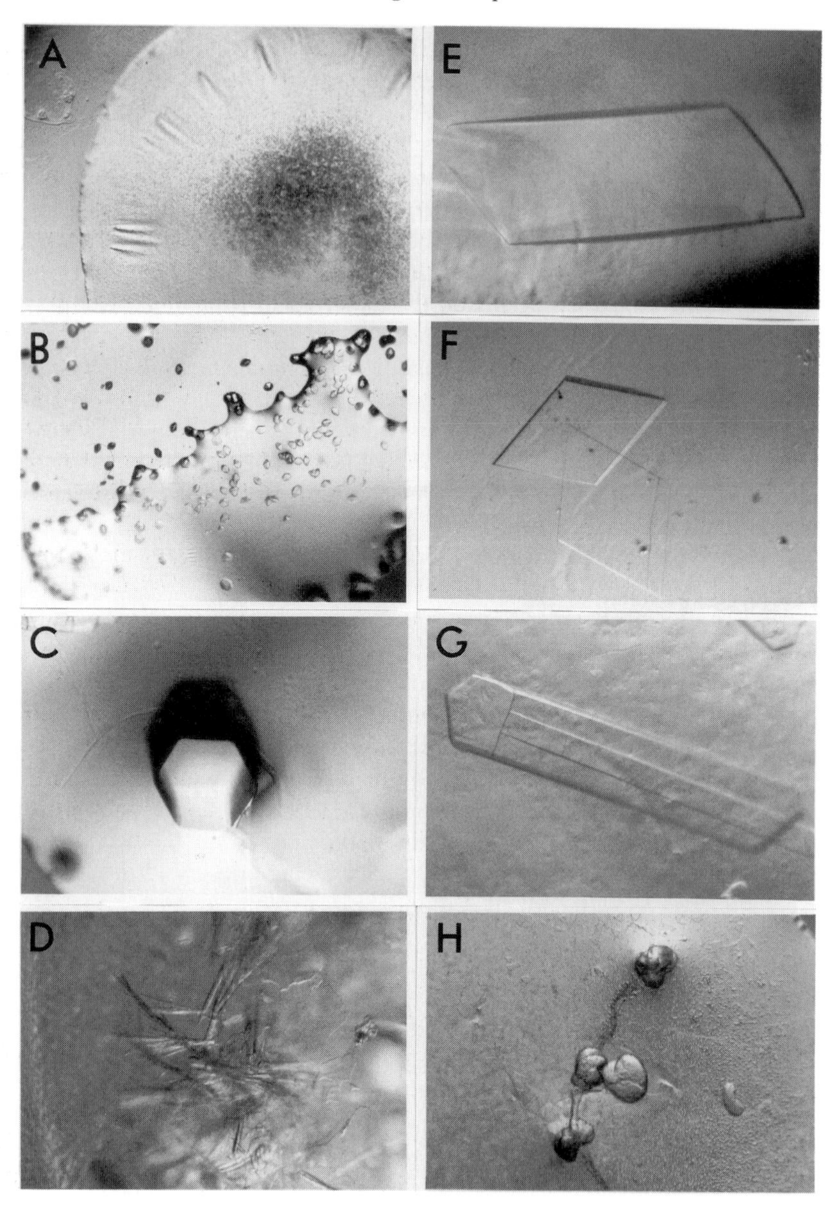

Figure 5. Photomicrographs of the results of crystallization experiments mentioned in the text. (a) The initial precipitate obtained in the crystallization of the complex of anti-peptide Fab′ B1312 with the 19 residue peptide corresponding to the C-helix of myohemerythrin. (b) When this precipitate was streaked onto drops under similar condition to those that yielded the precipitate crystal were obtained. (c) These crystals were used as macroseeding experiments to yield X-ray quality crystals. (d) Cross-seeding between anti-hemagglutinin Fab 26/9 and Fab 21/8. (e) Crystals obtained by spontaneous nucleation from 21/8 under the optimized conditions for the cross-seeded crystals.

were used to seed repeat experiments to dilute out the effect of the heterogeneous seeds. The concentration of PEG 10 000 required for optimal crystals of the complex between Fab 26/9 and the 13-mer peptide required a reduction to 12–15% and the pH range to 6.0–7.5 The crystals obtained from this second seeding are of X-ray quality diffracting to 2.5 Å resolution. The 13-mer–Fab complex crystals are morphologically similar to those of the complex with the 9-mer and belong to the same space group with a unit cell volume which is 3% larger. Seeds from Fab 26/9–13-mer complex crystals have been used to induce crystallization in solutions of Fab 21/8–13-mer mixtures, under identical crystallization conditions. The Fab 21/8–13-mer crystals first obtained by the streak seeding experiment (*Figure 5d*) were very thin needles. Optimization of the conditions resulted in better crystals, although seeding was still required. In the absence of seeding these refined conditions yielded large crystals, belonging to a different space group, with different cell dimensions (26/9–13-mer: $P2_12_12_1$, a = 73.6 Å, b = 104.4 Å, c = 67.0 Å, and 21/8-13-mer: P1, a = 76.4 Å, b = 63.8 Å, c = 76.5 Å, $\alpha = 96.6\,°$, $\beta = 124.5\,°$, $\gamma = 104.8\,°$ (*Figure 5e*). The two Fabs have heavy chains which belong to different classes; Fab 21/8 is derived from an IgG_{2b} while 26/9 is cleaved from an IgG_{2a}. To date no native Fab 21/8 crystals or complex crystals with the 9-mer have been obtained. Cross-seeding does not need to be carried out by streak seeding as, for example, large crystals from the chicken mitochondrial aspartate aminotransferase were used as seeds in the cross-seeding of the pig enzyme (4) (initial cross-seeded crystals of the chicken enzyme were badly twinned but were improved to X-ray quality in a second cycle of macro-seeding). However, streak seeding can substantially increase the speed with which crystals can be obtained.

4.1.2 Cross-seeding from native Fab to Fab–peptide complex

Anti-peptide Fab 50.1, that recognizes an epitope on the gp120 surface glycoprotein of HIV-1, is an example where native crystals were used to seed the Fab–peptide complexes. The native crystals can be grown in three different morphologies. Spontaneous nucleation has not been observed for any of the peptide complexes tested. The peptide lengths vary from 16–40 residues. Crystals of the Fab 13-mer complex have now been obtained by streak seeding with the native Fab crystals in 12–24% PEG 10 000, pH 5–8. In the Fab–peptide solution most of the protein is found partitioned in a gel

(f) Crystals of anti-HIV-1 Fab 50.1 complexed with a 13-residue peptide from the glycoprotein gp120 sequence of the MN isolate. The crystals were obtained by cross-seeding from native crystals (g) that grow spontaneously under similar conditions. (h) Epitaxial nucleation on a cellulose fibre. The first crystals for this Fab nucleated from a drop without fibres after a period of six months. The seeded crystals were obtained in less than a week. Notice the number of crystals nucleated on the fibre compared to the number nucleated separately.

phase covering the bottom of the drop. On addition of native seeds, crystals grow by acquiring protein from the gel phase surrounding the seeds. The morphology of the Fab–peptide complex crystals differs substantially from that of the native crystals (*Figure 5g*). It is also worth noting that while the crystals of the native Fab are very mosaic, data have been collected on well-ordered crystals of the seeded complex that diffract to better than 2.8 Å (*Figure 5f*).

4.1.3 Cross-seeding chemically modified proteins

As a last example we will consider cross-seeding to grow crystals of engineered proteins. The solubility of a protein can be easily altered when it is modified either chemically or by site-directed mutagenesis. Cross-seeding was used to produce crystals of a chemically modified subtilisin. Crystallization experiments with selenolsubtilisin, which results when serine 221, the active serine of the bacterial protease (26–28), is converted into a selenolcysteine (29), yielded no crystals spontaneously, even after extensive efforts were made to better purify the engineered enzyme (R. Syed, personnal communication). Crystals of the commercially available native subtilisin, from which the selenolsubtilisin was modified, were grown for the sole purpose of cross seeding from these to the modified enzyme. Since the quality of the starting material (without further purification) was low, the quality of the resulting crystals was rather poor. However, crystals of the selenolsubtiiisin obtained from the cross-seeding experiment were of better morphology and size than those of the original native enzyme. These crystals grew under conditions similar to those that gave crystals for the native subtilisin. Under optimized conditions it is now possible to obtain good quality crystals, which nucleate spontaneously, as long as preparations are of sufficient purity. The structure of selenolsubtilisin has been solved by molecular replacement with the selenol clearly defined in the electron-density map (R. Syed, personal communication). This is just one example where the streak-seeding technique has been applied when working with protein preparations that prove difficult to crystallize otherwise.

4.2 Epitaxial nucleation

Epitaxial nucleation is a particular instance of adhesion where the regularity of the surface facilitates nucleation. Many substrates mediate adhesion of proteins, and precautions need to be taken to avoid this interaction. Glass surfaces are generally siliconized, but even after this treatment crystals are still found to be preferentially attached to the siliconized surfaces. The strength of the interaction can be stronger than the forces that bond the crystalline lattice. In crystallization trials it is possible to find many instances in which crystals or microcrystals can be nucleated on cellulose fibres which are accidentally present in the protein–precipitant drop (*Figure 5h*). Because of the regularity of the fibres this can be considered a case of epitaxial nucleation. In most cases microcrystals are also observed spontaneously nucleating, under other conditions, away from foreign particles. The nucleation of crystals from aggregates and oils can also be considered a case of epitaxial nucleation. Here ordered surfaces may be present within the random aggregation of macromolecules which come out of solution as oils and precipitates. These surfaces may provide platforms suitable for macromolecular nucleation, and may possible support three-dimensional crystal growth. The streak-seeding technique can be used not only with small unusable crystals but also with any promising aggregate or precipitate, to test for the possibility that ordered planes within such aggregates may be able to stimulate the growth of crystals, or that such aggregates may be polycrystalline.

5. Crystallization of complexes

5.1 Considerations in the crystallization of complexes

When crystallizing a complex, e.g. a protein and its receptor, a protein and an effector, or an Fab–antigen complex, it is important to consider the heterogeneity of the resulting system. Both members of the complex will be somewhat heterogeneous, and the resulting mixture will be composed of complexed and uncomplexed molecules in different ratios depending on the molar ratio of the two molecules in the solution and the dissociation constant. To reduce the heterogeneity it is important to optimize the number of complexed versus the uncomplexed molecules. For protein complexes with small ligands it is possible to increase the number of complexed protein molecules by adding an excess of ligand. Theoretically, the excess ligand that is necessary to achieve the desired ratio of bound to unbound may be calculated from the dissociation constant if known. In practice it is best to set up experiments at different protein–ligand ratios, typically 1:1 to 1:20, since large excesses of ligand have been found to be unnecessary or to even inhibit crystal growth. Enzymes present another level of complexity when cocrystallization of enzyme substrate complexes is attempted. Catalysis of the substate into product will result in a mixture of free enzyme, enzyme–product complexes, and enzyme–substrate complexes. These crystallizations should be attempted by using non-productive substrate analogues, inhibitors that mimic the transition state analogue and undissociable end-products.

When the complex consists of proteins of comparable size the addition of an excess of one macromolecule does not decrease the amount of heterogeneity in the system. Only a limited number of such complexes have been crystallized to date, and so generalizations concerning the best ratios to use cannot yet be drawn. In the crystallization of the complex between the p55 IL-2 receptor (mol. wt = 44 000) and IL-2 (mol. wt = 14 000) it was found that crystals grew best when ligand and receptor were present in a 1:1 to 2:1 molar ratio (30). When considering the crystallization of such complexes we must take into consideration the theoretical possibility of binding an uncomplexed molecule to the lattice of the complex between the two molecules. In order for the crystal to continue its growth without defects the unbound molecule must either become complexed while still maintaining its lattice contacts, or else it must break all lattice bonds and diffuse away from the crystal surface to be replaced by a complexed molecule. The energy involved in each of the lattice interactions that must be broken will determine the inhibitory effect of the uncomplexed molecule with respect to the growth of the complex crystal. Assuming that the number of lattice bonds is proportional to the surface area of the molecule, the strength of each bond is roughly equal to one another, and that the molecules are roughly spherical in shape, we reach the conclusion that the number of bonds to be broken to dislodge the uncomplexed molecule is proportional to mol. $wt^{2/3}$. Hence a molar excess of the smaller molecule is expected to favour the crystallization of the complex. For this to be true crystallization must proceed slowly enough for the crystal not to incorporate the unbound molecule as a defect. If we consider the deleterious effect of incorporating a defect into the lattice the absence of the larger of the two molecules will carry a greater penalty. In the initial search for complex crystals between two proteins a 1:1 ratio of the two molecules or a small excess of the large molecule should be used for this reason. In cases where the affinity between the molecules in the complex is not high, and the off rate is substantial, we must consider the relative solubility of the complex versus the two uncomplexed molecules. In the case where the complex is less soluble than either of the two native molecules the situation is relatively simple, and in order to obtain a higher ratio of complexed to uncomplexed molecules it is important to use high protein concentrations of both molecules. If either or both molecules are less soluble than the complex, lower protein concentrations should be used as the precipitated native proteins will not be available to form crystals of the complex. For screening purposes the use of lower concentrations will allow for a larger number of trials to be attempted. When producing crystals for crystallographic work higher protein concentrations may be used, combined with seeding, early in the equilibration phase of the crystallization.

5.2 Use of streak seeding in protein–complex crystallization

The optimal ratio of the two molecules in the complex for the nucleation of crystals is not necessarily the optimal ratio for crystal growth. By using streak seeding, with different ratios of the two molecules, it is easy to determine the optimal ratio and concentrations of the two molecules experimentally by analysing the response along the streak line. While initial success in obtaining microcrystals, microcrystalline aggregates or even crystals can be achieved with less than absolutely pure macromolecules, as determined by SDS–

PAGE and isoelectric focussing (IEF) gels, it is important, as in all crystallizations, to attempt to obtain higher purity of the two molecules in order to grow X-ray quality crystals. The use of affinity columns as in the IL-2R-IL2 work (30) ensures that the only native molecules that can form complexes are present in the crystallization trials. However, it is also important that the subsequent elution conditions do not affect adversely the molecules being crystallized. When either or both molecules in the complex also crystallize in their native state, testing for their ability to crystallize by themselves will provide information on the effects of the affinity column and on their purity. This was important in the crystallization of the complex between Fab U/108 (mol. wt = 55 000) and thaumatin (mol. wt = 22 000 (31)]. Complex crystals with the Fab were obtained in our laboratory only from batches of thaumatin that gave good native crystals. In the work with the enzyme glycinamide ribonucleotide transformylase (GAR-Tfase) crystals were obtained both of the native enzyme and its complex with a powerful inhibitor, each in a different crystal form. With the inhibitor bound the enzyme becomes more soluble and crystallizes spontaneously from 20% PEG 6 000, while the native enzyme forms a precipitate at this concentration. The native crystallizes readily with either ammonium sulphate or phosphate (32). Initial results have shown that the precipitate of the native in PEG can be converted to the crystalline state when seeded with complex crystal seeds. Similarly, crystals can be grown for the complex in high salt by streak seeding with native crystal seeds from phosphate or sulphate.

5.3 Analytical techniques for determination of crystal content

5.3.1 Protein–protein complexes

SDS–PAGE (33) can be used to determine whether crystals of a putative complex contain both the protein and the Fab. Crystals must be separated from the mother liquor, from which they have been growing, and must be washed to remove material that may be deposited on the crystal surface rather than being an integral part of the crystalline lattice. The procedure described for the handling of crystals for macroseeding may be used here. The crystals are subsequently dissolved in distilled water typically with a final volume of a few microliters. The SDS–PAGE gel is then run on this protein sample. Crystals that do not dissolve under these conditions may have to be dissolved under more acidic, more basic, or higher-salt conditions, or by adding SDS solution directly to the crystals. When using high salt to dissolve the crystals, the salt may have to be dialysed out before running the polyacrylamide gel.

5.3.2 Fab–peptide complexes

The above technique cannot be applied when the Fab ligand is a small peptide.

In this case a native-polyacrylamide gel electrophoretic analysis (native-PAGE) (34, 35) using a PhastGel Gradient 8–25 (Pharmacia) of the native and complexed Fabs in solution, and on the dissolved crystals of the putative complexes, is the method of choice in the determination of the molecular composition of the crystals. The bound peptide modifies the mobility of the Fab, and usually shows as an identifiable shift in the position and distributions of the protein bands.

5.3.3 Chemical reactions in the crystal

The channels in protein crystals are typically large enough to allow for the diffusion of many small molecules throughout the lattice. Heavy-atom derivatization relies on this fact. If a chemical reaction can be done on the compounds which are presumed bound to the protein in the complex, so that a colour or fluorescence can be developed, the reaction can be tried on the crystals. In the crystallization of steroid complexes of the anti-progesterone Fab DB3 (36, 37), for example, the presence of the steroids needed to be verified before undertaking extensive data collection. Progesterone has a free ketone group at position 3 on the A ring. Reaction with 2,4-dinitrophenyl-hydrazine was used to detect the free steroidal ketone. The reagent was diffused into the crystals for 5 min. Acidification with dilute HCl then resulted in the dissolving of the outside of the crystals leaving a brownish-red precipitate, demonstrating the presence of the steroid. As a control, the same reaction was repeated with native DB3 crystals which did not dissolve, but developed a light yellow colour, while crystals of an unrelated F_{ab} remained clear. From these results subsequent determination of the content of the native preparations indicated that approximately 10–20% progesterone was present. Crystals of true native have been grown from steroid-free preparations by growing the antibody in cell culture.

6. Concluding remarks

The application of seeding methods in macromolecular crystallization has proven invaluable for obtaining high-resolution X-ray quality crystals when conventional methods have failed. It provides a means of analysing many conditions without requiring large amounts of protein. Streak seeding is particularly valuable as it provides a fast method of analytical seeding with easy visualization of the results. Cross-seeding is a powerful tool to crystallize a given protein with seeds from a related protein. The application of micro and macroseeding methods can result in the production of large single crystals for X-ray structure determination. Without such methods many of the projects that are currently being tackled in our laboratory would not have been viable.

Acknowledgements

We would like to thank Dr Richard Lerner for the initial anti-peptide antibody hybridomas produced in his laboratory by Dr Henry Niman (anti-HA) and Dr Terry Fieser (anti-Mhr Fab′ B13I2). We appreciate the excellent work of Gail Fieser and Warren Densley for the present production and purification of Fab 26/9 and Fab 21/8, and Rod Samodal for Fab′ B13I2. We thank Dr Robyn Stanfield for her crystallization data for Fab′ B13I2 and Dr Rashid Syed for unpublished results on selenolsubtilisin. We thank our collaborators Dr Steven Benkovic for the GAR-Tfase and Dr Sung Hou Kim for Fab U/108 and Thaumatin. We are grateful to A1 Profy and RepliGen Corp. for help and collaboration with the anti-HIV-1 antibody 50.1 and peptides from the gp120 surface glycoprotein. This work was supported by National Institutes of Health Grants AI-23498, GM-38795, and GM-38419 (to I.A.W.). This is publication #6569-MB from The Scripps Research Institute.

References

1. Wilson, I. A., Rini, J. M., Fremont, D. H., and Stura, E. A. (1991). *Methods in Enzymology*, **203**, 153.
2. Carter, C. W., Jr. and Carter, C. W. (1979). *J. Biol. Chem.*, **254**, 12219.
3. Betts, L., Frick, L., Wolfenden, R., and Carter, C. W., Jr. (1989). *J. Biol. Chem.*, **264**, 6737.
4. Eichele, G., Ford, G. C., and Jansonius, J. N. (1979). *J. Mol. Biol.*, **135**, 513.
5. McPherson, A. (1982). *Preparation and analysis of protein crystals*. John Wiley and Sons, New York.
6. Blundell, T. L. and Johnson, L. N. (1976). *Protein crystallography*. Academic Press, New York.
7. Arakawa, T. and Timasheff, S. N. (1985). *Methods in Enzymology*, **114**, 49.
8. Feher, G. and Kam, Z. (1985). *Methods in Enzymology*, **114**, 77.
9. McPherson, A. (1985). *Methods in Enzymology*, **114**, 112.
10. Thaller, C., Weaver, L. H., Eichele, G., Karlsson, R., and Jansonius, J. (1981). *J. Mol. Biol.*, **147**, 465.
11. Thaller, C., Eichele, G., Weaver, L. H., Wilson, E., Karlsson, R., and Jansonius, J. N. (1985). *Methods in Enzymology*, **114**, 132.
12. Stura, E. A. and Wilson, I. A. (1991). *J. Crystal Growth*, **110**, 270.
13. Stura, E. A. and Wilson, I. A. (1990). *Methods, A Companion to Methods in Enzymology*, **1**, 38.
14. Fieser, T. M., Tainer, J. A., Geysen, H. M., Houghten, R. A., and Lerner, R. A. (1987). *Procl. Natl. Acad. Sci. USA*, **84**, 8568.
15. Stura, E. A., Stanfield, R. L., Fieser, T. M., Balderas, R. S., Smith, L. R., Lerner, R. A., and Wilson, I. A. (1989). *J. Biol. Chem.*, **262**, 15721.
16. Stanfield, R. L., Fieser, T. M., Lerner, R. A., and Wilson, I. A. (1990). *Science*, **248**, 712.
17. McPherson, A. (1989). *Scientific American*, **260**, (3), 62.

18. Schiffer, M., Chang, C.-H., and Stevens, F. J. (1985). *J. Mol. Biol.*, **186**, 475.
19. Schulze-Gahmen, U., Rini, J. M., Arevalo, J. H., Stura, E. A., Kenten, J. H., and Wilson, I. A. (1988). *J. Biol. Chem.*, **263**, 17100.
20. Wilson, I. A., Bergmann, K. F., and Stura, E. A. (1986). In *Vaccines '86'* (ed. R. M. Channock, R. A. Lerner, and F. Brown), pp. 33–37. Cold Spring Harbor Laboratory, Cold Spring Harbor, NY.
21. Wilson, I. A., Skehel, J. J., and Wiley, D. C. (1981). *Nature*, **289**, 366.
22. Niman, H. L., Houghten, R. A., Walker, L. E., Reisfeld, R. A., Wilson, I. A., Hogle, J. M., and Lerner, R. A. (1983). *Proc. Natl. Acad. Sci. USA*, **80**, 4949.
23. Wilson, I. A., Niman, H. L., Houghten, R. A., Cherenson, A. R., Connolly, M. L., and Lerner, R. A. (1984). *Cell*, **37**, 767.
24. Houghten, R. A. (1985). *Proc. Natl. Acad. Sci. USA*, **82**, 5131.
25. Houghten, R. A., Hoffman, S. R., and Niman, H. L. (1986). In *Vaccines '86'* (ed. R. M., Channock, R. A. Lerner, and F. Brown), pp. 21–25. Cold Spring Harbor Laboratory, Cold Spring Harbor, NY.
26. Markland, F. S., Jr. and Smith, E. (1971). In *The enzymes*, vol. (3rd edn), (ed. P. D. Boyer), pp. 561–608. Academic Press, New York.
27. Kraut, J. (1971). In *The enzymes*, vol. 3 (3rd edn), (ed. P. D. Boyer), pp. 547–560. Academic Press, New York.
28. Neidhart, D. J. and Petsko, G. A. (1988). *Protein Eng.*, **2**, 271.
29. Wu, Z.-P. and Hilvert, D. (1989). *J. Am. Chem. Soc.*, **111**, 4513.
30. Lambert, G., Stura, E. A., and Wilson, I. A. (1989). *J. Biol. Chem.*, **264**, 12730.
31. DeVos, A. M., Hatada, M, Van der Wel, H., Krabbendam, H., Peerdermann, A. F., and Kim, S.-H. (1985). *Proc. Natl. Acad. Sci. USA*, **82**, 1406.
32. Stura, E. A., Johnson, D. L., Inglese, J., Smith, J. M., Benkovic, S. J., and Wilson, I. A. (1989). *J. Biol. Chem.*, **264**, 9703.
33. Laemmli, U. K. (1970). *Nature*, **227**, 680.
34. Andrews, A. T. (1981). In *Electrophoresis. Theory, techniques and biochemical and clinical applications*, (ed. A. R. Peacocke and W. F. Harrington), pp. 1–91. Clarendon Press, Oxford.
35. Hames, B. D. (1981). In *Gel electrophoresis. A practical approach*, (ed. B. D. Hames and D. Rickwood). IRL Press Limited, London.
36. Stura, E. A., Feinstein, A., and Wilson, I. A. (1987). *J. Mol. Biol.*, **193**, 229.
37. Stura, E. A., Arevalo, J. H., Feinstein, A., Heap, R. B., Taussig, M. J., and Wilson, I. A. (1987). *Immunology*, **62**, 511.

6

Crystallization in gels and related methods

M. C. ROBERT, K. PROVOST, and F. LEFAUCHEUX

1. Introduction

Gel growth is a particular case of solution growth and consequently is ruled by the same parameters; however, some of the parameters can be varied much more conveniently by gelling the growth medium.

After a short description of gel structures allowing identification of the parameters involved, typical techniques used for macromolecule crystal growth are reviewed in order to explain how the techniques are tailored for gel growth. Practical advice for the preparation of different gel-growth experiments is given together with comments about the expected results. To some extent, gel growth has features common with growth under reduced gravity or hypergravity. These related techniques are considered as well.

2. Basic considerations

The gels used for crystal growth are hydrogels; they are two-component media with the growth solution soaking a microporous flexible polymer network (*Figure 1*). The gelation process corresponds to the setting of a polymeric cluster stretching over the whole volume of solution. This process is either reversible for physical gels (like gelatine and agarose gels) which are obtained by temperature decrease, or irreversible for chemical gels (such as silica or polyacrylamide gels) which are obtained by the formation of strong bonds. Although a universal 'best' gel cannot be recommended (1), silica gels and agarose gels have proved their efficiency for growing macromolecular compounds (2, 3).

2.1 Preparation and structure of gels

2.1.1 Silica gels

The silica gel presents several advantages: it is stable, usable over a large range of temperature (0–60 °C), and compatible with many substances due to

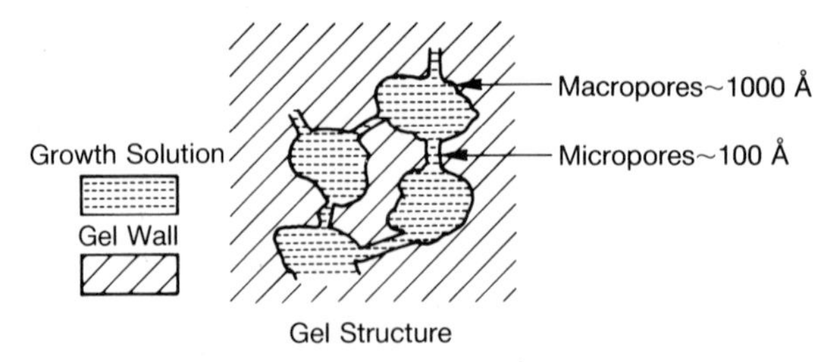

Figure 1. Schematic drawing of a gel growth medium.

having a chemical composition close to glass. The gel is easy to prepare by hydrolysis of siloxanes like tetramethoxysilane $Si(OCH_3)_4$ (TMOS) or tetraethoxysilane $Si(OC_2H_5)_4$ (TEOS), which lead to the formation of Si–O bonds. In the case of TMOS, the reaction is written:

$$-\overset{|}{\underset{|}{Si}}-OCH_3 + HOH + CH_3O-\overset{|}{\underset{|}{Si}}- \rightarrow -\overset{|}{\underset{|}{Si}}-O-\overset{|}{\underset{|}{Si}}- + 2\,CH_3OH$$

The polymerization reaction proceeds with the formation of loops and spherical particles which can either grow or aggregate according to the ionic strength, pH, and chemical composition of the gelling medium (4). The gel walls are formed by a three-dimensional arrangement of particle chains. Electron microscopy images and light-scattering experiments (5) allow estimation of the size of the largest gel pores, which can vary from 1000 nm for a 2% (v/v) TMOS gel to 250 nm for a 5% (v/v) TMOS gel. These macropores are interconnected through a dense system of micropores.

2.1.2 Agarose gels

The agarose gel is widely used for biological macromolecule separation. Its structure and gelling process are not fully understood, however. It is extracted from seaweed and consists of polysaccharide chains carrying different substituents (O-methyl, O-sulphate . . .). The gelation process proceeds through the association of chains in double helices and subsequent aggregation of several helices to form interconnected fibres (6). This fibrous network leaves large voids through which very large molecules can diffuse. The final structure depends on the characteristics of the gelling medium, as for silica gels, but also on the agarose source which can contain various types and amounts of chemical substituents; e.g. the gelling temperature is related to the methyl group content.

2.2. Gel properties related to crystal growth

2.2.1 Diffusion of solutes

Entrapping of growth solution by the gel network prevents the onset of natural convection which are density-gradient induced movements of macroscopic volumes of liquid. These gradients are unavoidable in growth solutions because they correspond to driving forces for mass transfer to the growing interface. In gel media, mass transfer proceeds by diffusion through the gel pores which provides a regular and somewhat adjustable supply of solute. The diffusion coefficients are not significantly different for light gels (TMOS gels $< 1\%$ v/v or agarose gels $< 0.2\%$ w/v) when compared to free solution: 11×10^{-7} cm^{2}s^{-1} in solution instead of 9×10^{-7} cm^{2}s^{-1} in 2% (v/v) TMOS gel for hen egg white (HEW) lysozyme. This decrease might be more pronounced for larger macromolecules and it could be interesting to measure the diffusion coefficients in a wide range of molecular weight.

2.2.2 Suspension of crystals

Solid particles such as growing crystals are also entrapped by the gel network so that they do not form sediments and can grow free from strain exerted by walls, holders, or other crystals. The only strain comes from the gel itself, but due to the low elastic limit (2) the gel generally fissures all around a growing crystal forming a cusp-like cavity (7). For fragile crystals, very light gels are recommended in order to minimize the gel resistance.

2.2.3 Influence of gel on crystal nucleation

Different types of crystal nucleation (see Chapter 10 for definitions) must be considered:

i. Heterogeneous nucleation

When the gelling process is achieved in the growth solution, heterogeneities such as dust, walls, or gas bubbles that serve as active sites of initiation of the polymerization reaction or cross-linkings of polymers, are embedded in a gel layer, thus becoming inefficient in generating crystal nuclei. This effect of gel is beneficial because foreign particles inducing strains can lead to unusable crystals.

ii. Secondary nucleation

This concerns nuclei generated from particles pulled out of a primary crystal by fluid shear forces (convection) or by impact of the crystal against the growth cell wall or neighbouring crystals. As solid and fluid movements are prevented in gels, such parasitic crystals cannot appear.

iii. Homogeneous nucleation

The solubility curves which correspond to an equilibrium state between crystal and mother solutions are not changed by working in gel medium (for the same solution composition). The critical supersaturation (minimum supersaturation required to get a nucleus) can be significantly changed, however. The nucleation rate may also be different; e.g. the number of nucleated crystals decreases by increasing the silica gel content (2). It is the reverse with agarose gels; the number of crystals increases by increasing the agarose content (3). These effects are not quite well understood especially in the latter case but it is of practical interest to have nucleation inhibiting or nucleation promoting gels at one's disposal.

2.2.4 Crystal growth parameters influenced by the presence of gel

Taking into account the above considerations, one can select from a list of parameters influencing biological macromolecule crystal growth (see *Table 1*, Chapter 1, and reference 8) those which are influenced by the presence of a gel structure:

- parameters related to supply of reactants:
  concentration of macromolecules or of crystallizing agent,
  pH variation,
  level of reducing agent or oxydant
- parameters related to mechanical behaviour of solid or liquid phases:
  gravity,
  density gradient,
  convection,
  vibration,
  sound (acoustic waves)

One cannot reasonably expect that gel can improve crystal growth with regard to biochemical parameters like purity or degree of denaturation of biological macromolecules (detrimental reactions between protein and gel polymer can even be feared). However, one can expect an indirect effect on the ageing of proteins since the growth time can be reduced; this requires the application of a very high supersaturation which would be impossible under regular growth conditions in free solution but which can be accommodated in a gel.

2.2.5 Crystal quality of gel-grown crystal

Characterization of the crystalline quality of gel-grown crystals of biological macromolecules is still in progress but it has been demonstrated for mineral crystals and for low molecular weight organic compounds (9) that the growth defect density (dislocations, growth bands) is drastically reduced by working in gel media. These studies have shown that the concentration of reactants must be increased with respect to those used in a convective medium, to avoid the formation of inclusions (lack of matter) in order to counterbalance the slowness of diffusive mass transfer.

The impurity content of gel-grown crystals is not lower than in solution but the impurities are distributed much more homogeneously (10). The resulting strain is reduced: it has been shown therefore that in the case of low molecular weight materials, it is possible to overdope such crystals.

3. Practical considerations

Following preliminary crystallizations in solution, which have led to crystals which are too small (due to an excessive nucleation), defective, and unusable for X-ray diffraction, a gel experiment might be undertaken. Too limited nucleation also poses some problems because it requires a very high initial supersaturation which is incompatible with regular further growth. In such cases, and for the above mentioned reasons, a gel experiment can be attempted in parallel with classical crystallization. This means that the same chemical parameters would be chosen (buffer, crystallizing agent, and so on) and the same growth method. Then, by examining the results of these first tests, one would be able to improve the growth conditions (changes of concentration or modification of the growth techniques).

Among crystal growth methods of biological macromolecules (see Chapter 4), liquid–liquid diffusion and vapour phase diffusion techniques can be transposed in agarose gels. Silica gels are not very well adapted to vapour phase diffusion because dehydration of the gel is not homogeneous and it leads to a dried skin on its surface.

The first task is to attempt to gelify the different solutions of the growth medium, first separately according to the protocols indicated below in order to find what path can be used for gel growth; i.e. one should test the possibility of gelifying in the presence of crystallizing agent (salts, PEG, MPD, and so on) at different concentrations.

3.1 Preparation of silica gels

Silica gels are prepared from siloxanes. TMOS and TEOS are liquids very soluble in alcohols (ethanol or methanol) but not directly in water. Two steps of the gelling process have to be considered separately: hydrolysis and polymerization. *Figure 2* shows the different steps of the gel preparation.

3.1.1 Hydrolysis

The hydrolysis reaction consumes water and releases alcohol (see Section 2.2.1); for a given TMOS gel of percentage p, this corresponds to $(2.5 \times p)$ cm^3 water and $(10.9 \times p)$ cm^3 methanol per litre of solution. This has to be taken in account for solubility considerations. Hydrolysis occurs through vigorous stirring of siloxane droplets in water; this emulsifying provides a large contact surface where hydrolysis reactions can take place. First the mixture looks like an oily emulsion in water, then, when no parasitic reaction occurs, it becomes clear and homogeneous. This stirring precludes a direct dissolution of siloxane in protein solutions. TMOS (or TEOS) is preferably added and homogenized, e.g. in the buffer solution prior to dilution with protein or salt solution (see *Figure 2*). This homogenization must be achieved as quickly as possible because it competes with the polymerization step which

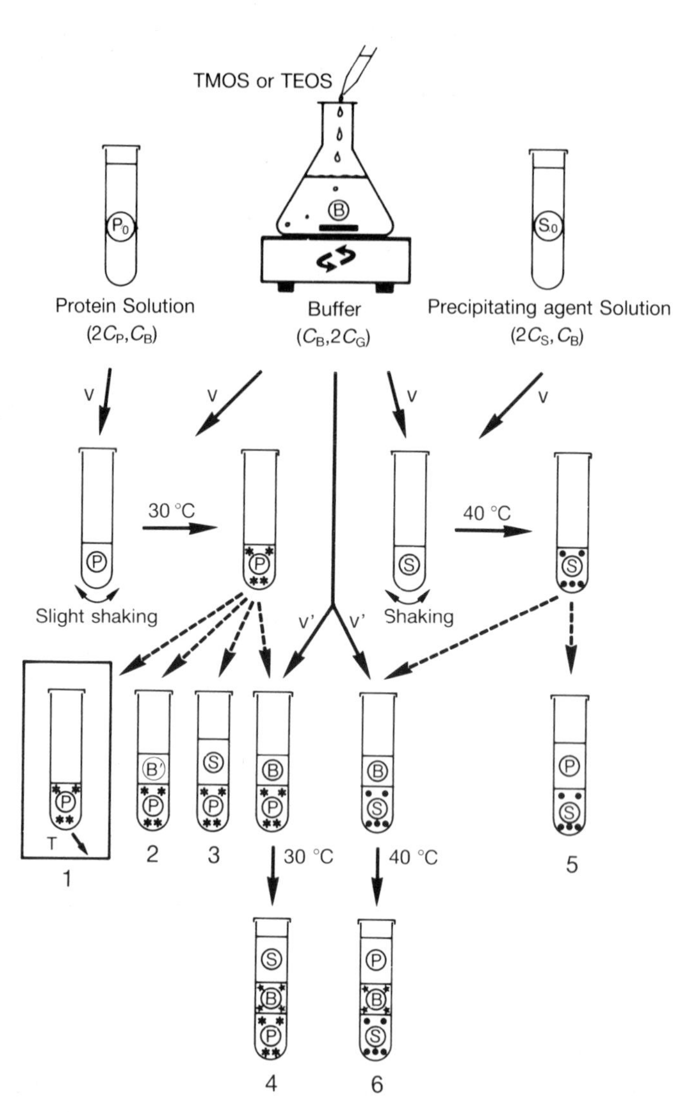

Figure 2. Operational sequence for preparing a gelled protein solution P or a gelled crystallizing agent solution S by emulsifying of TMOS or TEOS in buffer solution B. The final concentration in P are C_P (protein) and C_B (buffer) for cases 1, 2, 3, and 4. The final concentrations in S are C_S (crystallizing agent) and C_B (buffer) for cases 5 and 6. Gelling is speeded up by low temperature thermal treatment (30–40 °C). Growth occurs by temperature variation (case 1) or by pouring a liquid layer of reactant (buffer B′ in case 2, crystallizing agent in case 3, protein in case 5). Cases 4 and 6 are identical to cases 3 and 5 for the external layers with a gelled buffer layer inserted in-between.

begins as soon as monomers are available in the medium. Polymerization building the gel network must occur in a very quiescent medium. The homogenization is all the more rapid as the pH is low; e.g. this step which is rapid with a mixture TMOS-acetate buffer at pH 4.7 is much too slow with TEOS. In this case, one can first mix TEOS with one part of 0.1 M acetic acid solution (the mixture becomes rapidly clear) and add an equal part of 0.1 M sodium acetate.

As a rule, when mixing is difficult in a given buffer, one can attempt to dissolve separately the siloxane in one of the constituents of the buffer before completing with the other constituents. The polymerization reactions being very sensitive to temperature, it is possible to delay this step by lowering the temperature in a water-ice bath during mixing.

Siloxanes are corrosive liquids. Careful protection of skin and eyes are recommended. All vessels in contact with them must be thoroughly rinsed with alcohol prior to water cleaning.

3.1.2 Polymerization

Only clear transparent and homogeneous siloxane-solution mixtures can lead to usable gels. These mixtures are poured in clean, dried crystallization containers (small test tubes, capillaries, dialysis button, and so on) and allowed to gelify without mechanical disturbances. Surface dehydration of gels must be avoided either by sealing them with a minimum air volume enclosed or by closing them in a vessel containing a reservoir of solution giving the suitable vapour pressure of water.

The gelation time depends on many parameters: nature and concentration of species in solution, pH, temperature. In the case of thermally stable solutions one can shorten this time by increasing temperature. Practical considerations require gelation times less than a couple of days. One estimates that the gel is set when it resists pouring. It must stick to the crystal growth cell walls. It can look somewhat opalescent but without heterogeneities such as fissures.

3.2 Use of silica gels

Since the solubility of proteins varies as a function of many parameters (see *Table 1* of Chapter 1), different crystal growth protocols numbered 1–6 in *Figure 2*, are discussed below.

- Temperature.
  The gelled protein solution P is inserted in an incubator whose temperature is continuously lowered (or increased) (see *Figure 2*, 1).
- pH.
  A convenient buffer solution (B′) is layered over the gelled protein solution P (see *Figure 2*, 2).

- Crystallizing agent.
  (a) A solution of the crystallizing agent (S) is layered over the gelled protein solution P (see *Figure 2*, 3).
  (b) If the protein solubility varies strongly with the crystallizing agent concentration, insert a gelled buffer layer (B) between the gelled protein solution and the crystallizing agent solution to avoid 'plugging' of the interface by a dense precipitate (see *Figure 2*, 4).
- If the protein solution cannot be properly gelified, then pour it on a gelled crystallizing agent solution layer (see *Figure 2*, 5). In this case, if a too rapid variation of solubility occurs, a gelled buffer solution is inserted between the gelled crystallizing agent solution layer and the protein solution layer (see *Figure 2*, 6). The crystals will grow preferentially in the protein solution layer and some advantages of gel are lost, but others are preserved, e.g. the progressive supply of salt or the soft interface as crystal holder.

  From these basic configurations, many other possibilities can be imagined in order to increase the supersaturation. One can combine two preceding configurations, e.g. temperature variation and addition of crystallizing agent (1 and 3). It is also possible to presaturate the protein layer by adding a small amount of crystallizing agent in the protein solution layer just before gelling it, or to gel the three layers as shown in *Figure 2*, 4, or 6.

One can also play with the geometrical parameters, volumes of the different layers, section of the tube, but due to the low value of protein diffusion coefficients it is suggested to keep the diffusion distances as small as possible (isometric volumes). This is illustrated in *Figure 3* of reference 2: when a concentration equal to 1 (arbitrary unit) is set at the boundary of a gel layer, six hours later the concentration is only 1/100 at 5 mm away from this boundary for a diffusion coefficient of 10^{-6} cm^2s^{-1}.

Dialysis techniques can be directly extrapolated from these different configurations by setting a dialysis membrane on the solution to be gelled so that, after gelling, the gel does not present discontinuities at the level of the membrane. One can also use dialysis membranes to introduce macromolecules in a buffer gel previously set: the protein solution is simply percolated through a gel column supported by a dialysis membrane. This avoids a dilution of mother material.

Figure 3 shows examples of configurations 3, 5 and 6 (of *Figure 2*), in the case of (HEW) lysozyme crystals grown in acetate buffer solution using sodium chloride as crystallizing agent. The gelled protein solution layer is rather opalescent (*Figure 3a*), but some crystals growing in the bulk far from the interface can be seen. The gelled salt solution (*Figure 3b*), or the gelled buffer solution (*Figure 3c*), is clear and transparent: in the last two cases most of crystals have nucleated in the protein solution layer close to the interface.

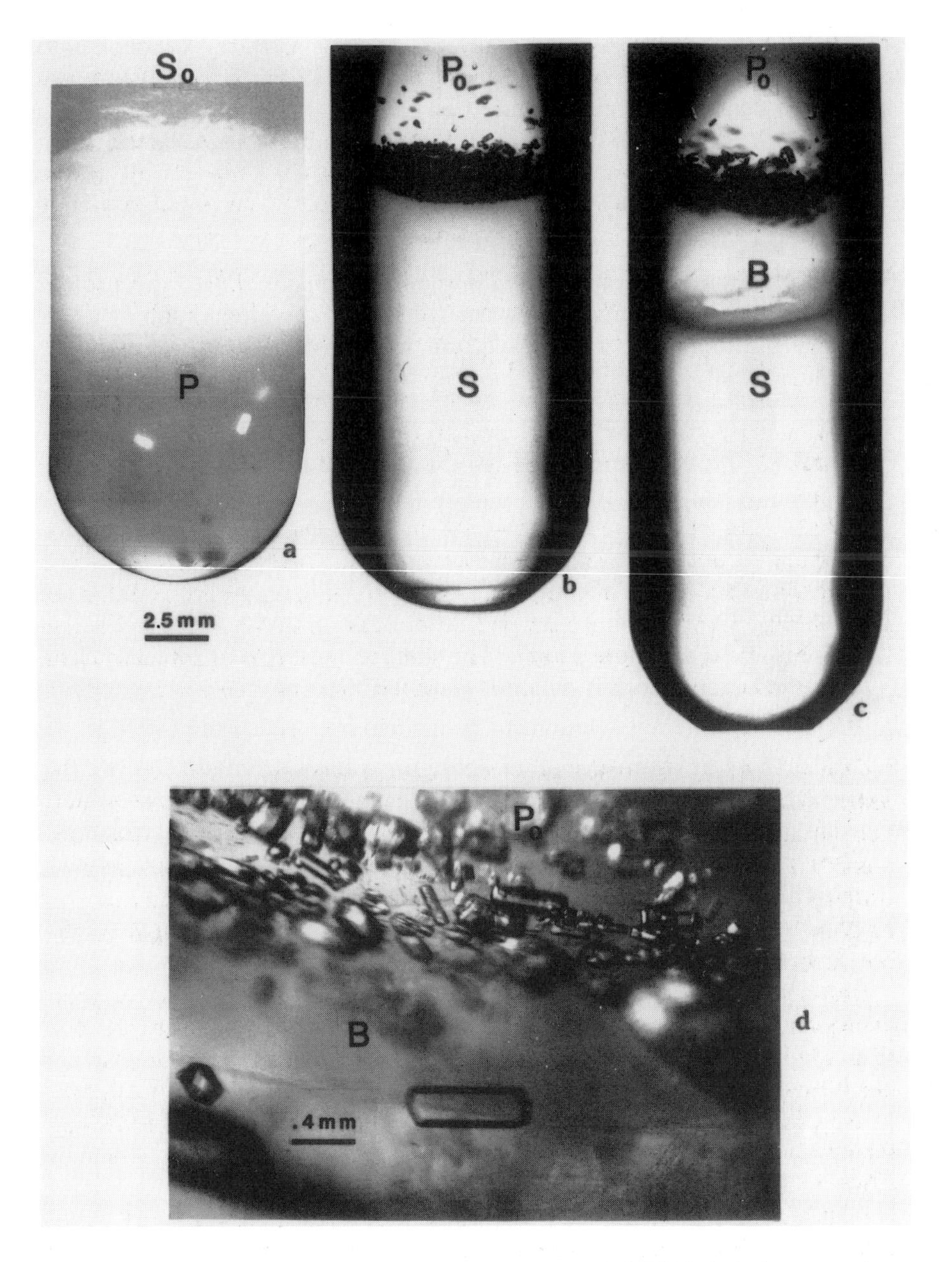

Figure 3. HEW lysozyme crystals grown by silica gel (TEOS) techniques. a, b, c correspond respectively to growth configurations 3, 5, 6 shown in *Figure 2*. B = acetate buffer at pH 4.7, S = NaCl. (a) TEOS = 1%(v/v), C_P = 7 mg/ml, C_{S0} = 1.4 M; (b) TEOS = 2% (v/v), C_{P0} = 13 mg/ml, C_S = 1.4 M; (c) TEOS = 2% (v/v), C_{P0} = 13 mg/ml, C_B = 0.1 M, C_S = 1.4 M. d: enlarged view of the interface in c of layer B and layer P_o.

Figure 3d shows an enlarged view of layer B-layer P_0 interface. Although both sides of this interface experience the same crystallization conditions, the nucleation density is strongly reduced on the gelled side and larger crystals are obtained there. This illustrates the nucleation inhibiting effect of silica gels.

3.3 Preparation of agarose gels

Select an agarose of purity and gelling point compatible with the protein under study. As an example, *Protocol 1* describes the preparation of a 1% (w/v) agarose solution in water made from an agarose of 36 °C gelling point. It is schematically illustrated in *Figure 4*.

Protocol 1. Preparation of a 1% (w/v) agarose solution

1. Add 1 g of agarose powder to 100 ml of cold water.
2. Raise the temperature progressively to 80 °C while the solution is continuously stirred at low rate (about 150 r.p.m.).
3. Maintain this temperature for 2 or 3 h.
4. Raise to 95 °C for about 30 min. The solution must appear perfectly clear and transparent. Keep it on a hot plate (80 °C).
5. Preheat the different solutions to be gelified in a water bath (40 °C).
6. Add a part of agarose mother solution to each solution. Due to the surface tension of this liquid, which is quite different from water, sampling with an automatic pipette with a plastic cone can be erroneous. Weight an identical sample to know the right agarose content. Mixing must lead to a homogeneous solution.
7. Pour the solution in a dried crystallization vessel and allow to set quiescently until the temperature decreases below the gelling point.

3.4 Use of agarose gels

Liquid–liquid diffusion techniques as previously described for silica gels can be applied to agarose gels. However, these gels are especially convenient for vapour diffusion techniques. An example of gelled droplet preparation is given in *Protocol 2* and illustrated in *Figure 4*.

Protocol 2. Gelled droplet preparation

1. Prepare samples of agarose mother solution, water, buffer, and crystallizing agent solutions, the respective volumes of which are calculated to give the right final concentrations.

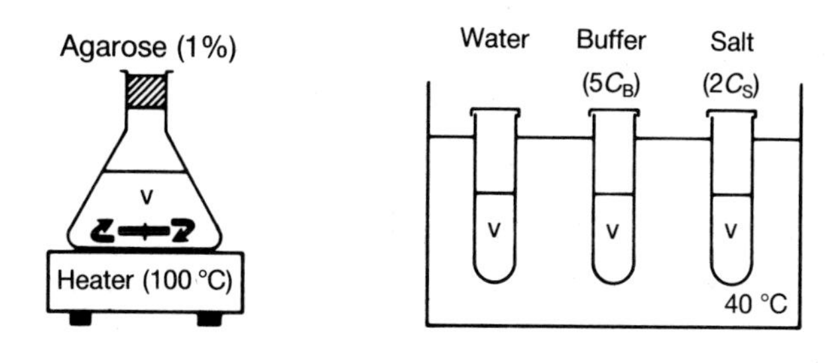

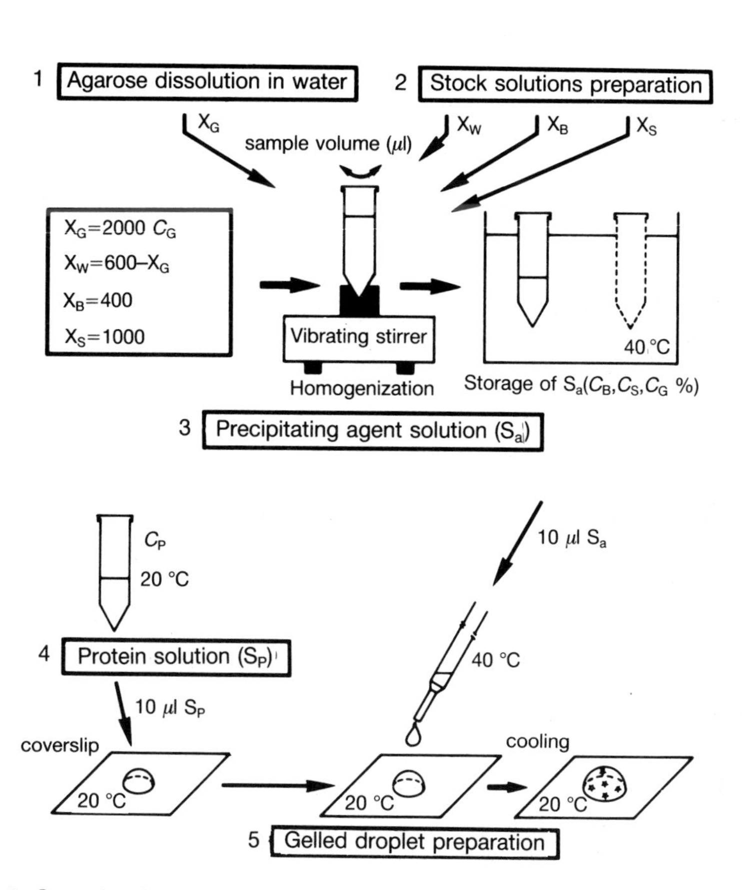

Figure 4. Operational sequence for a 1% (w/v) agarose solution preparation and subsequent dilutions with buffer and crystallizing agent to get 2 ml of solution S_A. The gelled drop is formed by coalescence of a cold drop of protein solution and a warm drop of S_A. After vapour equilibration with a reservoir of salt at concentration C_S, the concentrations in the drop will be C_P (protein), C_B (buffer), C_S (crystallizing agent), and C_G [% (w/v) agarose].

Protocol 2. *Continued*

2. Pour in a 1.5 ml Eppendorf tube and thoroughly mix with a vibrating stirrer. Keep this tube in a water bath. Prepare series of such tubes containing different % (w/v) of agarose.

3. A warm droplet of agarose–crystallizing agent solution is allowed to coalesce with a protein droplet set on a coverslip (eventually siliconized) immediately beforehand.

4. Invert the coverslip and suspend it over a reservoir filled with a solution (gelled or free) of crystallizing agent. Use of gelled droplets makes this handling easy and allows for larger volume of droplet.

In order to test the efficiency of gel and to optimize the gel content, it is useful to prepare a series of droplets from the same stock solutions but with different gel contents. In particular a reference experiment (gel free solution) is necessary.

Figure 5 shows three droplets of HEW lysozyme solutions containing respectively 0%, 0.05%, and 0.1% w/v of agarose with a constant concentration of protein (40 mg/ml) and of sodium chloride (1.2 M). Mechanical and thermal shocks undergone during crystal growth have been especially harmful for the gel-free experiment (*Figure 5a*) leading to numerous secondary nuclei and formation of defective crystals. This is not the case for the gelled droplets in which crystals, nucleated in the bulk, have been protected by their gel shell. These pictures show that the number of crystals increases with the agarose content. Studies of series of such experiments show that the number of optically defective crystals drops to zero for an agarose content above 0.1% w/v (3).

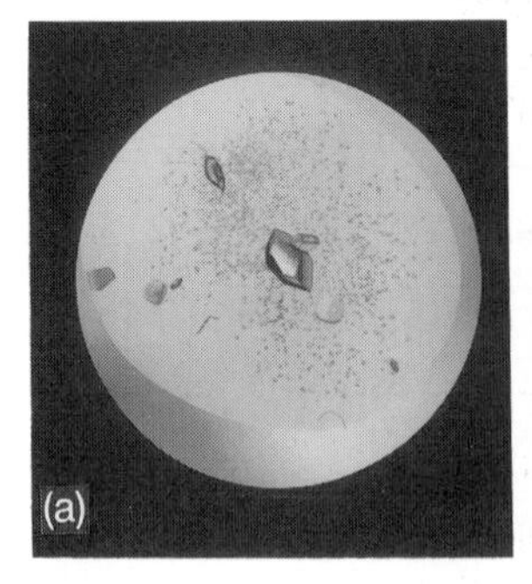

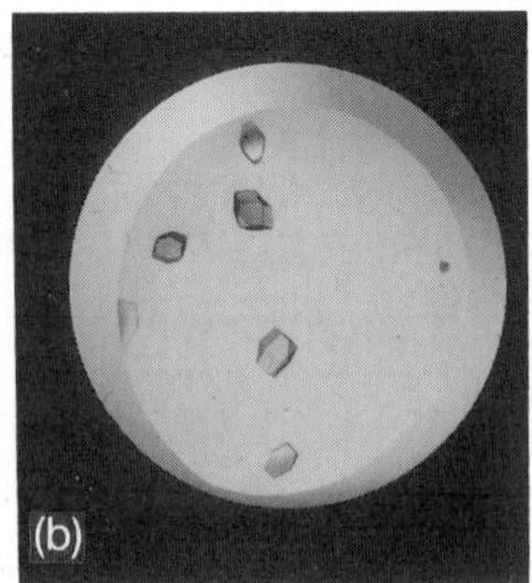

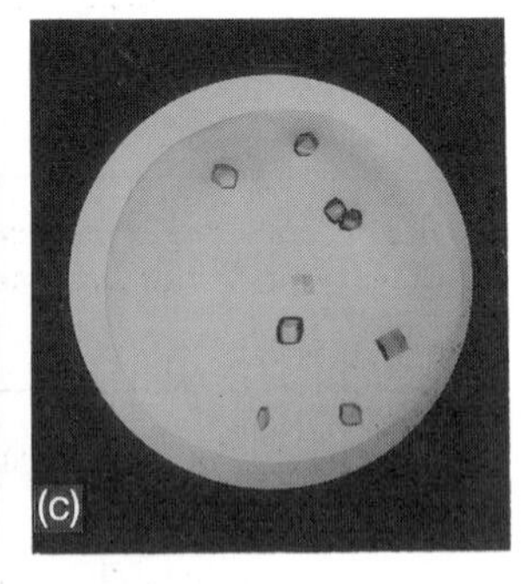

Figure 5. HEW lysozyme crystals growing in hanging drops prepared according to the method presented in *Figure 4* with B = acetate buffer at pH 4.7 and S = NaCl. In the three cases: C_P = 40 mg/ml, C_B = 0.1 M, C_S = 1.2 M; C_G = 0% (a), 0.05% (b), 0.1% (c).

4. Removal of crystals out of gels

The gels do not generally stick to the crystal surface because the growing crystal fissures the gel around it (see Section 2.2.2). Cusp-like cavities are often visible to the naked eye around macroscopic crystals. Such cavities exist but it is difficult to find evidence of microscopic crystals because they correspond to a thin film of liquid. A cusp is schematically drawn in *Figure 6*. Removal of crystals out of a gel is performed as described in *Protocol 3*.

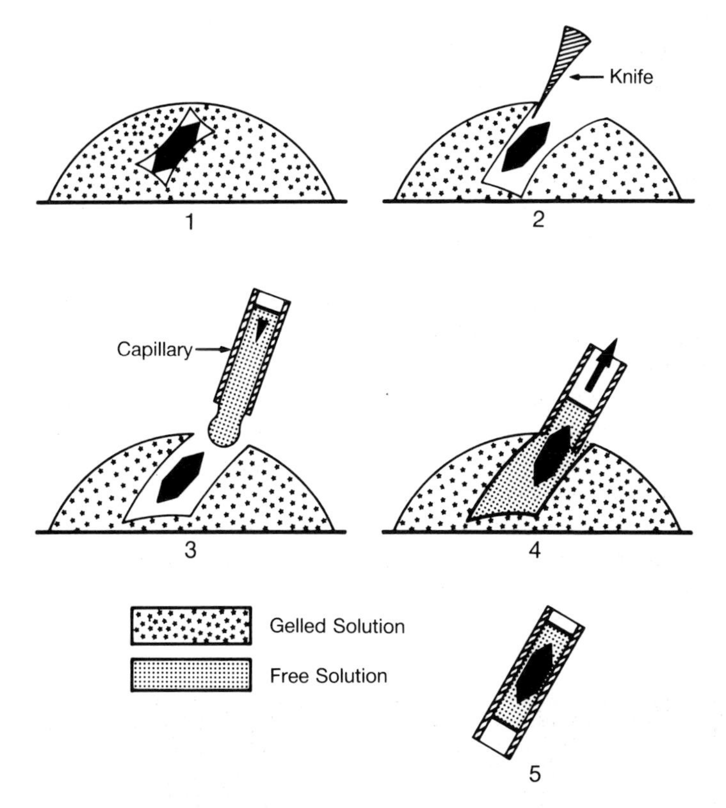

Figure 6. Schematic representation of operations required to remove gel-growth crystals out of the growth medium.

Protocol 3. Removal of crystals from a gel solution[a]

1. Extract a gel shell surrounding the grown crystal from the tube and set it on a cover slip (for hanging drop, take its cover slip).
2. Gently open the cusp (under the microscope) by pulling aside the gel parts with a surgical blade, without touching the crystal.

Protocol 3. *Continued*

3. Fill the enlarged cavity with a solution taken off the reference gel free droplet and wash crystal.
4. Suck this solution and the crystal into the capillary as for conventional crystal mounting (see also Chapter 12).

[a] Very light gels are somewhat thixotropic and can be directly sucked like liquids.

---

5. Related methods

5.1 Growth under reduced gravity

The behaviour of liquid and solid phases is completely changed by reducing the intensity of the gravity vector *g* (11). In the absence of gravity (zero *g* environment), the natural convections are prevented and solid particles floating in liquids do not sedimentate. Considering solution crystal growth, such a situation provides nearly all the advantages of gel growth. It avoids drawbacks such as a possible contamination or problems encountered to succeed in gelling the growth medium. As for gel crystal growth, mass transfer proceeds by diffusion and the crystals do not sedimentate.

Concerning nucleation, the different cases are:

(a) Heterogeneous nucleation is not affected by suppressing gravity forces.

(b) Secondary nucleation, which is due to gravity-induced movements of solid and liquid phases, is suppressed.

(c) Homogeneous nucleation is probably affected in the sense that the nucleation could occur at higher supersaturation (12). Previous space experiments suggest that large clusters of solute could survive in a reduced-gravity environment (13). Further experiments are planned to test this hypothesis.

In the present state of our knowledge, the influence of gravity field on biochemical parameters (see *Table 1*, Chapter 1) can be considered as rather speculative. The condition 'zero gravity' has been used to simplify the discussion but this condition is not at all realistic from an experimental point of view: drop towers, planes in parabolic flights, rockets, and space vehicles provide only reduced gravity levels (for example 10^{-6} *g* is assimilated to microgravity) with superimposed perturbations of *g* (*g*-jitter); this makes the analysis of results rather complicated. Gravity-level recordings during space experiments are needed to estimate a possible degradation of capabilities due to a spread of residual convections, or possible drifts of growing crystals.

Protein crystal growth experiments in space are either academic or application experiments. The first group corresponds to growth of model crystals for which satisfying samples are already obtained on earth. The purpose of these experiments is to test the capabilities of experimental facilities and the characteristics of space samples. Series of such experiments have shown that space samples have some structural pecularities whose origin is not yet well understood (14). The second group corresponds to crystal growth of highly interesting substances for which growth problems have not been solved by working under normal gravity. In a few cases positive results have been obtained, but it is not possible to generalize without having found an explanation for this improvement (15, 16).

Scientists interested in space experiments should first contact their national space agency about flight opportunities (experiment duration in crystal growth is too long to be achieved in drop towers or parabolic flight). Space instrumentation has been developed over the last few years and different prototypes corresponding to every classical growth technique are now available. The preparative work includes the same purification steps of starting material as for conventional techniques. The idea that 'purer crystals can be grown in space' is not founded. Compatibility tests between the chemical and biochemical reactants and materials entering in the growth cell composition must be made in order to see if denaturation of some substances is to be feared. Use of space vehicles implies waiting periods before the crystallization cell activation: experiments must request late access to their instrument and to very well controlled storage conditions. The delay for the recovery of the resulting crystals must be minimal.

The physico-chemical parameters such as the nature of buffer, crystallizing agent, or temperature, must be selected from series of experiments designed to find conditions which provide crystals even if these crystals are too small or defective. Conditions which do not provide crystals at all on earth should not be proposed for a space experiment. Starting from these conditions one can prepare a series of space experiments preferably with increased concentrations of protein and crystallizing agent in order to increase the supersaturation. The chances of succeeding are greatly improved when simulation experiments using gel media can be undertaken in parallel. The method consists in carrying out a series of gel experiments with the same configuration and the same physico-chemical parameters except for the gel content. By extrapolating the results obtained for decreasing gel contents, one can have an idea of what would be obtained in gel-free solution under reduced gravity. Mathematical simulations of mass transfer can also be made but this method can hardly take the nucleation step into account.

Critical analysis of space experiment results considered in the frame of earth simulation experiment results might either indicate how to improve the growth conditions or lead to the conclusion that the growth problem is not relevant for space.

5.2 Growth under hypergravity

A supersaturated state can be obtained by setting a concentration gradient in an ultracentrifuge. This approach is quite opposite to that indicated above:

- Sedimentation effects are reinforced which implies the use of this technique with mechanically resistant crystals or crystals whose density is close to that of their mother liquor.
- Mass transfer proceeds by convection. Hydrodynamical studies show that different convective regimes can be obtained according to the intensity of the gravity vector. Some of these regimes are very stable. Theoretical studies have been made in the case of melt growth of InSb showing that a more homogeneous composition can be obtained for these regimes (17).

Although no systematic studies have been performed for protein crystal growth, one study (18) examines the crystallization of catalase; using a centrifuge at 15 000 r.p.m., for 180 h in a fixed angle rotor, led to well-diffracting crystals of size $1.5 \times 1. \times 0.4$ mm^3. Main parameters which influence the experiment are the concentrations of protein and crystallizing agent, the type of rotor, and the rate of rotation.

It seems that the technique is better suited for large molecular weight biological macromolecules ($>$ 100 000) of diluted concentration. Further studies should prove the efficiency of this technique.

6. Conclusion

This chapter must be considered as a guideline allowing selection of a technique and adapting it to one's problem. Gel growth and related techniques consitute new trends in the protein crystallization field so that further progress is essentially expected from increasing the number of potential users.

References

1. Henisch, H. K. (1988). *Crystals in gels and Liesegang rings.* Cambridge University Press, Cambridge, USA.
2. Robert, M. C. and Lefaucheux, F. (1988). *J. Cryst. Growth*, **90**, 358.
3. Provost, K. and Robert, M. C. (1991). *J. Cryst. Growth*, **110**, 258.
4. Iler, R. K. (1979). *The chemistry of silica.* Wiley Interscience, New York.
5. Cabane, B., Dubois, M., Lefaucheux, F., and Robert, M. C. (1990). *J. Non Cryst. Solids*, **119**, 121.
6. Arnott, S., Fulmer, A., Scott, W. E., Dea, I. C. M., Moorhousee, B., and Rees, D. A. (1974). *J. Mol. Biol.*, **90**, 269.
7. Lefaucheux, F., Robert, M. C., Gits, S., Bernard, Y., and Gauthier-Manuel, B. (1986). *Rev. Int. Hautes Temp. Refract.*, **23**, 57.

8. McPherson, A. (1985). *Methods in Enzymology*, **114**, 112.
9. Andreazza, P., Lefaucheux, F., Robert, M. C., Josse, D., and Zyss, J. (1990). *J. Appl. Phys.*, **68**, 8.
10. Leon, S., Lefaucheux, F., Robert, M. C., Malgrange, C., and Lorin, J. C. (1989). *J. Cryst. Growth*, **97**, 631.
11. Authier, A., Benz, K. W., Wallrafen, F., and Robert, M. C. (1987). In *Fluid sciences and material sciences in space* (ed. H. U. Walter), pp. 405–46. Springer Verlag, NYC.
12. Robert, M. C. and Lefaucheux, F. (1987). *Proceedings 6th European symposium on material sciences under microgravity conditions* (Bordeaux 1986), pp. 77 *ESA SP256*.
13. Littke, W. and John, C. (1986). *J. Cryst. Growth*, **76**, 663.
14. De Lucas, L. J., *et al.* (1989). *Science*, **246**, 651.
15. Borisova, S. N., Trakhanov, I. G., and Kharitonenkov, I. G. (1990). *Acta Cryst. A*, **Vol A 46**, Suppl. Collected Abstracts 15th Congress of Intern. Union of Crystallography, C436 M.S. 13.01.02.
16. Bugg, C. E. and De Luca, L. J. (1990). *Acta Cryst. A*, **Vol A 46**, Suppl. Collected Abstracts 15th Congress of Intern. Union of Crystallography, C436 M.S. 13.01.01.
17. Muller, R. and Neumann, G., (1982). *J. Cryst. Growth*, **59**, 548.
18. Barynin, V. V. and Melik-Adamyan, V. R. (1982). *Sov. Phys. Crystallogr.*, **27**, 588.

7

Crystallization of nucleic acids and co-crystallization of proteins and nucleic acids

A.-C. DOCK-BREGEON and D. MORAS

1. Introduction

At a first glance crystallizing nucleic acids poses the same problems as crystallizing proteins since the variables under investigation are alike. Some differences can be underlined, however. Most of the papers dealing with crystallization of biological macromolecules describe protein crystal growth. Thus it seems important to add a contribution for nucleic acid crystallization. As seen from crystallization data banks and from the Protein Data Base, it is obvious that these banks contain a large majority of proteins; this is largely due to the difficulty in obtaining large single crystals of nucleic acids.

Among natural RNAs, only transfer RNAs and 5S RNA have yet been crystallized. These are also the smaller ones, with 75 to 120 nucleotides. Techniques have now been developed with other systems too, like ribozymes, fragments of messenger RNA, or viral tRNA-like RNAs. The great advances in oligonucleotide chemical synthesis during the past decade has provided the opportunity of making large amounts of pure desoxyribonucleotides of any desired sequence. This has led to the crystallization of oligonucleotide fragments, and was followed by the co-crystallization of complexes of proteins and synthetic oligodeoxynucleotides.

In the case of complexes of proteins with RNAs, the main difficulty was in purifying large quantities of homogeneous biological material with well-defined physicochemical properties. This prevented a good survey of crystallization conditions; the problem has now been overcome for some cases and problems of large complexity, such as the crystallization of ribosomes, have been addressed. Crystallizations of viruses will also be briefly described.

2. Preparation of samples

2.1 5S RNAs

The 5S RNA is a constituent of the large subunit of the ribosome and can be prepared, starting from the ribosomes, by a phenol extraction. It is separated from tRNAs and high-molecular-weight ribosomal RNAs by molecular sieving (1, 2). Crystals have been obtained with 5S RNA from *Thermus thermophilus* (3) and *E. coli* (4).

2.2 Sources of transfer RNAs

For the yeast *Saccharomyces cerevisiae* and the bacteria *E. coli* bulk tRNA is commercialized. The purification of a single species of a yeast tRNA is better done starting from the bulk tRNA from brewer's yeast since the structural integrity of the -CCA at 3′-end of these molecules has been demonstrated, while it was not the case for tRNAs from baker's yeast (5). Other sources need an extraction from the cells with phenol. Thermophilic and halophilic bacteria provide molecules of great interest for crystallization since these are more stable at a higher temperature or higher ionic strength (e.g. see the case of ribosomes).

Advances in genetics allows one to turn more readily to cloning (see Chapter 2), when a specific tRNA is chosen. This procedure was used for *E. coli* $tRNA^{Gln}$ which was crystallized with its cognate synthetase, glutaminyl-tRNA synthetase (6). Cloning allows overproduction of a single isoacceptor and then easier purification. A problem of undermodification of tRNAs may be encountered with this procedure, since the modifying enzymes could be in limited amount when a specific tRNA is overproduced. However, the example of $tRNA^{Gln}$ shows that the undermodification is limited enough and does not affect the crystallization of the complex.

Another way of producing large amounts of RNA is by *in vitro* transcription system with the use of highly active phage polymerases (7). The most used is T7 RNA polymerase, which is cloned and overproduced (8). The yield of *in vitro* transcription can reach several hundred moles of transcripts per mole of template. When used for tRNA production, the template is made according to the strategy described in *Protocol 1*.

Protocol 1. *In vitro* synthesis of an RNA molecule[a]

1. Construction of an insert ending by restriction sites and containing the T7 promoter and the tRNA sequence. This is made by ligation of synthetic DNA oligomers chosen to hybridize unambigously in tandem so as to give the correct, double-stranded sequence.

Protocol 1. *Continued*

2. Insertion of this 'pseudo gene' into a plasmid, digested with the adequate restriction enzymes.
3. Production of the plasmid by cell culture.
4. Extraction of the DNA.
5. Linearization of the template at the restriction site.
6. Transcription of the template in an appropriate medium containing the polymerase and the nucleotide monomers.

[a] The detailed experimental procedures are described in references 7, 9, and 10.

Limitations of the *in vitro* synthesis of tRNA are:

- The yield depends on the sequence of the 5′-end of the RNA-product, i.e. the +1–+6 region. The polymerase works more efficiently when residue 1 is a guanine (9).
- The *in vitro* transcription method produces tRNAs which lack the modified bases and start with 5′-triphosphate residues (or a mixture of 5′-mono and triphosphates when transcription is carried out in the presence of appropriate monophosphate nucleotides). These molecules seem to be correctly recognized by their cognate synthetases but could have less stable tertiary structures (10, 11).
- A major problem is transcription termination, since polymerase sometimes adds one or two extra nucleotides at the 3′-end (7).

Advantages are:

- When the synthetic gene is available, synthesis of RNA is rapid.
- Any wild-type sequence or any engineered variant can be prepared (provided the presence of a good 5′-start for the polymerase).

2.3 Purification of transfer RNAs

In the case of natural tRNAs the purification problem is rather complicated since the cellular extracts contain about 60 different species with similar structures.

2.3.1 Countercurrent fractionation

In countercurrent fractionation (12) tRNAs are distributed between an aqueous and an organic phase. Selectivity relies upon differential solubility of tRNAs in the two solvants. This method allows to handle very large quantities of bulk tRNA, typically 5 g. The enrichment may also be good: a tRNA at 2% in the starting material can be obtained at 20%. The major problem with countercurrent separation is the need of the machine, a complicated assembly of several hundred tubes in which the solvents distribute; this cannot be afforded reasonably for the purification of a unique species. When available it is, however, a very convenient first step in a purification procedure.

2.3.2 Adsorption chromatography

i. Anion exchange chromatography

Different classical supports interacting with the negatively charged tRNAs have been used as DEAE-Sephadex (13) or hydroxyapatite (14). Ionic interactions take place between phosphates of the tRNA and positively charged groups of the DEAE matrix, or calcium ions of the hydroxyapatite crystals. Additional weak interactions are especially important on hydroxyapatite. They are tuned by the tRNA structures and are sensitive to the presence of Mg^{2+} ions or urea, to the pH, and temperature. The resolution of such columns is limited since electrostatic forces, related to the number of accessible phosphates, are poor discriminators of tRNA species. However, the use of HPLC systems should favour a renewed interest of such columns since these, fo classical use for protein purification also, are commercially available (15).

ii. Mixed-mode chromatography

BD-cellulose: this most popular matrix is a DEAE-cellulose modified by the addition of benzoyl groups (16). The tRNAs are sorted by the interaction between phosphates and DEAE groups, and also by hydrophobic interaction between the accessible bases and the benzoyl moieties. Advantages of BD-cellulose chromatography are:

- The additional mode of binding is much more specific than the pure electrostatic one.
- The interactions depend also on the conformation of the molecules and can be tuned through variations of the pH or magnesium concentration.

The disadvantage of using BD-cellulose is:

- Up to now, the resolution obtained with the commercially available resin is unfortunately much lower than that reported in the literature, and this type of chromatography is to be combined with others to purify one specific species.

Of better resolution is the use of BD-cellulose with aminoacyl–tRNAs (in the case of aromatic amino acids) or with derivatized aminoacyl–tRNA (17, 18). *Reverse-phase chromatography (RPC)*: Much more resolutive are the reverse-phase systems, where inert supports are coated with quaternary ammonium derivatives of high molecular weight not miscible with water. The tRNAs distribute between an aqueous mobile phase and an organic phase, where phosphates interact with the positively charged quaternary amine moieties. The tRNA molecules are eluted by raising the sodium chloride concentration (19). Advantage of using RPC is:

- Excellent resolution: it is of great use for separating isoaccepting species of tRNAs.

The disadvantages of RPC are:

- The difficulty of finding a good combination of inert support and coating. This is illustrated by the publication of a series of improvements toward the best-known RPC 5 (20). The unavailability of the commercial support prompted new modifications (21, 22). An adaptation to HPLC has also been reported (23).
- Another problem can arise from the leakage of the coating, which implies the need to re-coat the inert support after several chromatographies. Some traces of the quaternary ammonium may be present in the tRNA sample; this possible contamination is not to be neglected in crystallization experiments.

An answer to the last problem is given by the covalent binding of the quaternary ammonium derivatives to the support (24). This seems to be also the direction chosen nowadays by the manufacturers, especially for HPLC matrices.

Hydrophobic interaction chromatography (HIC): It was first used with a Sepharose-4B support and a reverse gradient of ammonium sulphate (25) (see also Chapter 2). Advantages of HIC are:

- High resolution.
- This method allows processing of large quantities of material: according to Holmes *et al.* (25), 1 ml of Sepharose-4B could adsorb 8 mg of unfractionated tRNA. In fact even larger amounts of tRNA can be handled but at the cost of decreased resolution: in this case the hydrophobic binding is overwhelmed by precipitation of the tRNAs on the support, and the tRNAs elute according to their solubility.
- Adaptation to HPLC: HIC has recently been improved with the *n*-alkylated silica supports available for medium-and high-pressure chromatography. A very resolutive HPLC system has been reported, using C4-bonded silica gel for the separation of tRNAs from various organism (26).

The disadvantage of HIC is:

- The presence of the salt: getting rid of ammonium sulphate is necessary after the chromatography, especially if an ethanol precipitation is the following step. It can be done either by dialysis or by buffer exchange in a concentration set-up.

2.3.3 Design of a purification protocol

The purification of a single species results from the combination of these different methods. For crystallization, the need of large quantities of pure material ($\geqslant$ 10 mg) directs the choice of the first steps to the methods allowing the handling of the largest quantities of material: countercurrent distribution, precipitation on an hydrophobic matrix, and elution in ammonium sulphate solutions of decreasing concentration, or BD-cellulose (about 2 mg of tRNA can be bound on 1 ml of packed resin). The last steps will be the most resolutive (RPC, HIC).

The procedure is much simplified when overproducing strains are used. This is illustrated in *Table 1*, where the purification of one tRNA species from commercial bulk tRNA is compared to the a purification from an overproducing strain.

Table 1. Comparison of purification protocols of yeast $tRNA^{Asp}$ (starting from commercial bulk tRNA) and *E. coli* $tRNA^{Asp}$ (from an overproducing strain).

Yeast $tRNA^{Asp}$	***E. coli* $tRNA^{Asp}$**
• Preparation of bulk tRNA	
commercial tRNA (contains about 2.5% $tRNA^{Asp}$)	cell culture phenol extraction deacylation DNA elimination (gel filtration or DEAE–Sephacel) tRNA obtained contains about 40% $tRNA^{Asp}$)
• Chromatographic steps	
HIC BD-cellulose Porex 5C4 (HPLC)	DEAE (HPLC) Porex 5C4 (HPLC)
• Elapsed time	
2 weeks	1 week

For tRNA molecules resulting from *in vitro* transcription, the desired molecule is the major species in the reaction tube. It is separated from the template and the nucleotide monomers by gel filtration. The major problem is the separation of the correct product from elongated molecules since polymerase sometimes adds one or two nucleotides at the 3′-end (7, 9). The transcripts are usually purified by electrophoresis in polyacrylamide gels, but HPLC chromatography should also give good results.

2.3.4 General handling

i. Care of degradations

Beside the problem of sensitivity towards ribonucleases which requires work in sterile conditions, RNAs are sensitive to alkaline hydrolysis (27) and therefore alkaline pH should be strictly avoided. The cleavage of the polyribonucleotide chain is also favoured by some metal ions, of which the most effective is lead (28). A chelating agent, like EDTA, is generally introduced into the buffers, at a concentration in the order of 0.1–0.5 mM, to complex the traces of heavy metals. At the end of the purification the tRNA preparation must be checked for its integrity. The simplest method is an electrophoresis in a denaturing polyacrylamide–urea gel.

ii. Measure of tRNA concentration

Concentrations are obtained from optical density measurements at 260 nm with $A_{260nm} = \varepsilon \times c \times l$, where $\varepsilon = 25$ ml/mg, l is the optical path in cm, and c the concentration in mg/ml.

iii. Concentration of tRNAs

Two ways, described in *Protocol 2* and *Protocol 3*, can be used to concentrate tRNAs.

Protocol 2. Concentration of tRNA by precipitation with ethanol

1. Prepare the tRNA solution. It should contain Mg^{2+} ions ($\geqslant$ 2 mM) and a Na salt such as Na acetate ($\geqslant$ 10 mM, generally at pH 6.0). For good recovery the tRNA solution should be $\geqslant$ 0.1 mg/ml. If not, raise the Na acetate concentration to $\geqslant$ 100 mM. Take care that the solution does not contain too much salt (i.e. after a chromatography, dialuse first in water).
2. Add two or three volumes of ethanol (best quality). The precipitate forms.
3. Leave to precipitate completely at −20 °C (2 h or more) or at −80 °C (20 min or more).
4. Centrifuge at the lowest possible temperature (10 min at $\geqslant$ 4000*g* are sufficient).
5. Dry the pellet under vacuum in the presence of solid KOH.
6. Dissolve the pellet in the desired amount of buffer.

Protocol 3. Concentration of tRNA on a membrane

1. Prepare a set-up of the type Amicon (for large volumes) or Centricon (for volumes of some ml). Use membrane of correct cut-off (usually 10 000).
2. Concentrate by pushing the solvent through the membrane under nitrogen pressure (for the Amicon set-up) or by centrifugation (for Centricon).
3. Recover the solution when the desired volume is obtained.

This method can be used to change the solvent. When the tRNA is concentrated to a small volume, dilute it in the new solvent and concentrate again. Repeat several times.

iv. Last step of the preparation of a tRNA and storage

In order to gain more homogeneity, the tRNA solution to be crystallized is dialysed thoroughly in a buffer at low concentration of Mg^{2+} (e.g. 2 mM $MgCl_2$ and 10 mM Na cacodylate, pH 6.0). The same result can be obtained by buffer exchange. Samples of tRNA can be stored:

- Frozen in such solution at −20 °C or −18 °C.
- In dry form at 4 °C or −20 °C. For this purpose it is precipitated by ethanol (without adding anything else other than ethanol to a concentrated solution), then recovered by centrifugation in microtubes (Eppendorf style) in aliquots of a few milligrams. The pellets are dried. When needed, a sample is dissolved in water or buffer.

2.4 Oligonucleotides

2.4.1 Oligodeoxyribonucleotides

The method of phosphotriester and phosphoramidites in solid phase (29, 30) has been automated, and DNA-synthesizers are commercially available. At present it is straightforward to prepare DNA molecules and most molecular biology institutes provide this facility. Using a 10 μmol solid support resin, about 20 mg of a 20-mer oligonucleotide can be synthesized.

After cleavage from the resin by ammonia treatment, the final product is generally purified by HPLC. The expected product is contaminated by shorter molecules that result from incomplete reaction within a cycle of synthesis when the previous product was not fully deprotected. Other contamination results from side-reactions, like de-purination that may occur during the acid treatment for the removal of the 5′-protecting group at the end of each cycle. The purification can be performed before or after the complete de-blocking of the protecting groups by ammonia treatment (31).

The purification procedure calls for the different properties of oligonucleotides, in a way similar to that discussed for tRNAs. The resins are generally adapted to HPLC. These are ion-exchangers or reverse-phase chromatography columns. They are generally made of silica particles modified with functional groups in C4, C8, or C18. They can be used in two different ways, according to the counter-ion choice. With ammonium ions the retention of oligonucleotides is mainly due to hydrophobicity and this method is used to separate oligonucleotides of similar length but of different sequences. With triethylammonium ions, the ion-pairing phenomena comes into play; the oligonucleotides are adsorbed to the stationary phase via their counter-ions. In this case the strength of the interaction is dependent on the hydrophobicity of the counter ion, and also on the length of the oligonucleotide, since the strength of the interaction is proportional to the charge of the oligonucleotide (31). In both cases, the oligonucleotides are eluted by increasing concentrations of acetonitrile. An advantage of the second method is that triethylammonium salts are volatile, and so the product is easy to recover by lyophilization.

2.4.2 Oligoribonucleotides

Oligoribonucleotides can be produced using the *in vitro* transcription method. The procedure is simplified when short molecules (12–35 nucleotides) are synthesized, since in this case the template can be synthesized directly. The 17-nucleotides promoter of T7 polymerase is double-stranded, and the sequence to transcribe is single stranded (9), so two oligodeoxynucleotides are needed. A large-scale transcription reaction can produce up to 20 mg of transcripts from 200 nM DNA-template. Here again, the sequence of the 5′-end is limited to a guanine when large amounts of product are awaited. Gel electrophoresis or HPLC are used for the final purification.

Another way of producing short RNA fragments is chemical synthesis (30). RNA is, however, more difficult to make than DNA, because of the additional group to protect the 2′-hydroxyle of the ribose. Several protecting groups have been designed and chemical synthesis in solution has produced about 10 mg of the tetradecamer $U(UA)_6A$, which was crystallized (32). The solid-phase method has also been developed and is now automated (33). Commercial DNA-synthesizers can also be used with RNA monomers with the same facilities as for DNA synthesis, but giving somewhat lower yields. The great advantage of the chemical synthesis upon the enzymatic one is that modified nucleosides, or even deoxynucleotides, can be introduced at a specific position (38), and also that any sequence can be designed. It appears as the method of choice for short oligoribonucleotides.

3. Crystallization of nucleic acids

Several examples (arbitrarily chosen) of crystallization of oligonucleotides are given in *Table 2*. More complete data can be obtained from the review in reference 35. For tRNAs a compilation of crystallization conditions is given in reference 36.

3.1 General features

3.1.1 Crystallizing agents

The more widely used are alcohols, and specially MPD, which is not volatile and therefore easy to handle. It is used in the range of 10% (v/v) in the case of tRNAs, and 30% in the case of oligonucleotides. Isopropanol has also given good results with tRNAs, and especially with $tRNA^{Phe}$ (37). In the field of tRNAs, another successful precipitant is ammonium sulphate. It gave good results with yeast $tRNA^{fMet}$ (38) and yeast $tRNA^{Asp}$ (39). PEG precipitate

Table 2. Some examples of crystallization conditions of oligonucleotides

Sequence	Precipitant	Temp (° C)	Concentrations oligo.	buffer[a]	spermine	Mg^{2+} [b]	Crystals	Ref
Z-form:								
CGCGCG	isopropanol 5%		2.0 mM	30 mM pH 7.0	10 mM	15.0 mM	$P2_12_12_1$ 0.9 Å	48
m^5CGTAm^5CG	MPD 8% vs. 50%		4.0 mM	30 mM pH 7.0	7.0 mM	10.0mM	$P2_12_12_1$ 1.2 Å	40
(^{5}BrCG)$_3$	MPD 10% vs. 60%	18 or 37	0.5 mM	20 mM pH 6.5		NaCl 200 mM	$P2_12_12_1$ 1.4 Å	56
B-form:								
CGCATATATGCG	MPD 10% vs. 40%		0.5 mM	–	0.4 mM	22.0 mM $Mg(Ac)_2$	$P2_12_12_1$ 2.2 Å	75
5′-CGCAAAAAAGCG GCGTTTTTTCGC-5′	MPD 5% vs. 30–45%	4	0.2 mM	–	0.5 mM	10.0 mM $Mg(Ac)_2$	$P2_12_12_1$ 2.5 Å	76
CCAAGATTGG (G.A mismatch)	MPD 45%	4	3.0 mM	–	–	0.7 M	C2	77
5′-ACCGGCGCCACA TGGCCGCGGTGT-5′	MPD 40%	4	1.0 mM	50 mM pH 6.0	1.2 mM	18.0 mM $Mg(Ac)_2$	R3 2.8 Å	59

A-form:								
GGCCGGCC	MPD 30%		1.2 mM	25 mM pH 7.0	0.6 mM	3.0 mM	$P4_32_12$ 2.25 Å	78
GGGGCTCC (G.T mismatch)	slow evaporation	4	3.0 mM	50 mM pH 6.5	–	28.0 mM	$P6_l$	79
CTCTAGAG	MPD 7% vs. 50%	18	1.2 mM	60 mM pH 6.8	1.0 mM	25.0 mM	$P4_12_12$ 2.15 Å	80
RNA–DNA hybrid:								
r(GCG)d(TATACGC)	MPD 40%		1.5 mM	30 mM pH 6.0	8.0 mM	15.0 mM	$P2_12_12_1$ 1.9 Å	81
RNA								
$rU(UA)_6A$	MPD 35%	35	4.0 mM	40 mM pH 6.5	–	0.4 M	$P2_12_12_1$ 2.25 Å	32

[a] When the crystallization medium is buffered, the buffer is always sodium cacodylate.
[b] Most often, $MgCl_2$. In other cases, the salt is specified.

tRNAs at concentrations of a few percent for a medium-sized PEG (M_r 4000–8000); e.g. a different crystal form of yeast $tRNA^{Asp}$ could be obtained with PEG (36). Crystals of the Z-DNA hexamer $d(CG)_3$ were obtained with a smaller-sized PEG 600 (40). The precipitants PEG or MPD mixed with NaCl, NH_4Cl, or $(NH_4)_2SO_4$ are also interesting possibilities: the salt acts as an electrostatic shield and modulates the interaction between tRNA and additives.

3.1.2 Temperature

The temperature range of nucleic acids stability allows examination of a large range of temperatures, from 4 °C (usual cold-room temperature) to 30 °C or 35 °C (in a bacteriologic incubator or an oven). The 35 °C assays bring new paths towards crystallization especially when mixed precipitants are tried since it modifies the phase partitions.

3.1.3 pH and buffers

In nucleic acid crystallization, the pH appears to play a smaller role than in protein crystallization where the overall charge of a protein, and then its capacity of packing in a certain way, may be tuned through pH variations. The situation is quite different with nucleic acids, which are negatively-charged polyelectrolytes. At pH 5.0–4.0 cytidines are protonated, and such a pH range can therefore promote crystallization when there is an accessible cytidine, by introducing a potential additional interaction. A too-low pH could, however, induce local structural artefacts. Taking pH into account is also of importance for mismatched oligonucleotides. In the case of RNA the problem of degradations forbid the use of alkaline pH. The buffer is often Na cacodylate (pH range 6.0–7.0), which has the advantage of preventing bacterial growth (which may be a problem in PEG) and which has a rather constant pH over a large temperature range. In $(NH_4)_2SO_4$, the buffer concentration must be high enough to maintain pH against variations due to ammoniac evaporation (i.e. 100–300 mM). In low ionic strength media (PEG or MPD) the buffer itself can introduce an electrostatic shield, and variations of the buffer concentration may modulate the electrostatic interactions between molecules and additives.

3.1.4 Nucleic acid concentration

For tRNAs, the crystallization is generally tried in the order of 5–20 mg/ml, i.e. 0.2–0.8 mM. Higher molar concentrations are generally used for oligonucleotides, as indicated in *Table 2*.

3.2 Additives

Nucleic acids are polyelectrolytes and therefore the counter-ions are important additives of the crystallization medium. Two families of cations are generally used, polyamines and divalent cations. Their role in the crystallization differs subtly and parallels their structural effects.

3.2.1 Polyamines

Polyamines are widely distributed in biological systems. They are involved in DNA condensation as well as in protein synthesis (41, 42). Some examples of polyamines are given in *Table 3*. For crystallization, spermine is most generally used. It is a linear molecule with four positive charges at neutral pH. The popularity of spermine comes from its key role in the production of the first crystals of a tRNA (43). After the crystallization of yeast $tRNA^{Phe}$, with crystals diffracting to high resolution (44–46), spermine was systematically tried with nucleic acids, including oligonucleotides. Spermidine, which is an asymmetric molecule bearing three positive charges at neutral pH, was also reported to promote crystal growth, but with less success.

Spermine binds in the grooves of nucleic acids. The refinement at 2.5 Å resolution of the structure of yeast $tRNA^{Phe}$ has shown two spermine molecules (46). One spermine molecule is coiled in the deep major groove of the anticodon arm of the tRNA, at the junction of D- and T-stems. It is hydrogen bonded to four different phosphates on both sides of the groove. The second spermine molecule interposes a string of positive charges between the extended polynucleotide chain of the variable region and the P9–P10 sharp turn. Spermine molecules have also been identified in crystals of the Z-DNA structures of the hexamer $d(CG)_3$ (48), the tetramer $d(CG)_2$ (49), and the hexamer $d(m^5CG)_3$ (50). Interestingly, no spermine was identified in crystals of A-type oligonucleotides.

3.2.2 Divalent cations

Divalent cations, and especially Mg^{2+} ions, have been known for a long time to be involved in the stabilization of nucleic acid structures and to play an important role in their functions. The crystallographic structures of tRNAs have given a first insight into the structural effect of Mg^{2+} on the conformation of tRNA. Preferential Mg^{2+} sites are located mostly in the non-helical regions of the tRNA molecule and appear to stabilize loops and bends of the tertiary structure. Some of these Mg^{2+} sites are of interest of crystallization, since they are bridging tRNA molecules and therefore seem to stabilize the crystal packing [e.g. one Mg^{2+} in the D-loop of the refinement by Holbrook *et al.* (51)].

Table 3. Polyamines

putrescine	$H_2N–(CH_2)_4–NH_2$
cadaverine	$H_2N–(CH_2)_5–NH_2$
spermidine	$H_2N–(CH_2)_3–NH–(CH_2)_4NH_2$
thermine	$H_2N–(CH_2)_3–NH–(CH_2)_3–NH–(CH_2)_3–NH_2$
spermine	$H_2N–(CH_2)_3–NH–(CH_2)_4–NH–(CH_2)_3–NH_3$

The structures of oligonucleotides in the A- or B- helical forms, refined to better resolution than tRNAs, have brought little additional information about the preferred co-ordination of Mg^{2+} ions. The Z-structures, on the contrary, give generally more details about ion binding, a consequence of their better resolution. Examples of Mg^{2+} binding to Z-DNA can be found in the structure of d(m^5CGTAm^5CG) (40). One Mg^{2+} ion is surrounded with six oxygen atoms, one of which is a phosphate oxygen of the backbone and the others are in water molecules. Other interesting examples of Mg^{2+} sites are found in the structure of $d(CG)_3$ (52) or d(CGTACGTACG) (53) where intermolecular Mg^{2+} sites are described. The presence of such sites confers probably, with the hydrogen-bonding possibilities, an increased stability to the crystal packing, and may explain why in general Z-DNA crystals diffract to higher resolution.

Other divalents cations can be used instead of Mg^{2+}, or in addition to it. For tRNAs crystallization, different divalent cations have been tried, like manganese, calcium, cobalt, nickel, barium, mercury. Care must be taken, however, since some metal ions may induce hydrolysis of the phosphodiester bonds in RNA molecules, especially lead (28). The crystallographic structures of mono- or dinucleotides give an insight on the mode of binding of several ions to nucleic acids, e.g. calcium binding to ApA (54). These ions provide sometimes interesting stabilization of local structures or new packing possibilities.

More complex ions can also be tried, of which cobalt hexammine is an interesting case. It stabilizes Z-DNA with an efficiency that is five orders of magnitude greater than Mg^{2+}. Cobalt hexammine favoured the crystallization of $d(CG)_3$ (52) and d(CGTACGTACG) (53) in the Z-form. Cobalt hexammine was also identified as an helix-stabilizing agent in the case of $tRNA^{Phe}$ (55).

3.2.3 Monovalent ions

The example of the cluster of ions in d(m^5CGTAm^5CG) (40) has shown that Na^+ ions can also play a role in helix stabilization, and therefore can favour the crystallization of nucleic acids, in addition to Mg^{2+}. Na^+ only, without Mg^{2+}, was used for the crystallization of $d(Br^5CG)_3$ in the Z-form (56), and the structure shows how Na^+ can bridge two neighbouring molecules in the crystal. Compared to Mg^{2+}, the octahedral co-ordination of Na^+ is less precisely defined and in certain cases can be accommodated more easily.

3.2.4 Concentration of the counterions

A very important parameter is the relative concentration of spermine and Mg^{2+}, as well as the ratio spermine/, or Mg^{2+}/nucleic acid molecule. For magnesium a 'rule of thumb' for the first trials is 0.5–1.0 Mg^{2+} per phosphate (e.g. for a tRNA at 0.4 mM corresponding to a phosphate concentration of 30 mM, try Mg^{2+} concentrations of 15 mM and 30 mM). Smaller or larger concentrations may be tried if the results are disappointing; e.g. in this range of Mg^{2+} concentration the tetradecamer $U(UA)_6A$ did crystallize readily but the crystals showed very poor diffraction (with maximal resolution of about 7 Å). The best crystals were obtained at a Mg^{2+} concentration of 400 mM. For spermine the 'rule of thumb' is one spermine molecule for 10–12 base pairs (e.g. for a tRNA at 0.4 mM the spermine concentration to try is 3 mM). Since spermine and Mg^{2+} act as counterions, the ionic strength of the medium has to be taken into account: in $(NH_4)_2SO_4$ solutions or when monovalent salts are added, the concentrations of Mg^{2+} and spermine have to be somewhat higher than in PEG or MPD. The relative concentration of spermine and Mg^{2+} is also to be considered; at higher Mg^{2+} concentration, higher spermine concentrations can be tested, since Mg^{2+} brings its own shielding effect. An excess of spermine, especially at low ionic strength, very often produces crystals which do not diffract. Some assays without spermine should also be tried. *Table 2* shows some examples of crystals obtained in the absence of spermine.

3.3 Crystal packing

Among the oligonucleotides that have been crystallized only a few different packing modes have been observed. This is partly due to the systematic exploitation of successful crystallization attempts, by introducing variants into the crystallizing matrix made by the parts of the oligonucleotide that are involved in packing interactions. The lattices are commonly built upon the aromatic base-to-base stacking interaction of one molecule upon the other; e.g. the Z-DNA hexamer orthorhombic crystals [e.g. $d(CG)_3$] are built according to this packing mode, with the DNA molecules forming infinite helices in the crystal. The stacking of the end base-pairs is very similar to that found inside the sequence. Additional hydrogen bonds between the backbones of the neighbouring molecules bring more stability to the lattice (57). The first family of B-DNA structures were derived from the dodecamer d(CGCGAATTCGCG), and all contained the terminal CG sequence which was involved in the packing arrangement through overlapping and hydrogen bonding between the terminal bases in a sequence specific way (58). A new family may now root from the structure of the *NarI* site (59) which provides a striking example of a direct interaction between one phosphate and the amino group of cytidines of a symmetrical molecule. This new packing is also sequence-dependent but in a completely different way.

The A-DNA structures, although crystallizing in four several space groups, show also clear similarities of packing interactions. The common packing motif is the stacking of a terminal base-pair upon the minor groove of a symmetrical molecule. The RNA oligomer $U(UA)_6A$ follows the same type of interaction. The existence of 2′-hydroxyl groups allows the formation of additional intermolecular hydrogen bonds (32). In tRNA crystals also the stacking interactions are important (60).

Formation of crystals of poor quality is a problem often encountered with nucleic acids. This may be due to the geometry of the helices, which can pack easily despite rotational disorder. The answer is to play with additives, temperature, pH, trying to find a way of introducing some structural change or to bind additional small molecules which could act as a lever, promoting lattice buiding.

4. Co-crystallization of nucleic acids and proteins

Most fundamental biological mechanisms and particularly those regarding the storage and expression of the genetic message involve interactions between proteins and nucleic acids. This promoted a need for three-dimensional structural knowledge. When the nucleic acid moiety of the complex of interest are small ligand-like nucleotides (e.g. ATP, GTP, or NAD), or short oligonucleotides, it is sometimes possible to diffuse it into a receptor crystal [e.g. dT_4 in Kleenow fragment of DNA polymerase I from *E. coli* (61)]. But with larger substrates which cannot penetrate, or when much conformation changes occur, co-crystallization is a necessity.

4.1 General features of protein and nucleic acid co-crystallization

4.1.1 Co-crystallization or crystallization of pre-existing complexes

When dealing with complexes, a variety of situations can be encountered and crystallization strategy must adapt for each particular problem. Some complexes can be isolated and purified from natural sources (e.g. viruses, ribosomes, nucleosome) while many other nucleic acid protein complexes are transient and an independent purification of each component is necessary.

The heterologous nature of the particles introduces additional problems. Protein nucleic acid recognition involves specific interactions between macromolecules of very different electrostatic properties, and for a given protein the binding areas are adapted to this complementarily. In order not to be a competitive site, the other part of the molecule will act as a repellent to the nucleic-acid substrate. In known crystal structures of complexes the crystal packing is built upon contacts between macromolecules of the same type, i.e. protein–protein or nucleic acid–nucleic acid interactions.

4.1.2 Stability

Difficulties may arise regarding the stability of the complex particle in the crystallization conditions (pH, ionic strength, . . .) and it is important to ascertain the physical existence of the complex in such conditions. A K_d vaue of 10^{-5} can be considered as an upper limit of stability for a crystallizable complex. The time-scale of crystallization experiments creates another problem when nucleic acids are substrates of enzymatic reactions. If RNA is the substrate it is now possible to chemically synthesize a mixed nucleic acid with the reactive ribonucleotide being replaced by a deoxyribonucleotide (34). When DNA is the substrate various solutions have been found, like pH changes or removal of the cations necessary for the enzymatic reactions [e.g. omit Mg^{2+} in the co-crystallization of *Eco RI* (62), *Eco RV* (63) or Kleenow fragment (61) with their DNA substrates]. As with DNAse I, in some special cases the cleaved oligonucleotide was co-crystallized with the enzyme (64).

4.1.3 Homogeneity of samples

A problem which is specific to natural samples containing large nucleic acids like nucleosome and ribosomes is the inhomogeneity of the nucleic acid part of the samples. Even if the size of the nucleic acid component can be defined with some accuracy, the random dispersion of the nucleotide sequence is a major problem. In the case of nucleosome core particles, this problem was one of the major limitation to obtaining high-resolution diffracting crystals. A solution can be found with the development of chemical synthesis or *in vitro* genetic engineering techniques, which enables a large-scale preparation of long oligonucleotide sequences, up to a hundred nucleotides.

When looking at the successful crystallization attempts in the past ten years, one is impressed by the improvement of the quality of the crystals obtained for complexes with synthetic oligonucleotides of small or average size. The best diffracting crystals can now pass the limit of 2 Å resolution. Many DNA protein complex crystals diffract to better than 3 Å resolution. These include enzymes with their substrates like *Eco Ri*, *Eco RV*, DNAse I or repressors like trp, cro, phage λ and others.

4.1.4 Stoichiometry

A slight excess of substrate is a general trend of all experiments. In one example where stoichiometry was well analysed, that of the complex between yeast $tRNA^{Asp}$ and aspartyl–tRNA synthetase, a variation of the tRNA concentration around the stoichiometric value (2:1) was the main cause of polymorphism (65). In some cases, however, a larger excess of DNA concentration over protein concentration was used; e.g. the crystallization of λ repressor fragment (1–92) with a 20-mer operator was done at a molar excess of two DNA duplexes for one protein dimer. The correct stoechiometry, one DNA duplex per protein dimer, was found in the crystals (66).

4.1.5 Purity

This is of general concern. Since we are dealing with two molecules the problem is more crucial here. An illustration of the importance of the nucleic acid purity is given with crystallization of the operator binding domain of the λ repressor with the λ operator site. Much better crystals were obtained when the synthetic operator was further purified with HPLC (66). For the crystallization of yeast aspartyl–tRNA synthetase with $tRNA^{Asp}$, improvement of the protein purification protocol produced a new and better-diffracting crystal form (67). Impurities can also affect the complex stability.

4.1.6 Crystallizing agents

High salt conditions were long believed to be unfavourable to the stability of nucleic acid protein complexes. This was the main reason for the success of crystallization attempts with alcohols, and among them MPD is the most popular (in the range 15–25%). However, many successful attempts were made with $(NH_4)_2SO_4$ at high concentration (in the range of 2M). This can be explained by a screening effect of ammonium and/or sulphate ions which hamper unspecific contact and prevent aggregation. It is an experimental observation that although high NaCl concentrations are disruptive of complexes, $(NH_4)_2SO_4$ or ammonium citrate do not have such disruptive effects. Mixtures of $(NH_4)_2SO_4$ and PEGs were also successful in some cases.

4.1.7 pH

Slightly acidic or neutral pH seem the best bet although attempts at slightly basic pH (7.5–8) are not uncommon.

4.1.8 Additives

For this part we enter in more specific problems linked to the nature of the systems investigated. $MgCl_2$ and $CaCl_2$ are the most common additive salts used. Phosphate salts have to be avoided for two reasons: they very often lead to insoluble compounds and they act as competitors for nucleic acid binding sides. When existing, cofactors (like ATP, GTP, L-tryptophan) should always be used as an important variable in crystallization screenings.

4.2 Complexes of proteins and synthetic oligodeoxynucleotides

A compilation of crystallization conditions which gave crystals (diffracting to high resolution) of protein–DNA complexes is given in *Table 4*. Crystals have been obtained with salts, alcohols, or PEG as precipitants.

The main problem of this type of binary complexes is the choice of the best sequence and the optimal length of the nucleic acid substrate. Two major constraints have to be taken into account: the biological relevance of the sequence and the stability of the duplex. Beside that, no general rule can be applied. An effect of the number of base pairs (which should have been a multiple of seven) was thought to be important after the co-crystallization of the DNA-binding domain of phage 434 repressor and its operator (68). Later examples were no longer in this line. Some co-crystallizations were made with

Table 4. Crystallization conditions of DNA–protein complexes (updated summer 1990)

Protein and DNA / Precipitant[a]	Temp	Concentrations protein[b]	Concentrations DNA[c]	buffer	additives	Crystals	Ref
● phage 434 repressor (fragment 1–69) + operator (14 bp, symmetric, blunt ends)							
$(NH_4)_2SO_4$ 1.3 M	4 °C	0.5 mM (2:1)	0.25 mM	Na phosphate 5 mM pH 4.7		I422 3.2–4.5 Å	68
● phage 434 repressor (fragment 1–69) + operator (20-mer, asymmetric, complementary overhangs of 1 nt.)							
PEG 3000 12–14%	4 °C	2 mM (2:1)	1.0 mM		NaCl 100 mM $MgCl_2$ 120 mM	$P2_12_12_1$ 2.5 Å	82
● phage 434 Cro repressor + operator (14 bp, symmetric, blunt ends)							
$(NH_4)_2SO_4$ 2.7 M	20 °C	(2:1)	1–2 mM	Tris–HCl 10 mM pH 8.0	EDTA 1 mM DTT 0.1 mM NaCl 0.7 M $[Co(NH_4)_6]Cl_3$ 10 mM $MgCl_2$ 1 mM	C2 3.0–5.0 Å	83
● phage λ repressor (fragment 1–92) + operator (20-mer, asymmetric, complementary overhangs of 1 nt.)							
PEG 400 10 vs. 20%	20 °C	0.91 mM (1:1)	0.91 mM	BTP 15 mM pH 7.0	NaN_3 1 mM	$P2_1$ 2.5 Å	66
● phage λ Cro repressor + operator (17 bp, asymmetric, blunt ends)							
NaCl 0.1 vs. 3.5–4.0 M or slow evaporation		2.5 mg/ml	2.5 mg/ml	Na caco. 20 mM pH 6.9		$P6_2(P6_4)$ 3.7 Å	69
● *E. coli* trp repressor + Trp + operator (18 bp + overhanging 5′-T)							
MPD	20 °C	0.4 mM	0.6–0.8 mM	Na caco.	L-Trp 2 mM		84
(1) 20% vs. 40%				10 mM pH 7.2	$CaCl_2$ 11 mM	$P2_1$ 2.5 Å	
(2) 15–20% vs. 30–40%				10 mM pH 7.2	$CaCl_2$ 18 mM	P1 3.0 Å	
(3) 25% vs. 50%			pH 5.0	100 mM	$MgCl_2$ 6 mM	$P3_121$ $(P3_221)$ 3.2 Å	

Table 4. (*continued*)

Protein and DNA Precipitant[a]	Temp	Concentrations protein[b]	Concentrations DNA[c]	buffer	additives	Crystals	Ref
• *E. coli* CAP protein + cAMP + DNA binding site (30 bp + overhanging 5′-G)							
PEG 3350 5–10%		4–6 mg/ml	1.5 molar xcess	MES 50 mM pH 5–6	NaCl 0.2 M $CaCl_2$ 0.1 M cAMP 2 mM spermine 2 mM 0.3% *n*-octylglucoside 0.02% NaN_3 DTT 2 mM	$C222_1$ 3.0 Å	70
• *E. coli* Kleenov fragment of Pol I + DNA substrate (8 bp + 3 bases, single stranded 5′ overhang)							
$(NH_4)_2SO_4$ 1.5 M				citrate 400 mM pH 5.6	EDTA		61
• *E. coli* restriction endonuclease *Eco RI* + DNA substrate (13-mer with overhanging 5′-T)							
PEG 400 8% vs. 16%	4 °C	2.7 mg/ml (2:9)	2.8 mg/ml	BTP 40 mM pH 7.4	NH_4acetate 0.5 M dioxane 15%	P321 2.6 Å	62
• *E. coli* restriction endonuclease *Eco RV* + DNA fragments							
PEG 4000						$P2_12_12_1$ 2.0 Å	63
• *E. coli* endonuclease *HhaII* + DNA substrate (7 bp)							
PEG 4000 5% vs. 10%	20 °C	20–25 mg/ml	2.2 mg/ml	K phosphate 10 mM pH 7.2	NaCl 0.2 M EDTA 1 mM DTT 0.1 mM	C2 4.0 Å	85
• Bovine pancreatic DNase I + DNA substrate (8-mer, nicked).							
PEG 600	4 °C				EDTA 15 mM	$C222_1$ 2.0 Å	64

[a] The concentration in the reservoir is given, or initial concentrations in the form: C(drop) versus C(reservoir).
[b] The molar ratio of protein monomers versus DNA duplex is given in brackets.
[c] When several DNA duplexes were tried, the conditions indicated are those producing the best crystals.

blunt-ended oligomers [e.g. phage λ cro repressor with a 17-mer operator (69) or phage 434 repressor with a 14-mer operator (66)]. Others underline the importance of overhanging nucleotides. These could reinforce the end-to-end stacking of DNA duplexes which seem to be a common mode of packing. Clearly, there is no generally applicable rationale that specifies the optimal length and terminal structure of the oligonucleotides to be used in crystallizing protein–DNA complexes. The principal limitation of the choice seems to be the production of the oligonucleotides, especially if the required sequence is large. This problem has been nicely overcome in the crystallization of CAP protein complexed with DNA (70). Ten oligonucleotides, up to 20 nucleotides in length, were synthesized. These are able to self-hybridize and were mixed to generate 19 different double-stranded segments (of 28–36 base-pairs) with symmetric overhangs of zero, one, or two bases. Crystallization conditions were examined with 26 different DNA segments, 28 or more base pairs in length, that explored a variety of sequences (symmetric or not), length, and extended 5′ or 3′-termini. Crystals of variable quality were produced, one of them diffracting to 3.0 Å resolution.

4.3 Complexes of proteins and RNAs

Complexes between tRNAs and their cognate aminoacyl-tRNA synthetases are the only examples of successful co-crystallization of proteins and RNAs. The stability range of this type of complex is not very high: K_d values are in the range 10^{-6}–10^{-9}. Well-characterized crystals, which led to high-resolution structure determination, were obtained only in two cases; the *E. coli* glutamine system (6) and the aspartic acid system from yeast (67).

The main limitation in the field has been the poor supply of biological material with reliable physicochemical integrity. This prevented a good survey of crystallization conditions. In the case of the yeast aspartic system, the first attempt of a large-scale purification of the synthetase and the tRNA was set up from wild-type yeast cells and commercial bulk tRNA (65). However, the method necessitated three weeks of hard work with rather poor yield of intact enzyme due to proteolysis. An improvement of the purification procedure reduced the time-scale to three days with concomittant increase of the yield. Together with revised conditions of crystallization, this improvement resulted in much better crystals diffracting to 2.7 Å resolution (67). In the case of the *E. coli* glutamine system, the problem of sample quantities was overcome by cloning and construction of overproducing strains for both the enzyme and the tRNA (6).

For crystallization, most of the problems and constraints with tRNA synthetase systems are similar to those described in the general part. *Table 5* shows that both attempts were successful in high salt conditions. Spermine, which was a very important additive for obtaining good diffracting crystals of free tRNA, is not necessary to obtain co-crystals of the complex.

Table 5. Crystallization conditions of RNA–protein complexes

Complex	Con. of tRNA and enzyme	Temp	pH	Buffer	Precipitant	Additives	Ref
GlnRS[a]:tRNA^{Gln} (*E.coli*) 1:1	(1:1 complex) 10 mg/ml	17 °C	6.8–7.0	Pipes 80 mM	Na citrate 44–64%	$MgCl_2$ 20 mM ATP 4 mM 2-mercaptoethanol 20 mM	6
			7.0–7.5	Pipes 80 mM	ammonium sulphate 1.8–2.0 M	$MgSO_4$ 20mM ATP 8 mM NaN_3 0.02%	
AspRS[b]:tRNA^{Asp} (yeast) 1:2	(enzyme) 10mg/ml (80 μM) tRNA^{Asp} 4.8mg/ml (190 μM)	4 °C	7.5	Tris-maleate 40 mM	ammonium sulphate 1.0 vs. 2.4 M	$MgCl_2$ 5 mM	67

[a] glutaminyl–tRNA synthetase; [b] aspartyl–tRNA synthetase.

4.4 Ribosomes and their subunits

Protein biosynthesis takes place on the large ribonucleoprotein particles called ribosomes. These organelles are made of two subunits which associate upon biosynthesis initiation to form a full particle. Although the first observation of crystalline material was made *in vivo* as part of a mechanism of hibernation in a variety of lizards, the only real successful attempts to grow large three-dimensional crystals of ribosomes was achieved with bacterial particles. In bacteria the smallest subunit (30S) has a molecular weight of 700kD and contains about 20 proteins and one RNA chain (16S). The large subunit (50S) weights 1.6 million daltons and consists of about 35 different proteins and two RNA chains (23S and 5S). The full particle (70S) is the operating unit. Although the size of ribosomes is comparable to that of viruses, the lack of internal symmetry transforms the problem into a formidable challenge.

Crystallization properties and biological activities of the particles are strongly correlated, i.e. inactive particles do not crystallize. Conversely, redissolved crystals are active. In the case of 70S ribosomes from *T. thermophilus*, the best crystals are grown from the material obtained after dissolution of previously formed egg-like crystals (71).

When many crystalline forms of particles and subunits of different sources have been grown, as shown in *Table 6*, the best three-dimensional crystals available are obtained with 50S particles from *Halobacterium marismortui* and with 30S and, more recently, 70S particles of *T. thermophilus*. The quality of the crystals depends on the procedure use for the preparation of the ribosomal subunits and of the strain of a given bacterial species.

Since high salt conditions are disruptive for ribosomes from eubacteria, crystallization conditions were searched mostly with volatile organic solvents. Initially, the crystallization droplets contain no precipitant or a very small quantity of it. In order to reduce the rate of crystal growth and to avoid some technical difficulties linked to the use of volatile solvents, crystallization assays are often performed directly in X-rays capillaries (72). In contrast, halophilic ribosomal particles are stable at high salt concentrations. The growth solution mimics to some extent the natural environment within the halobacteria and contains KCl, NH_4Cl and $MgCl_2$. In these cases PEG is often the crystallization agent. To improve the crystalline order, the concentration of the salts are finely adjusted. Better crystals are obtained when the salt concentrations were reduced to the minimum needed for storage without loss of activity (73). More particularly the equilibrium between Mg^{2+} and monovalent ions is delicate. Three-dimensional crystals of 50S ribosomal particles from *Bacillus stearothermophilus* grow in relatively low Mg^{2+} concentrations, whereas the production of two-dimensional sheets occur at high levels of Mg^{2+}. Similarly, for 50S subunits of *H. marismortui*, the lower the Mg^{2+} concentration is, the thicker the crystals are (72).

Table 6. Crystallization conditions for ribosomes and ribosomal particles

Particle	Source	Temp	pH	Precipitant[a]	Salts and additives	Max. resol.	Ref
30S	*E. coli*	4 °C	8.2	ethyl butanol 30%			73
	T. therm.	4 °C	8.2	ethyl butanol 20% ethanol 10%			73
	H. maris.	19 °C	5.7	PEG 5%	KCl 2.0 M NH_4Cl 0.5 M $MgCl_2$ 0.1 M		73
50S	*B. stearo.* (I)	4 °C	6.6	PEG 2.5%	KCl 0.5 M $(NH_4)_2SO_4$ 0.18 M $MgCl_2$ 0.03 M		73
	(II)	4 °C	6.4	PEG 2.5%	$(NH_4)_2SO_4$ 0.18 M $MgCl_2$ 0.02 M		73
	(III)	4 °C	8.4	methanol (1%) to 12% ethylene glycol 12%	NaCl 0.5 M spermine (10 mM)	10–13 Å	72
	H. maris.	19 °C	5.0–5.6	PEG (4–5) to 9%	KCl (1.2) 1.7 M NH_4Cl 0.5 M $MgCl_2$ (0.05–0.10) 0.10 M spermidine (10 mM)	5.5 Å	72
70S	*B. stearo.*	4 °C	6.6	PEG 2.5%	KCl 0.5 M $(NH_4)_2SO_4$ 0.18 M $MgCl_2$ 30 mM		73
	T. thermo.	4 °C	7.5	MPD 15%	KCl 0.2 M NH_4Cl 75 mM $MgCl_2$ 25 mM	20 Å	71

[a] Numbers in brackets correspond to initial concentration in the drop, when specified in the referenced article. Otherwise, the concentrations indicated correspond to those of the reservoir solutions.

A major problem of ribosomes crystals is the poor limit of resolution, which could be due either to radiation damage and crystal degradation, or to intrinsic disorder of the molecules in the crystal. Data collection at very low temperatures (−150 °C) showed no improvement of the maximum resolution but increased the lifetime of the crystals.

4.5 Viruses

In this large family of particles, proteins form the protecting shell which encapsidates the genetic material (RNA or DNA). The quaternary structure of viruses is dominated by the nature of protein–protein interactions. Two main shapes are observed: helical rods as in TMV and filamentous bacteriophage, and isometric capsids in spherical viruses. Crystals of viruses were among the first obtained and now many viruses have been crystallized. The list of those for which the three-dimensional structure is known is given in *Table 7*. Details of the crystallization conditions can be found in this table as well as in the attached references.

The importance of the external capside and hence the non-effect of RNA or DNA on crystal formation is nicely demonstrated with cowpea mosaic virus (CPMV). The genome of this virus consists of two RNA molecules, RNA1 (5.9 kb) and RNA2 (3.5 kb), which are encapsidated in separate particles. Empty capsids are also formed *in vivo*. All three components are of the same size and appear to have identical surfaces. Isomorphous crystals were obtained with each of the isolated components or with a mixture of the three components, and the same ratio of components was found in the crystals and in the crystallizing solution (74).

Since protein–protein interactions are dominant in viruses' crystal packings the crystallization conditions resemble those used for proteins; e.g. the pH range is larger than for nucleic acids (i.e. 4.9–8.0) and limited only by the stability of the capsid. PEG (2–3%) alone or mixed with $(NH_4)_2SO_4$ in the 0.5 M range is the most widely used crystallizing agent. Crystallization is usually done at room temperature (20 °C).

References

1. Monier, R. (1971). In *Procedure in nucleic acids research*, (ed. G. L. Cantoni and D. R. Davies), Vol. 2, pp. 618–28. Harper and Row, New York.
2. Kime, M. J. and Moore, P. B. (1983). *Biochemistry*, **22**, 2615.
3. Morikawa, K., Kawakami, M., and Takemura, S. (1982). *FEBS Lett.*, **145**, 194.
4. Abdel-Meguid, S. S., Moore, P. B., and Steiz, T. A. (1983). *J. Mol. Biol.*, **171**, 207.
5. Giegé, R. and Ebel, J. P. (1968). *Biochim. Biophys. Acta*, **161**, 125.
6. Perona J. J., Swanson, R., Steiz, T. A., and Söll, D. (1988). *J. Mol. Biol.*, **202**, 121.

Table 7. Crystallization conditions of some isometric viruses

Virus[a]	Concentration	Temp	pH	Buffer	Precipitant and additives	Ref
CPMV–hexag.	20 mg/ml	20 °C	4.9	Na citrate 0.6 M		74
–cubic	35 mg/ml	20 °C	7.0	K phosphate 50 mM	$(NH_4)_2SO_4$ 0.4 M PEG 8000 2%	
BPMV	15 mg/ml	20 °C	7.0	Na phosphate 20 mM	PEG 8000 2%	74
BBV	8 mg/ml	20 °C	6.9–7.2	Na phosphate 50 mM	Ammonium sulphate 0.55 M PEG 8000 1% (accelerates)	74
FHV	10–20 mg/ml	20 °C	6.0	Na citrate 25 mM	PEG 8000 2.5% 0.1% ß-octylglucoside	74
NDV	10 mg/ml	20 °C	7.2	Na phosphate 50 mM	PEG 8000 2.5% 1 mM NaN_3	74
NßV	8 mg/ml	20 °C	7.6	Tris 0.17 M	Ca nitrate 1.0 M PEG 8000 2.25%	74
TBSV	30 mg/ml	4 °C			$(NH_4)_2SO_4$ 0.5 M	86
STNV	10–12 mg/ml or 7–8 mg/ml + 0.4% PEG 6000		6.5	Na phosphate 50 mM	Mg^{2+} 1 mM	87
rhinovirus 14	5 mg/ml	20 °C	7.2	Tris	$CaCl_2$ 10 mM PEG 800 0.25 or 0.5%	88
mengo virus	5 mg/ml		7.4	phosphate	PEG 8000 2.8%	89

[a] CPMV, cowpea mosaic virus; BPMV, beanpod mottle virus; BBV, black beetle virus; FHV, flockhouse virus; NDV, nodamura virus; NßV, nudaurelia capensis ß virus; TBSV, tomato bushy stunt virus; STNV, satellite tobacco necrosis virus.

7. Lowary, P., Sampson, J., Milligan, J., Groebe, D., and Uhlenbeck, O. C. (1986). In *Structure and dynamics of RNA*, (ed. P. H. Van Knippenberg and C. W. Hilbers). NATO ASI Series, Series A, Life Sciences, Vol. 110, pp. 69–76. Plenum, New York.
8. Davanloo, P., Rosenberg, A. H., Dunn, J. J., and Studier, W. (1984). *Biochemistry*, **81**, 2035.
9. Milligan, J. F., Groebe, D. R., Witherell, G. W., and Uhlenbeck, O. C. (1987). *Nucl. Acids Res.*, **15**, 8783.
10. Sampson, J. R. and Uhlenbeck, O. C. (1988). *Proc. Natl. Acad. Sci. USA*, **85**, 1033.
11. Perret, V., Garcia, A., Grosjean, H., Ebel, J. P., Florentz, C., and Giegé, R. (1990). *Nature*, **344**, 787.
12. Doctor, B. P. (1971). In *Procedure in nucleic acids research*, (ed. G. L. Cantoni and D. R. Davies), Vol. 2, pp. 588–607. Harper and Row, New York.
13. Nishimura, S. (1971). In *Procedure in nucleic acids research*, (ed. G. L. Cantoni and D. R. Davies), Vol. 2, pp. 542–64. Harper and Row, New York.
14. Spencer, M., Neave, E. J., and Webb, N. L. (1978). *J. Chromatogr.*, **166**, 447.
15. Nishimura, S., Shindo-Okada, N., and Crain, P. F. (1987). *Methods in Enzymology*, **155**, 373.
16. Gillam, I., Millward, S., Blew, D., von Tigerstrom, M., Wimmer, E., and Tener, G. M. (1967). *Biochemistry*, **6**, 3043.
17. Roy, K. L., Bloom, A., and Söll, D. (1971). In *Procedure in nucleic acids research*, (ed. G. L. Cantoni and D. R. Davies), Vol. 2, pp. 524–41. Harper and Row, New York.
18. Gillam, I., Blew, D., Warrington, R. C., von Tigerstrom, M., and Tener, G. M. (1968). *Biochemistry*, **7**, 3459.
19. Kelmers, A. D., Novelli, G. D., and Stulberg, M. P. (1965). *J. Biol. Chem.*, **240**, 3979.
20. Weiss, J. F., Pearson, R. L., and Kelmers, A. D. (1968). *Biochemistry*, **7**, 3479.
21. Narihara, T., Fujita, Y., and Mizutani, T. (1982). *J. Chromatogr.*, **236**, 513.
22. Flanagan, J. M., Fujimura, R. K., and Jacobson, K. B. (1986). *Anal. Biochem.*, **153**, 299.
23. Bischoff, R. and McLaughlin, L. W. (1985). *Anal. Biochem.*, **151**, 526.
24. Singhal, R. P., Griffin, G. D., and Novelli, G. D. (1976). *Biochemistry*, **15**, 5083.
25. Holmes, W. M., Hurd, R. E., Reid, B. R., Rimerman, R. A., and Hatfield, G. W. (1975). *Proc. Nat. Acad. Sci. USA*, **72**, 1068.
26. Dudock, B. S. (1987). In *Molecular biology of RNA. New perspectives.* (ed. M. Inouye and B. Dudock), pp. 321–29. Academic Press Inc., New York and London.
27. Brown, D. M. (1974). In *Basic principles in nucleic acids chemistry* (ed. P. O. P. Ts'O), Vol. 2, pp. 1–90. Academic Press Inc., New York and London.
28. Werner, W., Krebs, W., Keith, G., and Dirheimer, G. (1976). *Biochim. Biophys. Acta*, **432**, 161.
29. Gait, M. J., Matthes, H. W., Singh, M., Sproat, B. S., and Timas, R. (1982). *Nucl. Acids Res.*, **10**, 6243.
30. Ohtsuka, E., Ikehara, M., and Söll, D. (1982). *Nucl. Acids Res.*, **10**, 6553.
31. Pingoud, A., Fliess, A., and Pingoud, A. (1989). In *HPLC of macromolecules, a practical approach*, (ed. R. W. A. Oliver), pp. 183–208. IRL Press, Oxford.

32. Dock-Bregeon, A. C., Chevrier, B., Podjarny, A., Johnson, J., de Bear, J. S., Gough, G. R., Gilham, P. T., and Moras, D. (1989). *J. Mol. Biol.*, **209**, 459.
33. Ogilvie, K. K., Usman, N., Nicoghosian, K., and Cedergren, R. J. (1988). *Proc. Natl. Acad. Sci. USA*, **85**, 5764.
34. Perreault, J. P., Wu, T., Cousineau, B., Ogilvie, K. K., and Cedergren, R. (1990). *Nature*, **344**, 565.
35. Kennard, O. and Hunter, W. N. (1989). *Quarterly Rev. Biophys.*, **22**, 327.
36. Dock, A. C., Lorber, B., Moras, D., Pixa, G., Thierry, J.-C., and Giegé, R. (1984). *Biochimie*, **66**, 179.
37. Kim, S. H. Quigley, G., Suddath, F. L., McPherson, A., Sneden, D., Kim, J. J., Weinzierl, J., and Rich, A. (1973). *J. Mol. Biol.*, **75**, 421–28.
38. Johnson, C. D., Adolplh, K., Rosa, J. J., Hall, M. D., and Sigler, P. B. (1970). *Nature*, **226**, 1246.
39. Giegé, R., Moras, D., and Thierry, J. C. (1977). *J. Mol. Biol.*, **115**, 91.
40. Wang, A. H. J., Hakoshima, T., van der Marel, G., van Boom, J. H., and Rich, A. (1984). *Cell*, **37**, 321.
41. Tabor, C. W. and Tabor, H. (1976). *Annu. Rev. Biochem.*, **45**, 285.
42. Sakai, T. T. and Cohen, S. S. (1976). *Prog. Nucl. Ac. Res. Mol. Biol.*, **17**, 15.
43. Young, J. D., Bock, R. M., Nishimura, S., Ishikura, H., Yamada, Y., RajBandhary, U. L., Labanauskas, M., and Connors, P. G. (1969). *Science*, **166**, 1527.
44. Kim, S. H., Quigley, G., Suddath, F. L., and Rich, A. (1971). *Proc. Natl. Acad. Sci. USA*, **68**, 841.
45. Ichikawa, T. and Sundaralingam, M. (1972). *Nature New Biol.*, **236**, 174.
46. Ladner, J. E., Finch, J. T., Klug, A., and Clark, B. F. C. (1972). *J. Mol. Biol.*, **72**, 99.
47. Quigley, G. J., Teeter, M., and Rich, A. (1978). *Proc. Natl. Acad. Sci. USA*, **75**, 64.
48. Wang, A. H. J., Quigley, G. J., Kolpak, F. J., Crawford, J. L., van Boom, J. H. van der Marel, G., and Rich, A. (1979). *Nature*, **282**, 680.
49. Crawford, J. L., Kolpak, F. J., Wang, A. H. J., Quigley, G. J., van Boom, J. H., van der Marel, G., and Rich, A. (1980). *Procl. Natl. Acad. Sci. USA*, **77**, 4016.
50. Fujii, S., Wang, A. H. J., van der Marel, G., van Boom, J. H., and Rich, A. (1982). *Nucl. Acids Res.*, **10**, 7879.
51. Holbrook, S. R., Sussman, J. L., Warrant, W. R., and Kim, S. H. (1978). *J. Mol. Biol.*, **123**, 631.
52. Gessner, R. V., Quigley, G. J., Wang, A. H. J., van der Marel, G., van Boom, J. H., and Rich, A. (1985). *Biochemistry*, **24**, 237.
53. Brennan, R. G., Westhof, E., and Sundaralingam, M. (1986). *J. Biomol. Struc. Dyn.*, **3**, 649.
54. Einspahr, H., Cook, W. J., and Bugg, C. E. (1981). *Biochemistry*, **20**, 5788.
55. Hingerty, B. E., Brown, R. S., and Klug, A. (1982). *Biochim. Biophys. Acta*, **697**, 78.
56. Chevrier, B., Dock, A.-C., Hartmann, B., Leng, M., Moras, D., Thuong, M. Y., and Westhof, E. (1986). *J. Mol. Biol.*, **188**, 707.
57. Wang, A. H. J. and Teng, M. K. (1988). *J. Crystal Growth*, **90**, 295.

58. Dickerson, R. E., Goodsell, D. S., Kopka, M. L., and Pjura, P. E. (1987). *J. Biomol. Struc. Dyn.*, **5**, 557.
59. Timsit, Y., Westhof, E., Fuchs, R. P. P., and Moras, D. (1989). *Nature*, **341**, 459.
60. Moras, D. and Bergdoll, M. (1988). *J. Crystal Growth*, **90**, 283.
61. Steitz, T. A., Beese, L., Freemont, P. S. Friedman, J. M., and Sanderson, M. R. (1987). *Cold Spring Harbor Symp. Quant. Biol.*, **52**, 465.
62. Grable, J., Frederick, C. A., Samudzi, C., Jen-Jacobson, L., Lesser, D., Greene, P., Boyer, H. W., Itakura, K., and Rosenberg, J. M. (1984). *J. Biomol. Structure and Dynamics*, **1**, 1149.
63. Winckler, F. K., Brown, R. S., Leonard, K., and Berriman, J. (1987). In *Crystallography in molecular biology* (ed. D. Moras, J. Drenth, B. Strandberg, D. Suk, and K. Wilson). NATO ASI Series, Series A: Life Sciences, Vol. 126, pp. 345–52. Plenum, New York.
64. Suck, D., Lahm, A., and Oefner, C. (1988). *Nature*, **332**, 464.
65. Lorber, B., Giegé, R., Ebel, J.-P., Berthet, C., Thierry, J.-C., and Moras, D. (1983). *J. Biol. Chem.*, **258**, 8429.
66. Jordan, S. R., Whitcombe, T. V., Berg, J. M., and Pabo, C. O. (1985). *Science*, **230**, 1383.
67. Ruff, M., Cavarelli, J., Mikol, V., Lorber, B., Mitschler, A., Giegé, R., Thierry, J.-C., and Moras, D. (1988). *J. Mol. Biol.*, **201**, 235.
68. Anderson, J., Ptashne, M., and Harrison, S. C. (1984). *Proc. Natl. Acad. Sci. USA*, **81**, 1307.
69. Brennan, R. G., Takeda, Y., Kim, J., Anderson, W. F., and Matthews, B. W. (1986). *J. Mol. Biol.*, **188**, 115.
70. Schultz, S. C., Shields, G. C., and Steitz, T. A. (1990). *J. Mol. Biol.*, **213**, 159.
71. Trakhanov, S., Yusupov, M., Shirokov, V., Garber, M., Mitschler, A., Ruff, M., Thierry, J.-C., and Moras, D. (1989). *J. Mol. Biol.*, **209**, 327.
72. Yonath, A., Frolow, F., Shoham, M., Müssig, J., Makowski, I., Glotz, C., Jahn, W., Weinstein, S., and Wittmann, H. G. (1988). *J. Crystal Growth*, **90**, 231.
73. Glotz, C., Müssig, J., Gewitz, H. S., Makowski, I., Arad, T., Yonath, A., and Wittmann, H. G. (1987). *Biochem. Internat.*, **15**, 953.
74. Sehnke, P. C., Harrington, M., Hosur, M. V., Li, Y. R., Usha, R., Tucker, R. C., Bomu, W, Stauffacher, C., and Johnson, J. J. (1988). *J. Crystal Growth*, **90**, 222.
75. Yoon, C., Privé, G. G., Goodsell, D. S., and Dickerson, R. (1988). *Proc. Natl. Acad. Sci. USA*, **85**, 6332.
76. Nelson, H. C. M., Finch, J. T., Luisi, B. F., and Klug, A. (1987). *Nature*, **330**, 221.
77. Privé, G. G., Heinemann, U., Chandrasegaran, S., Kan, L. S., Kopka, M., and Dickerson, R. (1987). *Science*, **238**, 498.
78. Wang, A. H. J., Fujii, S., van Boom, J. H., and Rich, A. (1982). *Proc. Natl. Acad. Sci.*, **79**, 3968.
79. Brown, T., Kennard, O., Kneale, G., and Rabinovitch, D. (1985). *Nature*, **315**, 604.
80. Hunter, W. N., Langlois D'Estaintot, B., and Kennard, O. (1989). *Biochemistry*, **28**, 2444.
81. Wang, A. H. J., Fujii, S., van Boom, J. H., van der Marel, G., van Boeckel, S. A. A., and Rich, A. (1982). *Nature*, **299**, 601.

82. Aggarwal, A. K., Rodgers, D. W., Drottar, M., Ptashne, M., Harrison, S. (1988). *Science*, **242**, 899.
83. Wolberger, C. and Harrison, S. C. (1987). *J. Mol. Biol.*, **196**, 951.
84. Joachimiak, A., Marmorstein, R. Q., Schevitz, R. W., Mandecki, W., Fox, J. L., and Sigler, P. B. (1987). *J. Biol. Chem.*, **262**, 4917.
85. Chandrasegaran, S., Smith, H. O., Amsel, M. L., and Ysern, X. (1986). *Proteins*, **1**, 263.
86. Harrison, S. C. and Jack, X. (1975). *J. Mol. Biol.*, **97**, 173.
87. Jones, T. A. and Liljas, L. (1984). *J. Mol. Biol.*, **177**, 735.
88. Arnold, E., Erickson, J. W., Fout, S. G., Frankenberger, E. A., Hecht, H. J., Luo, M., Rossman, M. G., and Rueckert, R. R. (1984). *J. Mol. Biol.*, **177**, 417.
89. Luo, M., Vriend, G., Kamer, G., Minor, I., Arnold, E., Rossman, M. G. Boege, U., Scraba, D. G., Duke, G. M., and Palmenberg, A. C. (1987). *Science*, **235**, 182.

8

Crystallization of membrane proteins

F. REISS-HUSSON

1. Introduction

Crystallization of integral membrane proteins is one of the most recent developments in protein crystal growth: in 1980, for the first time, two membrane proteins were successfully crystallized, bacteriorhodopsin (1) and porin (2). Since then, a number of membrane proteins (about 20) yielded three-dimensional crystals. In a few cases, the quality of the crystals was sufficient for X-ray diffraction studies. The first atomic structure of a membrane protein, a photosynthetic bacterial reaction centre, was described in 1985 (3), followed by closely-related structures of reaction centres from other bacterial species (4–6). The structure of porin was recently determined (7). Crystallization of membrane proteins is now a very actively growing field, and has been discussed in several reviews (8, 9).

The major difficulty in the study of membrane proteins, which for years hampered their crystallization, comes from their peculiar solubility properties. These originate from their tight association with other membrane components, particularly lipids. Indeed integral membrane proteins contain hydrophobic regions buried in the lipid bilayer core, as well as hydrophilic regions with charged or polar residues more or less exposed at the external faces of the membrane. Disruption of the bilayer for isolating a membrane protein can be done in various ways: extraction with organic solvents, use of chaotropic agents, or solubilization by a detergent. The last method is the most frequently used, because the biological activity of the protein may be maintained if the detergent is chosen carefully. Up to now, successful crystallizations of membrane proteins were always performed starting from their detergent solubilized state. Thus this chapter will be restricted on specific aspects of three-dimensional crystallizations done in micellar solutions of detergent.

2. Crystallization principles

The general principles discussed in Chapter 4 for the crystallization of biological macromolecules apply for membrane proteins: the solution must be brought to supersaturation by modifying its physical parameters (concentrations of constituents, ionic strength, and so on), which may lead to nucleation. The main differences from the behaviour of soluble proteins stem from the following two points.

(a) The entity which is going to crystallize is the protein–detergent complex, not the protein alone. Yet most of the detergent found in the crystal is disordered. This has been demonstrated for two detergents (C_{12}DAO and C_8G respectively) associated with two bacterial reaction centres (10a, b). One or two ordered detergent molecules crystallize within the protein (5, 11). The amount of disordered detergent in the crystals is fairly high, however: about 200 molecules of detergent are associated to one reaction-centre protein, and they form a ring around the hydrophobic transmembrane α helices. The detergent ring is interconnected with its neighbours by bridges. Thus ribbon-like detergent structures are running throughout the crystal. These findings explain why the characteristics of the detergent molecules (such as their length) are so crucial in the crystallization process. Indeed, they should fit around the hydrophobic regions of the protein without hindering the inter-protein contacts.

(b) The solubility of the protein–detergent complex is governed not only by the protein properties, but also (and mainly) by those of the detergent micellar solution. Generally, as will be discussed below, this detergent is non-ionic; its micellar solution therefore only exists in a limited range of concentration and temperature, defined in a phase diagram (*Figure 1*). Outside of this range, the micellar solution may spontaneously break apart into two immiscible aqueous phases; one is enriched in detergent, the other one being essentially depleted in detergent. The temperatures and concentrations at which separation is observed define a curve, called consolution boundary; depending on the detergent, this boundary may be reached starting from the micellar solution either by increasing or by decreasing the temperature as shown in *Figure 1*. When phase separation takes place, the solubilized membrane protein generally partitions into the detergent-rich phase (12) (but exceptions are known for glycoproteins). Phase separation is a function of all constituents of the solution: detergent, protein, nature and concentration of salt, and concentration of a crystallizing agent like PEG. Phase separation seems to play a major role in the crystallization: it is quite often observed that crystallization takes place right before phase separation occurs. Choosing crystallization conditions close to the consolution boundary seems

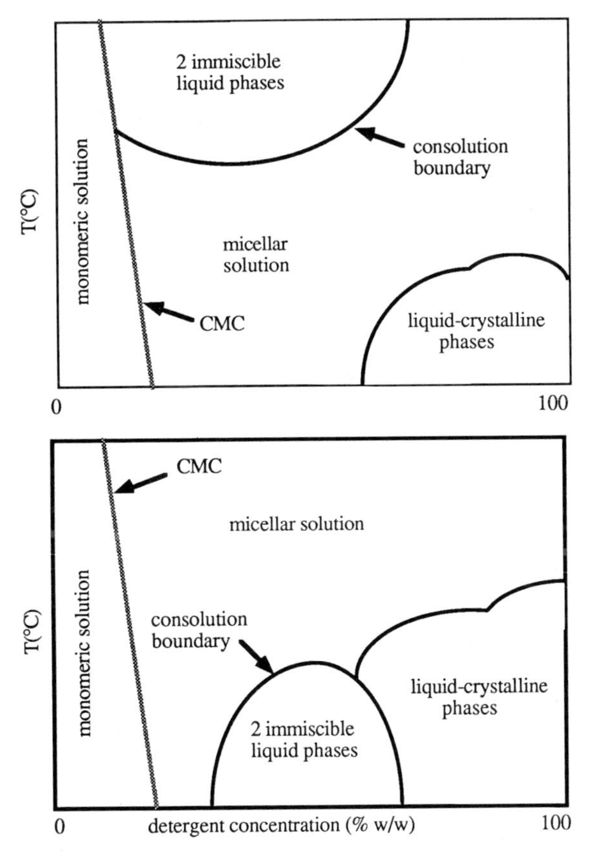

Figure 1. Schematic phase diagrams commonly observed for non-ionic detergent–water mixtures. Depending on the detergent, the region occupied by two immiscible liquid phases may be situated either above (see upper diagram) or below (see lower diagram) the consolution boundary.

equivalent to bringing the protein–detergent complexes into a supersaturated state.

Thus the detergent is playing a very important role in the crystallization and it is better to gain some knowledge of its properties before embarking on experiments.

3. Detergents for crystallization

The detergents which have already led to crystallization of membrane proteins belong to several chemical classes (see *Table 1*). Some have been specifically synthetized for this purpose. Most of them, however, were already used by the biochemist for solubilizing and purifying membrane proteins. As such they have been reviewed in a recent volume of this series

Table 1. Structure of detergents.

***n*-Alkyl-ß-glucosides (C_nG)**

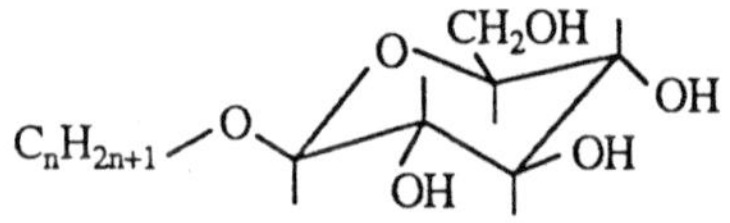

***n*-Alkyl-ß-thioglucosides**

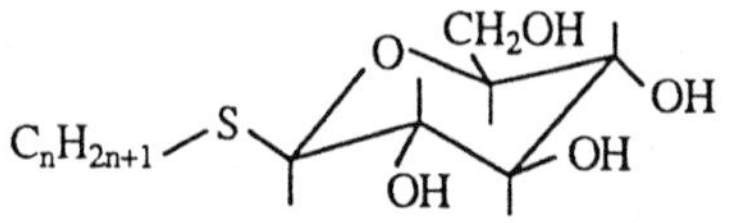

***n*-Alkyl-maltosides (C_nM)**

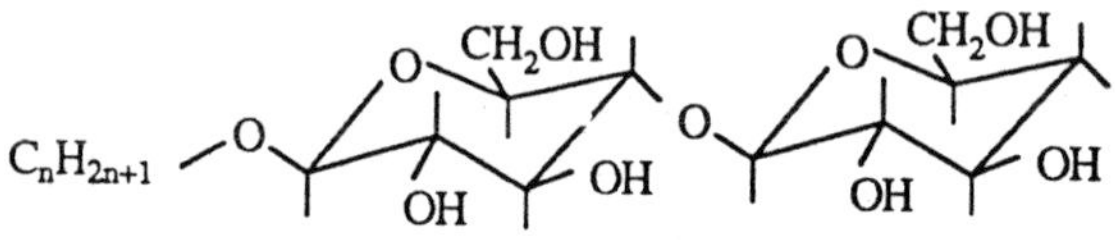

***n*-Alkyl-dimethylamineoxides (C_n-DAO)**

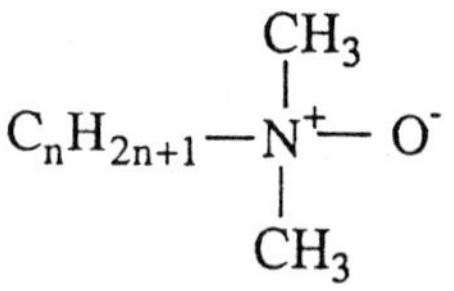

$$C_nH_{2n+1}-N^{+}(CH_3)_2-O^{-}$$

***n*-Alkyl-oligoethylene glycol-monoethers (C_nE_m)**

$$C_nH_{2n+1}-(CH_2-CH_2-O)_m-H$$

(13). With a few exceptions, they are either non-ionic, with uncharged polar groups, or zwitterionic at the pH used. Furthermore, they all are short aliphatic compounds, the length of their hydrophobic part not exceeding that of a normal C_{12} hydrocarbon chain. This last feature results in moderate or high critical micellar concentration (CMC) values (the CMC being the concentration limit between molecular and micellar solutions as shown in *Figure 1*). In micellar solution, micelles are in dynamic equilibrium with monomeric detergent still present in concentration equal to the CMC. Thus the higher the CMC, the larger this exchange. Such a mobile character might play a role in crystallization of the detergent–protein complex.

One further requirement of these detergents is purity and chemical homogeneity, which are important for:

- reproducibility of experiments: when a detergent is heterogeneous (e.g. it contains a mixture of various hydrocarbon chains) or impure (traces of fatty alcohol, and so on), its composition is badly defined and may vary from batch to batch. This may lead to painful situations; with a new detergent batch, crystals may no longer be obtained or still grow but with lower quality.

- quality of the crystals: e.g. trace impurities in C_8G have been reported to interfere with crystal growth of bacteriorhodopsin.

Thus it is advisable to pay attention to the problem of purity and to be able to check it. Thin layer chromatography (TLC) is one useful tool, and it is simple and fast to perform (as described in *Protocol 1*).

Protocol 1. Detergent TLC of *n*-alkyl-ß-glucosides and -maltosides[a]

1. Equilibrate in a TLC tank a silica gel plate (Merck 60, 0.25 mm thickness) with enough ethylacetate/methanol 4:1 (v:v) to wet the bottom of the plate over 5 mm.
2. Dissolve 1 mg detergent in 100 µl ethanol in an Eppendorf tube.
3. Spot with a microcap 10 µl of this solution at 1 cm of the bottom of the plate, and let it dry. Put the plate back in the tank and wait until the solvent front is at 1 cm of the top.
4. Take the plate out and let it dry. Spray under a hood with a 2N H_2SO_4 solution in water. Char in a 90 °C oven. Only one spot should be present.

[a] TLC of *n*-alkyl-oligoethyleneglycol monoethers can be performed with water-saturated methylethylketone as the solvent.

Trace impurities may, however, escape detection by TLC. Another parameter which is very sensitive to detergent composition is the CMC. Comparing the experimental value to those in the literature gives a clue to the purity of the sample. The CMC can be measured by changes in several properties of the solution (surface tension, drop size) or by spectral changes in absorption (14) or fluorescence (15) of solubilized dyes (see *Protocol 2*).

Protocol 2. Determination of the CMC by ANS fluorescence

1. Prepare a working solution of 10 µM ANS (8-anilino-1-naphtalenesulfonic acid Mg salt) in water by dilution of a stock 400 µM ANS solution. Fluorescence of 10 µM ANS is taken as the blank.
2. Prepare 500 µl of a stock detergent solution in 10 µM ANS at a detergent concentration about 100 times the expected CMC. Mix thoroughly. Fluorescence of this solution is taken as the 100% control.
3. Read the emission at 490 nm with excitation at 370 nm while titrating a 2 ml sample of 10 µM ANS with small aliquots of stock detergent solution (up to 50 µl).

Protocol 2. *Continued*

4. Plot the relative fluorescence versus detergent concentration. A steep increase in fluorescence indicates onset of micellization. Draw a straight line through the points in the steep increase region; its intersection with the X axis is the CMC.

Table 2 (16–18) summarizes useful properties of the most often used detergents. Some specific points will be discussed below.

Table 2. Properties of some detergents used for crystallization.

Detergent	mol. wt	CMC(mM)	ref	n[a]	Suppliers[b]
C_8G	292.4	23	16	78	various
C_9G	306.4	6.5	16		C,F
$C_{10}M$	482.6	2.2	16		C,F
$C_{12}M$	510.6	0.16	15	130	various
C_6DAO	145.4				F
C_8DAO	173.4				F
$C_{10}DAO$	201.4	10.4	8		F
$C_{12}DAO$	229.4	2.0	17	73	various
C_8E_4	306	8.5	18	82	B, K
C_8E_5	350	9.2	18		B, K
$C_{12}E_8$	518	0.071	18	120	C, K
$C_{12}E_9$	583	0.071	18		C, F

[a] number of monomers per micelle
[b] Suppliers: B, Bachem, C, Calbiochem, F, Fluka, K, Kohyo

3.1 *n*-Alkyl-ß-glucosides (C_nG)

The C_8 compound (C_8G) has been used in most crystallizations reported so far. It is also very useful in membrane protein biochemistry and generally behaves like a mild detergent (but not always: some membrane proteins are inactivated by C_8G). Its CMC is high and it forms small micelles in water. Therefore its dialysis is fast (half time of a few hours). Because of the possible hydrolysis of the ether linkage, C_8G (and other glucosides) must be stored dry and frozen, and freshly prepared solutions should be used. This instability may explain why various commercial brands were found to contain unidentified trace impurities which could be eliminated by chromatography on a mixed-bed strong ion exchanger (21) (see *Protocol 3*). This purification primarily removes ionic contaminants, but might also remove organic impurities. In our hands, C_8G from Bachem was satisfactory without repurification.

Protocol 3. Purification of C_8G[a]

1. Pour a column (10 × 150 mm) with a strong mixed bed exchanger in the H–OH form (Rexin I-300 or Biorad AG501X8). Wash with 120 ml ethanol then with 600 ml water. Stop the flow when water is draining to the gel surface.
2. Dissolve 5 g C_8G in 50 ml water and put the solution on the gel. Eluate at a flow rate of 0.2 ml/min. Then wash with water at the same rate.
3. Collect the first 100 ml of eluate. Lyophilize and store at −20 °C.

[a] This protocol may be scaled down for smaller detergent quantities.

Members of this class with C_7 and C_6 chain length are commercially available, but their CMC is much too high. The C_9 compound has been used for one crystallization (see *Table 3*). Compounds with longer chain are of limited use because of their poor solubility, not exceeding the CMC.

Table 3. Crystallization conditions for some membrane proteins

Protein	Detergent	Precipitant	Additive	Technique	Ref.
RC *Rps. viridis*	$C_{12}DAO$	AS	HPTH	v.d.	30
RC *Rb. spheroides* R26	C_8G	PEG/NaCl		v.d.	4
" " "	$C_{12}DAO$	PEG/NaCl	HPTH	v.d.	5
" " 241	C_8G	PEG/NaCl	HPTH	v.d.	5
" " Y	C_8G	PEG/NaCl		dial.	6
LH B800–850 *Rb. capsulatus*	$C_{12}DAO$	PEG/NaCl	HPTH	v.d.	31
LH B800–850 *Rps. acidophila*	C_8G, or $C_{12}DAO$	AS+phosph. " "	benzamidine	dial. "	32
RC-B875 *Rps. palustris*	$C_{12}DAO$	PEG		v.d.	33
Prostaglandin synthase	C_8G	PEG/NaCl		v.d.	9
Porin *E. coli*	C_8G, + C_8E_4, C_8E_5	PEG/NaCl		dial. or v.d.	20
OmpF *E. coli*	various	PEG/NaCl		free interf.	22
Porin *Rb. capsulatus*	C_8E_4	PEG/LiCl		v.d.	34
Maltoporin *E. coli*	$C_{10}M$ + $C_{12}E_9$	$PEG/MgCl_2$		dial.	35
PS1 RC	$C_{12}M$	PEG		v.d.	36
PS1 RC	$C_{12}M$	PEG/NaCl, or/$MgCl_2$		batch, v.d.	37
pea LHII	C_9G	phosphate		v.d.	8

Abbreviations: RC, reaction centre; LH, light-harvesting complex; PS1, photosystem. Detergent: see *Table 1*. Precipitant: AS, ammonium sulphate. Additive: HPTH, heptane-1,2-triol. Technique: v.d., vapour diffusion; dial, microdialysis; free interf., free interface diffusion.

3.2 *n*-Alkyl-thioglucosides

The use of these detergents has been reported once (22) and they could be assayed instead of the glucosides. The C_6, C_7, and C_8 compounds are commercially available. One of their advantages is their higher stability towards hydrolysis.

3.3 *n*-Alkyl-maltosides (C_nM)

The C_{12} and C_{10} compounds have been used in crystallizations. $C_{12}M$ has been used also for purification of membrane proteins. Its CMC is fairly low, and the size of the micelles in pure water is larger than for C_8G. Although it may be dialysed, it is a slow process (half-time for removal ~20 h). It is susceptible towards hydrolysis and should be stored frozen. Contamination by dodecanol decreases its solubility and may be detected by appearance of a white precipitate in solutions kept at 5 °C (16).

3.4 *n*-Alkyl-dimethylamineoxides (C_nDAO)

No hydrolysis is observed for this class of detergents; solutions are stable when kept at 5 °C. The C_{12} compound has been used successfully for the crystallization of bacterial reaction centres. It is zwitterionic at $pH > 3$ (and cationic at $pH < 3$). Its micellar size is similar to C_8G, but its CMC is lower, resulting in a slower dialysis rate. It is available from various sources, either as the pure C_{12} species, or as a cheaper mixture containing primarily the C_{12} species but also other chain lengths. Traces of H_2O_2 (left over from synthesis) may contaminate some batches. This can be removed by adding 10 μg of catalase per ml of 30% stock solution of $C_{12}DAO$ in water. Shorter chain analogs (C_6–C_{10}) are available. Some of them have been used for crystallization in mixtures with the C_{12} one.

3.5 *n*-Alkyl-oligoethylene glycol-monoethers (C_nE_m)

Besides pure $C_{12}E_8$ which has been used for biochemistry, pure compounds of shorter chain (C_5–C_8) are commercially available with a defined number of ethyleneglycol units ranging from 1–5. The C_8E_4, C_8E_5, and $C_{12}E_9$ detergents are suitable for crystallizations, either pure or mixed with C_8G. A cheaper polydisperse micture, C_8E_m, has also been mixed with $C_{12}M$ in some cases.

Aqueous solutions of these detergents are not stable. Peroxides and aldehydes are formed on storage under air, particularly in the light. Purification protocol involves treatment with a reducing agent ($SnCl_2$ or Na_2SO_3) followed by solvent extraction (23) (see *Protocol 4*).

Besides these five main classes, there are other detergents potentially useful for crystallization (such as the *n*-alkyl-glucamides), which have been less extensively used (8).

Protocol 4. Purification of *n*-alkyl-oligoethyleneglycol detergents

1. Stir for 2 h a 10% (w/v) detergent solution in distilled water with $SnCl_2$ [0.5% (w/v) final concentration].
2. Add NaCl to 10% (w/v) final concentration; then add an equal volume of dichloromethane and mix thoroughly. Let stand until two layers separate.
3. Discard the upper water layer and recover the lower organic layer which contains the detergent. Wash it with an equal volume of 1% (w/v) NaOH in 10% NaCl, then three times with 10% NaCl (the pH of the last water layer must be 7).
4. Dry the organic extract for 24 h over anhydrous Na_2SO_4, then in a flash evaporator at 40 °C.
5. Store the purified detergent at −20 °C.

New detergents related to those described above appear steadily on the market, such as *n*-heptylcarbamoyl-methyl-α-D-glucoside and *n*-methyl-*n*-decanoyl-maltosylamine (cheaper substituents for C_8G and $C_{12}M$, respectively). It is recommended that you perform your own investigation of chemical catalogues.

4. Purification of membrane proteins before crystallization

General methods for solubilization and purification of membrane proteins have been given in Chapter 4 of reference 13 and will not be detailed. This section focuses on several specific points: purity requirement, procedures for detergent exchange and for sample concentration.

4.1 Purity requirements

As for the crystallization of soluble proteins, the starting protein solution should be as pure and homogeneous as possible and the same precautions should be taken to avoid denaturation, proteolysis, and microheterogeneities (see Chapter 2).

Protein purity is usually checked by SDS–PAGE. However, one should be aware that contaminants may escape detection; e.g. lipopolysaccharides and lipoprotein contaminants of porin preparations are not stained by Coomassie Blue but only by silver staining (20).

Residual lipids represent another common source of impurity for membrane protein preparations. Indeed, they may withstand the solubilization by a mild detergent and remain associated with the detergent–protein complexes. Their non-specific, random binding prevented crystallization in several cases. Lipid content should therefore be checked by TLC (24) of organic solvent extracts (25) (see *Protocol 5*). Alternatively, phosphorus content of these extracts can be measured (26) as most residual lipids are phospholipids.

Protocol 5. Lipid analysis

1. Preparation of lipid extract.
  (a) Mix the protein solution (~1 mg/ml) with 20 volumes of hexane/isopropanol 3:2 (v/v).
  (b) Shake well; then centrifuge at low speed (5000 g) for 20 min.
  (c) Recover the supernatant. Repeat steps 1 and 2 on the pellet.
  (d) Dry the combined supernatants under a stream of N_2 or with a flash evaporator. Dissolve the residue in the minimal volume of chloroform.

2. TLC of the lipid extract.
The experimental procedure is similar to *Protocol 1*, except that the developing solvent is chloroform/methanol/water, 65/25/4 (v/v). The neutral lipids are running near the solvent front; the other lipids are fractionnated into various classes. They are identified with reference to published R_f values, or by using specific stains (24) rather than H_2SO_4 charring.

If present, lipids should be eliminated as much as possible; the ease of removal depends on the protein–detergent couple and there is no general recipe. Chromatographic techniques may be adequate even if they were not devised for this purpose. Ion-exchange chromatography has been reported in several cases to lower the lipid content, probably because of the extensive detergent washes of the adsorbed protein. Another chromatographic step which has been used for purification of membrane proteins prior to crystallization is chromatofocusing in presence of a detergent (9). Besides providing with a homogeneous preparation of defined isoelectric point, it may also result in lowering the phospholipid content. Purity of the detergent may play a role in elution by IEF, as reported in the case of bacteriorhodopsin which appeared heterogeneous when impure C_8G was present (21).

Homogeneity of the preparation requires also its monodispersity, i.e. all the protein–detergent complexes should have the same composition. This is verified by gel filtration experiments, e.g. on FPLC columns (Superose) or HPLC ones (such as the TSK-SW or TSK-PW gels). From these experiments one can estimate the size of the whole complex, including detergent. On the other hand, the amount of bound detergent may be determined by several techniques [see (27) for a review]. Combining these results allows determination of the number of protein units present per complex. The best crystals have been obtained so far in cases where this number is small: either 1 (bacterial reaction centres), or 3 (bacterial porins), or possibly 6 (bacterial light-harvesting complexes).

4.2 Detergent exchange

Purification of a membrane protein is often done in presence of a detergent, with crystallization performed with a different detergent. This may be a result of the cost of one detergent, or because solubilization and purification require a particular detergent, or the variation of the detergent nature during crystallization trials. In all cases, exchange of detergent, hereafter called detergent 1 and detergent 2, has to be performed; several methods may be used.

4.2.1 Dialysis

This is the simplest procedure but not applicable in all cases. The meaningful parameter of a dialysis membrane is its cut-off value. It must be low enough for retention of the protein–detergent complex. Thus for large complexes, highly permeable membranes can be used; e.g. for a complex of 100 kd, a Spectrapor 7 membrane with a cut-off value of 50 kd may be used. With such a pore size, exchange of detergents with CMC values higher than 1 mM is relatively rapid (a few days). On the other hand, for a small complex (e.g. 15 kd) a Spectrapor 1 membrane, (cut-off value 6 kd) should be chosen and only detergents of high CMC (10 mM or so) will exchange at an acceptable rate. Dialysis may be performed in two steps: e.g. we routinely exchange 0.1% (w/v) $C_{12}DAO$ for 0.8% (w/v) C_8G as follows:

(a) First dialyse for 48 h against a detergent-free buffer, with several changes of reservoir (Spectrapor semi-micro tubing, cut-off 12 000 kd). Removal of $C_{12}DAO$ results in increased turbidity of the sample.

(b) Transfer the bag in C_8G containing buffer for another 24 h. Loss of turbidity indicates re-dissolution of the protein.

From a practical point of view, dialysis may be performed either in the familiar closed bags, or for small volumes (< 500 μl) in microdialysis cells, either home-built (see Chapter 4) or commercially available (Pierce, Amicon). Some dialysis membranes, particularly those stored wet [e.g. Spectrapor (Amicon) type 7] are specially prone to fungi contamination. Before use, a good precaution is to boil membranes for 1 min in 1% (w/v) $NaHCO_3$, then to soak them 3 times in highly pure water (MilliQ grade) and to use them immediately afterwards.

4.2.2 Chromatography

The sample which contains detergent 1 is chromatographed on a column equilibrated with detergent 2, and eluted with detergent 2. This method is feasible with any type of detergent, with various chromatographic supports.

- Gel filtration is gentle and can be used with all detergents. However, usually it dilutes appreciably the sample.
- Ion-exchange chromatography (DEAE or CM exchangers or hydroxy-apatite) is restricted to non-ionic detergents; it has the advantage of concentrating the sample when elution is done by a steep salt increase.

If the presence of salt in the final sample is not wanted, a mixed column consisting of ion-exchanger superposed on gel filtration matrix (Sephadex G25) will exchange the detergent and de-salt the sample altogether (28). Microcolumns built with Pasteur pipettes are useful for such ion-exchange procedures.

4.2.3 Precipitation

The protein in the presence of detergent 1 is first precipitated with ethanol, which is a solvent of detergent 1; the precipitate is washed for eliminating detergent 1, and re-dissolved in detergent 2. This method has been used only for porins. It requires a very sturdy protein.

The efficiency of these procedures can be judged from the absence of detergent 1 in the final sample. Unfortunately, very few detergents exist in a labelled form and those are expensive. Colorimetric determination is possible for glucosides and maltosides, with reagents specific for reducing sugars (29). In other cases, TLC of extracts is the only method.

4.3 Sample concentration

A concentrated stock solution (often 10 mg protein/ml or more) is required for preparing the samples in crystallization trials. As said above (see Section 4.2), concentration may be achieved in the same time as detergent exchange by ion-exchange chromatography. Another very useful method is ultra-filtration. Here again a large number of devices is available. For final volumes of 1 ml and up, stirred cells equipped with Amicon YM or XM membranes (with a wide choice of cut-off values) are convenient, and may be operated under nitrogen pressure. The major drawback is foaming during stirring. For smaller volumes, a number of devices allow concentration down to 100 μl or so with low-speed centrifugation (Millipore, Amicon). Whatever the device, the cut-off value should be chosen as high as possible, taking into account the size of the protein–detergent complex; this avoids concentrating the detergent in solution in the same time as the protein–detergent complexes.

5. Crystallization protocols

For conducting crystallization trials on a membrane protein, several parameters have to be chosen (see *Table 1* of Chapter 1) to which should be added the nature and concentration of detergent. Some of these parameters have been discussed earlier with reference to soluble proteins (see Chapter 4). We will stress a few specific points based on published crystallization protocols (*Table 3*).

5.1 Detergent

The choice of detergent is still empirical. Some membrane proteins may crystallize with a wide variety of detergents. In one case, a systematic search has been done over 23 detergents with Omp F, an *E. coli* porin (22). Among them, 16 non-ionic detergents (from classes described in *Table 1*) could be used successfully. Omp F crystallized also in micellar solutions of short-chain lecithins (diC_6- or diC_7-glycerophosphatidylcholines) or lysolecithins (monoC_{14}- or monoC_{16}- glycerophosphatidylcholines) used as detergents. However, crystals could not be obtained with ionic detergents, nor with detergents derived from bile salts (cholate, CHAPS, or CHAPSO). A non-ionic, non-steroid polar group and a short alkyl chain seemed thus to be the only requirement, without narrow specificity.

On the other hand, the reverse situation may prevail, and strict requirements may exist for chain length or detergent type; e.g. a chloroplast light-harvesting complex, LHCII, crystallizes reliably with C_9G and poorly with C_8G (8). Therefore, for an unknown protein, screening should be done using at least two homologs of each detergent class. However, before beginning an extensive screening, one could try the most popular C_8G and $C_{12}DAO$. Indeed, they allowed a number of successful crystallizations: bacterial reaction centres, light-harvesting complexes, and *E. coli* porins gave crystals with one or both of them. From a certain point of view, these two detergents may be considered as equivalent: in two bacterial reaction centres crystals, the regions they occupy respectively around the hydrophobic α helices could be nearly superimposed (10a, b).

$C_{12}DAO$ should be tried alone and also in presence of an additive (see below, Section 7) which was required in some cases. C_8G has been used either alone, or mixed with low amounts of other short-chain detergents: these were, however, not essential for crystallization but improved crystal growth. Thus first trials may be done with pure C_8G only.

The initial concentration of detergent in the sample should be chosen only slightly higher than the CMC (see *Table 2*). Under this condition, the detergent is present either as monomers or as part of protein–detergent complexes, with very few pure detergent micelles. For porin crystallization, the optimal range of C_8G concentration is narrow: 8–9 mg/ml. Below the CMC (less than 7 mg/ml) or well above it (more than 10 mg/ml) growth rate and nucleation are excessive (20). The same optimal range was found for C_8G with bacterial reaction centres, and it does not seem to be very sensitive to the protein concentration.

5.2 Additives

For the crystallization of some membrane proteins, small molecules with amphiphilic character have been added (see *Table 3*). Most often used is heptane-1,2,3 triol (high melting point isomer); hexane-1,6 diol, benzamidine, glycerol, and triethylamine phosphate were also used. Their effects are various.

- They may be absolutely required: best example is heptane-1,2,3-triol, essential for crystallizing *Rhodopseudomonas viridis* reaction centre with C_{12}DAO.
- They improve crystal growth and quality, but crystallization still takes place in their absence: this is the case of heptane-1,2,3-triol for *Rhobacter sphaeroides* reaction centre with C_{12}DAO.
- In other cases their presence has no effect whatsoever: it is the case of *E. coli* porin with C_8G.

These 'additives' have been usually used at quite high molarities, when compared to those of detergent: in the cases cited above heptane-1,2,3-triol was present at 0.1–1M. Their mode of action is still poorly understood. One explanation, which has some experimental support, is they modify the micellar structure of the detergent by partitioning the micelles; they may also change the consolution boundaries and could bring them in a favourable temperature range.

Whatever the case, such additives may be tried when all previous trials done in their absence have failed. It is better to test them beforehand on the protein in solution since they may have a denaturing effect.

5.3 Crystallizing agent

In *Table 3* (30–37), one may notice that the conditions used so far to achieve supersaturation are not very diverse, with either the presence of PEG at a suitable concentration, or 'salting-out' due to high ionic strength.

5.3.1 PEG

PEG, which is a classical crystallizing agent for soluble proteins, probably acts in the same way on detergent solutions by competing with the micelle polar groups for water molecules and by modifying the structure of the solvent. This destabilizes the micelles and the protein–detergent complexes. Thus, when PEG is continuously added to a membrane protein in detergent solution the micellar solution is perturbed and one of two things may happen; protein and detergent will precipitate, or the solution spontaneously separates in two immiscible liquid phases, with the protein and most of the detergent in one phase and PEG in the other (see *Figure 1*). At a given temperature, the solubility limit of the protein is depending on all constituents of the solution

(PEG, detergent, protein, and salts). Crystals will eventually form when the system is slowly approaching this limit by increasing PEG concentration.

PEG is available in a variety of polymeric ranges. The optimal range of PEG concentration for crystallization of a given protein depends on PEG molecular weight. It may be very narrow (about 1.5% w/v), as observed for porin (20) and a bacterial reaction centre (38). Re-purification is strongly recommended before use (39, 40), and is described in Chapters 2 and 4.

Crystallization of soluble proteins with PEG is usually performed at low ionic strength. On the contrary, for membrane proteins salt is generally required along with PEG (see *Table 3*). Again, the optimal range of salt concentration may be narrow. Thus the two meaningful parameters are the concentrations of PEG and salt, for a given protein with a given detergent (pH and temperature being fixed). For C_8G at room temperature, it has been observed (9) that these conditions are quite similar for unrelated proteins like *E. coli* porin, *Rb. sphaeroides* reaction centre, and prostaglandin synthase, as if they were little influenced by the protein but mainly by the detergent. This observation, if it is generalized by further experiments, would greatly simplify the search for crystallization conditions with the system C_8G–PEG–NaCl.

No systematic comparison of the influence of various salts on the crystallization conditions of membrane proteins has been published. We have observed for the crystallization of a bacterial reaction centre that NaCl could be replaced by a number of other monovalent salts; their optimal concentrations were not identical and had to be optimized. However, there was no significant influence of the salt nature on crystal growth or characteristics (unpublished experiments).

5.3.2 'Salting-out'

Surprisingly, a few membrane proteins have been crystallized by high salt concentrations; ammonium sulphate and phosphate have been used in the presence of $C_{12}DAO$, C_8G, or C_9G (*Table 3*). These salts decrease the solubility of a membrane protein–detergent complex by a mechanism probably more complicated than for the salting-out of soluble proteins (see Chapter 9). Their presence may modify the interactions between water and the hydrophilic regions of the protein. More importantly, they induce a salting-out of the detergent itself: e.g. monovalent salts modify the upper consolution boundary of C_8E_5; the solubility shift is mainly determined by the anions and follows the Hofmeister series (41).

How does one choose between these various precipitating agents? The choice may be restricted by considering the stability of the protein in their presence: e.g. LHCII, a light-harvesting chloroplast membrane protein, is denatured by PEG and by ammonium sulphate; it was therefore crystallized in the presence of high phosphate concentrations, which do not affect it. The biological activity of the protein should therefore be checked in the presence of increasing amounts of these various precipitants before any crystallization trial. Then the useful range of precipitant concentrations is determined by measuring the lowest precipitant concentration leading to phase separation or precipitation, at given protein and detergent concentrations.

5.4 Optimization

Once crystals (most often microcrystals) have been observed in trial experiments, crystallization conditions have to be improved for crystal size and quality. The strategy is based on the same principles as for soluble proteins (see Chapter 4). Excessive nucleation, leading to a 'shower' of microcrystals, should be avoided; at the same time, growth rate should be kept low enough, as crystal defects are frequently observed when the rate is too high (hollow crystals may even be obtained). Practically, this implies repeating the trials with the different parameters (pH, concentrations, and so on) slightly modified around their initially positive values, over a fine grid. At this stage, use of an additive (see Section 5.2) or of a small amount of a second detergent may be included as a further variation. The crystal form of several membrane proteins has been shown to depend upon several parameters: type of detergent, pH, nature of the buffer, and ionic strength when PEG together with salt are present (20). By varying these parameters, it has been possible to select a form which grows better, or is more suitable for structure determination because of its symmetry or unit cell dimensions.

6. Experimental techniques

Crystallization of membrane proteins may be performed with all the experimental set-ups described in Chapter 4. Vapour diffusion and micro-dialysis have been more frequently used than batch crystallization and free liquid–liquid interface [which was only used for screening experiments (22)] (see *Table 3*).

Because of the wetting properties of detergents, their drops tend to spread when formed on a planar glass slide, and to fall out when the slide is inverted. Therefore, vapour diffusion with hanging drops is restricted to drop volumes less than 10 μl. With sitting drops formed on depression slides, there is no restriction in volume. Treatment of glass surface with silicone or silane derivatives is unnecessary.

Microdialysis is performed in capillaries or microtubes closed by dialysis membrane with sample volumes less than 150 μl, equilibrated against a reservoir. Choice of the cut-off value of membranes should take into account the molecular weights of the components of the sample (see Section 4.2.1). Depending on this cut-off value, and also on the thickness of the membrane, dialysis rate, and diffusable species may be controlled; e.g. PEG 4000 and 6000 diffuse (but slowly) through a membrane of cut-off value 25 000 daltons, together with water and salts, but not if a cut-off value of 2000 daltons is used.

We have observed that with these two membrane types, all other conditions being the same, crystallization of a bacterial reaction centre with PEG 4000 does not occur similarly (unpublished experiments).

Choosing between microdialysis and vapour diffusion is often a matter of personal preference. Cost of microdialysis is higher when the detergent is expensive, as detergent must be present in the microdialysis reservoir, but may be omitted from the vapour diffusion reservoir. One of the advantages of microdialysis method over vapour diffusion for screening experiments is the possibility of changing individual constituants of the mixture; furthermore, the detergent concentration may be kept constant throughout the crystallization process, by putting it at the same concentration in the sample and in the reservoir. Changing dialysis reservoir is also very easy.

The main disadvantage of microdialysis is the rapid equilibration between sample and reservoir (much faster than through vapour diffusion), which may be troublesome if growth rate has to be slowed down. In that case, double dialysis (see Chapter 4) is recommended.

7. Concluding remarks

At the present time a number of membrane proteins have been crystallized in presence of detergent, demonstrating the feasibility of the process, and providing guidelines for further experiments. Various detergents are suitable. Their role is important: their properties and their phase diagrams govern the crystallization conditions; they are still associated with the protein in the crystal lattice. The methodology requires (as for soluble proteins) a systematic search of the different parameters, especially the nature of the detergent. This adds one more factor to this empirical analysis. However, if this is carefully done on a highly-purified membrane protein, the chances of success should in principle be as high as for other macromolecules.

Few membrane proteins crystallized so far are suitable for high-resolution X-ray diffraction analysis. However, improvement of crystal quality should be possible, as for soluble proteins, by a careful control of protein purity and homogeneity, and optimization of crystal growth conditions. Very recently the resolution limit of a porin has been increased from 2.5 to 1.8 Å by modifying the purification protocol (42). This remarkable result should guide and encourage everybody working in this field.

References

1. Michel, H. and Oesterheld, M. (1980). *Proc. Natl. Acad. Sci. USA*, **77**, 1283.
2. Garavito, R. M. and Rosenbusch, J. P. (1980). *J. Cell Biol.*, **86**, 327.
3. Deisenhofer, J., Epp, O., Miki, K., Huber, R., and Michel, H. (1985). *Nature*, **318**, 681.

4. Chang, C. H., Schiffer, M., Tiede, D., Smith, U., and Norris, J. (1985). *J. Mol. Biol.*, **186**, 201.
5. Yeates, T. O., Komiya, H., Chirino, A., Rees, D. C., Allen, J. P., and Feher, G. (1987). *Proc. Natl. Acad. Sci. USA*, **85**, 7993.
6. Arnoux, B., Ducruix, A., Reiss-Husson, F., Lutz, M., Norris, J., Schiffer, M., and Chang, C. H. (1989). *FEBS Lett.*, **258**, 47.
7. Weiss, M. S., Kreusch, A., Schiltz, E., Nestel, U., Welte, W., Weckesser, J., and Schulz, G. E. (1991). *FEBS Lett.*, **280**, 379.
8. Kühlbrand, W., (1988). *Quarterly Rev. Biophys.*, **21**, 429.
9. Garavito, R. M. and Picot, D. (1990). *Methods, A Companion to Methods in Enzymology*, **1**, 57.
10. (a) Roth, M., Lewitt-Bentley, A., Michel, H., Deisenhofer, J., Huber, R., and Oesterhelt, D. (1989). *Nature*, **340**, 659.
10. (b) Roth, M., Arnoux, B., Ducruix, A., and Reiss-Husson, F. (1991). *Biochemistry*, **30**, 9403.
11. Deisenhofer, J. and Michel, H. (1989). *EMBO J.*, **8**, 2149.
12. Bordier, C. (1981). *J. Biol. Chem.*, **256**, 1604.
13. Findlay, J. B. C. (1989). In *Protein purification methods: a practical approach* (ed. Z. L. V. Harris and S. Angal), pp. 59–82. IRL Press, Oxford.
14. Rosenthal, K. S. and Koussaie, F. (1983). *Anal. Chem.*, **55**, 1115.
15. De Vendittis, E., Palumbo, G., Parlato, G., and Bocchini, V. (1981) *Anal. Biochem.*, **11**, 278.
16. Van Aken, T., Foscall-Van Aken, S., Castleman, S., and Freguson-Miller J. (1986). In *Methods in enzymology*, (ed. S. Fleischer and R. Fleischer), Vol. 125, pp. 27–35. Academic Press, London.
17. De Grip, W. J. and Bovee-Geurts R. (1979). *Chemistry and physics of lipids*, **23**, 321.
18. Hermann, K. W. (1962). *J. Phys. Chem.*, **66**, 295.I
19. Degiorgio, V. (1985). In *Physics of amphiphiles, micelles vesicles and microemulsions*, (ed. H. Corti), pp. 303–35. North Holland, Amsterdam.
20. Garavito, R. M. and Rosenbusch, J. P. (1986). In *Methods in Enzymology*, **125**, 309.
21. Lorber, M., Bishop, J. B., and DeLucas, L. J. (1990). *Biochim. Biophys. Acta*, **1023**, 254.
22. Eisele, J. L. and Rosenbusch, J. P. (1989). *J. Mol. Biol.*, **193**, 419.
23. Ashani, V. and Catravas, G. N. (1980). *Anal Biochem.*, **109**, 55.
24. Kates, M. (1972). In *Techniques in lipidology*, (ed. T. S. Work and E. Work), Laboratory Techniques in Biochemistry and Molecular Biology, Vol. 3, Part II. North Holland, Amsterdam.
25. Radin, N. S. (1981). *Methods in Enzymology*, **72**, 5.
26. Bartlett, G. R. (1959). *J. Biol. Chem.*, **234**, 466.
27. Moller, J., le Maire, M., and Andersen, J. P. (1986). In *Progress in protein lipid interactions*, (ed. A. Watts and J. J. De Pont), Vol. 2, pp. 147–96. Elsevier, Amsterdam.
28. Rivas, E., Pasdeloup, N., and le Maire, M. (1982). *Anal. Biochem.*, **123**, 194.
29. Rao, P. and Pattabiraman, T. N. (1989). *Anal. Biochem.*, **181**, 18.
30. Michel, H. (1982). *J. Mol. Biol.*, **158**, 567.

31. Welte, W., Wacker, T., Leis, M., Kreutz, W., Shinozawa, J., Gad'on, N., and Drews, G. (1985). *FEBS Lett.*, **182**, 260.
32. Papiz, M. Z., Hawthornwaite, A. M., Cogdell, R. J., Woolley, K. J., Wightman, P. A., Ferguson, L. A., and Lindsay, J. G. (1989). *J. Mol. Biol.*, **209**, 833.
33. Wacker, T., Gad'on, N., Becker, A., Mantele, W., Kreutz, W., Drews, G., and Welte, W. (1986). *FEBS Lett.*, **193**, 267.
34. Nestel, U., Wacker, T., Woitzik, D., Weckesser, J., Kreutz, W., and Welte, W. (1989). *FEBS Lett.*, **242**, 405.
35. Stauffer, K. A., Page, M. G. P., Hardmeyer, A., Keller, T. A., and Pauptit, R. A. (1990). *J. Mol. Biol.*, **211**, 297.
36. Ford, R. C., Picot, D., and Garavito, M. (1987). *EMBO J.*, **6**, 1581.
37. Witt, I., Witt, H. T., Di Fiore, D., Rogner, M., Hinzichs, W., Granzin, J., Betzel, C., and Dauter, Z. (1988). *Ber. Bunsenges. Phys. Chem.*, **92**, 1503.
38. Ducruix A., Arnoux, B., and Reiss-Husson, F. (1988). In *The photosynthetic bacterial reaction center: structure and dynamics*, (ed. J. Breton and A. Vermeglio), NATO ASI Series, Vol. 149, pp. 21–25. Plenum, New York.
39. Jurnak, F. (1986). *J. Crystal Growth*, **76**, 577.
40. Ray, W. J. and Puvathingal, J. (1985). *Anal. Biochem*, **146**, 307.
41. Weckström, K. and Zulauf, M. (1985). *J. Chem. Soc.*, Faraday Trans. I, **81**, 2947.
42. Kreusch, A., Weiss, M. S., Welte, W., Weckesser, J., and Schulz, G. E. (1991). *J. Mol. Biol.*, **217**, 9.

9

Phase diagrams

M. RIES-KAUTT and A. DUCRUIX

1. Introduction

Establishing a solubility diagram is one way of quantifying the influence of crystallization parameters (see Chapter 1, Table 1) on the solubility of biological macromolecules. In this chapter, theoretical concepts of solubility will be developed firstly, to reiterate how a compound dissolves in a solvent, and to point out what is particular to biological macromolecules especially concerning the type of molecular interactions which can be found in a macromolecular solution. Indeed, interactions between solvent or additives and a biological macromolecule, which are necessary for its solubilization, can compete with intramolecular interactions responsible for the tertiary structure. The concepts of salting-in and salting-out will be summarized.

Secondly, the methods for phase-diagram determinations will be explained in order to show the methods for establishing such diagrams. Usually rather large amounts of macromolecules are required. However, progress in:

(a) molecular biology, offering the possibility of overproducing large amounts of biological macromolecules for spectroscopic studies (X-ray, NMR), and

(b) scaling-down experimental procedures,

both make it possible to define at least some points on a diagram. Nevertheless, if complete solubility diagrams of your own protein cannot be established, it is worth showing crystal growth experiments on a schematic diagram.

Thirdly, the influence of the major crystallization parameters on protein solubility will be presented and illustrated with the data on hen egg-white (HEW) lysozyme. Finally, some practical applications will show interpretation of crystal growth problems by representing the supersaturation zones on theoretical diagrams.

2. Concepts of solubility

To crystallize a biological macromolecule in a given solvent, one must first bring it to a supersaturation state. A biological macromolecule follows the same thermodynamic laws as any other type of molecules (inorganic or small organic molecules) regarding supersaturation, nucleation, and crystal growth (1). Nevertheless, biological macromolecules present peculiarities because the interactions necessary to solubilize them in a solvent are similar, and may be competitive with the intermolecular interactions which are responsible of their tertiary structure.

2.1 General aspects

Solubilization of a solute (e.g. a biological macromolecule) in a solvent requires solvent–solute interactions, which must be similar to the interactions of solvent–solvent molecules and the solute–solute interactions of the compound to be dissolved. Solute–solvent interactions can be of different types: ion–dipole, dipole–dipole, hydrogen bonds, or complexation. Based on their dielectric constant, ε, solvents are classified as shown in *Table 1*.

Table 1. Polarity and dielectric constants (ε) of solvents at 20 °C

		ε > 15 **Dipolar**		ε < 15 **Non-polar**	
Protic		Aprotic			
(O-H, N-H . . .)					
water	80	acetonitrile	37	benzene	2.3
formic acid	58	acetone	21	*n*-hexane	1.9
methanol	34				
ethanol	25				

The dipolar solvents (ε > 15) bear a dipole between atoms of different electronegativity (e.g. C=O, C–Cl, and so on). Depending on the presence of labile protons (O–H, N–H, and so on), these dipolar solvents are subdivided into protic, and non-protic or aprotic. Non-polar solvents (ε < 15) are generally hydrocarbons.

Ionic species of solid electrolytes (e.g. salts) are dissociated in a polar solvent by ion–dipole interaction, and in addition by H-bond interaction in the case of protic solvents. On the contrary, non-electrolytes are dissolved in apolar solvents, unless the non-electrolyte solute presents the possibility of forming H-bonds in which case they are soluble in protic solvents (e.g. solubility of carboxylic acids in alcohol).

2.2 Solubility of biological macromolecules

In general, the solubility of a chemical compound in a solvent, at a given temperature, is defined as the amount of compound dissolved in a solution in equilibrium with an excess of undissolved compound; e.g. at 25 °C, ammonium sulphate dissolves until its concentration reaches 4.1 moles per litre of water, the excess remaining undissolved. More salt can be dissolved when raising temperature, and when temperature is brought back to 25 °C, the solution is *supersaturated*. The excess of salt will crystallize until its concentration reaches the solubility value at 25 °C (4.1 moles per litre of water).

In the case of biological macromolecules, the solubility is additionally defined by the characteristics of the solvent, as the solvent consists usually of water which contains various chemicals to bring the solution to a given pH (buffer), ionic strength (salts), and eventually other additives. These compounds can affect the solubility of macromolecules:

(a) directly, by interacting with the different functional groups of the protein (different net charge, binding of counterions, and so on) and modifying eventually the macromolecular conformation. Working in different conditions, of buffer and salts, is in fact like working with a different macromolecule from a physico-chemical point of view.

(b) indirectly, through modification of the structure and physico-chemical properties of the solvent; e.g. changing the pH modifies the net charge of the protein and thus the nature of the polyelectrolyte. Furthermore, if salt bridges are involved in the tertiary structure, these may be disrupted and the protein may change its conformation or even denature.

Finally, it should be mentioned that different conventions defining solubility exist in the literature: concentration values may be measured before complete equilibrium is reached, or in the presence of precipitate instead of crystals (2). In the present chapter, solubility is given as the concentration of a biological macromolecule solution in *equilibrium with crystals* of this macromolecule (3).

2.3 Perturbations of interactions

A biological macromolecule is a polymer of amino acids or nucleotides, which is folded into a tertiary or a quaternary structure:

- mainly by dipole–dipole interactions, H-bonds (e.g. C=O · · · H–N–), and van der Waals interactions;
- by some covalent bonds (S–S bridges);
- occasionally by salt bridges (e.g. $-COO^-$ · · · $^+H_3N-$) between charged residues.

'Water soluble' proteins contain mostly the hydrophobic side-chains in the core, exposing the hydrophilic side-chains on their surface. They are thus considered as polyelectrolytes able to dissolve in water (protic polar solvent) as described before. The special case of membrane proteins (see Chapter 8) should be mentioned here as they bear, at least on the surface embedded in the membrane, hydrophobic residues which interact in their natural medium with lipidic (apolar) compounds. In practice, detergents are added to the water solutions in order to induce hydrophobic interactions between the hydrophobic residues of the protein and the hydrophobic tail of the detergent. The hydrophilic head of the detergent then allows interactions with the solvent.

The stability of biological macromolecules in solution relies on the competition of solvent–solute interactions with the intramolecular interactions which are necessary to maintain the tertiary structure. The balance of interactions controlling the solubility and/or the conformation of a macromolecule can be modified by (4, 5):

(a) temperature. An increase of temperature increases the disorder of solvent molecules, but also allows macromolecule conformations of higher total free energy to form. In fact the modifications of solvent–solute interactions are directly linked with modifications of both chain–chain contacts and solvent structure.

(b) pH. Like temperature, pH changes affect both solute and solvent. Nevertheless, altering H^+ and OH^- concentrations of the solvent is minor compared to protonation or deprotonation of titrable groups of the protein.

(c) salts. Salts can act in different ways. (i) They are responsible of the ionic strength, and thus affect macromolecular electrostatic interactions by charge shielding. The repulsion between polyelectrolytes of same net charge is decreased. (ii) They may form direct electrostatic interactions with charged residues at the surface of proteins through non-specific monopole–monopole interactions; e.g. interactions of anions with lysine or arginine side chains, or of cations with glutamic or aspartic side chains. (iii) Salts may act by monopole–dipole interactions with dipolar groups of the macromolecule (peptide bonds, amino, hydroxyl, or carboxyl groups or amides). These interactions may lead to partial denaturation of the protein. (iv) Non-polar interactions may occur between solvent-exposed hydrophobic residues and the hydrophobic part of organic salts (i.e. carboxylates, sulphonates, or ammonium salt). These interactions are also involved in the solubilization of solvent-exposed hydrophobic residues by the hydrophobic tail of an ionic detergent. (v) Salts may associate with binding-sites. Ions can present a direct interaction with a specific part, of well-defined geometry, of the biological macromolecule. The characteristic is the specificity and not the strength of the interaction, because even a weak association ($K_d \approx 10$–10^2 mM) is considered as binding. The strength of a specific interaction is expressed by the association constant, K_a (M^{-1}), or the dissociation constant K_d (M, with $K_d = 1/K_a$). At very low levels of affinity, the specific and direct effect of site binding becomes difficult to distinguish from indirect effects of the additive on the structure of water.

(d) H-bond competitors. Molecules having a H-bonding potential (formamide, urea, guanidinium salts, and so on) compete at high concentration ($\geqslant$ 4 M) with H-bonds of water and the structural intramolecular H-bonds of the protein. In the latter case they act as 'denaturating' agents. On the contrary, they seem to stabilize hydrophobic bonds.

(e) hydrophobic additives. They may interact directly with hydrophobic parts of the protein, but also alter the solvent structure. Non-ionic detergents (6) act in this way.

(f) organic solvents. Addition of organic solvents implies a modification of the dielectric constant and consequently changes of the various interactions (intramolecular solute–solute, solvent–solute, and so on).

2.4 Salting-in and salting-out

The change of protein solubility when increasing salt concentrations has been studied in terms of salting-in and salting-out (3, 7–10).

2.4.1 Salting-in

The increase of protein solubility at low salt concentration (< 0.5 M) is called salting-in. This phenomenon is explained by non-specific electrostatic interactions following Debye–Hückel theory between the charged protein and the ionic species of the salt (7).

2.4.2 Salting-out

In the salting-out phenomenon, the protein at high ionic strength behaves as a neutral dipole and solubility is mainly governed by hydrophobic effects (7). Salting-out corresponds to a decrease of protein solubility at high salt concentrations according to the Cohn–Green empirical linear equation:

$$\log S = \beta - K_S m, \text{ or } \log S = \beta - K'_s\mu \qquad [1]$$

S is the protein solubility. β, the intercept at $m = 0$, is a constant at high salt concentration and is function of the net charge of the protein, thus strongly pH dependent (at the pI, β tends to 0). The magnitude of β, as well as charge distribution, varies with temperature. K_s or K'_s is the salting-out constant. It is independent of pH and temperature, but depends on the nature of the salt. m is used for the *molal* (3, 7) (g salt/1000g water) or *molar* (9) (mole salt/litre water) salt concentration. μ, the ionic strength, takes into account the valency Z_i of all ions present in the solution. It is calculated with molal (8, 10) or molar (3, 10) salt concentration, as follows:

$$\mu = 1/2\ \Sigma\ (m_i Z_i^2) \qquad [2]$$

Experimentally defined solubility curves rarely fit with the linear function as described above (11–14). The observation of non-linearity may be essentially due to following reasons:

(a) As the salt concentration decreases, solubility tends to the solubility value of the protein in the buffer. The curves of HEW lysozyme solubility in presence of various salts in 50 mM NaAcO (pH 4.5) (11) converge at low salt concentrations towards the HEW lysozyme solubility curve in Na AcO.

(b) At low salt concentrations higher amounts of protein are required for crystallization, so the solubility value can be affected by the presence of higher amounts of 'protein associated' salts in following ways: (i) At its isoionic point, protein has a zero net charge and only H^+ and OH^- ions are present as counterions. In contrast the isoelectric point (pI), where the protein net charge is also zero, depends on the nature of the salt in presence, and the counterions are cations and anions. Decreasing or increasing the pH of an isoionic protein solution is equivalent to neutralizing titrable groups of the protein by adding acid or base. This is illustrated by the following example: to bring an isoionic HEW lysozyme solution (pI ≈ 10–11) to pH 4.5, about 10 molar equivalents of HCl are required. This implies that a 200 mg/ml (≈ 14 mM) HEW lysozyme solution at pH 4.5 contains approximately 140 mM (14 mM × 10 equiv.) Cl ions. (ii) If the protein solution has not been properly dialyzed after purification, salts will remain with the protein. This problem often affects reproducibility of crystallization experiments with different batches of biological macromolecule preparations.

(c) When binding can no longer be neglected, compared to preferential exclusion.

2.4.3 Green's general equation

A compound added to the protein–water (solute–solvent) system can bind to the protein (preferential binding) or be excluded (preferential exclusion) depending on preferential protein–additive or protein–water interactions.

The net interaction of salting-out is preferential exclusion, even though molecules or additives may bind to the protein. A thermodynamic approach has been described by Arakawa and Timasheff (3, 15).

The variation of protein solubility over the whole range of salt concentration reflects the resultant effect of both electrostatic and hydrophobic interactions, the first being predominant at low salt concentration and the second at high salt concentration as shown in *Figure 1*. This is expressed by the more complete equation (8):

$$\log S = \log S_0 + k_i \sqrt{C} - k_0 C \qquad [3]$$

where k_i and k_0 are the salting-in and salting-out constants, respectively; S_0 the solubility of the biological macromolecule in pure water; and C the molar salt concentration.

Though the relations [1] and [3] proved to be useful, it has been observed very often that experimental data on solubility are not a linear function of salt concentrations (11–14).

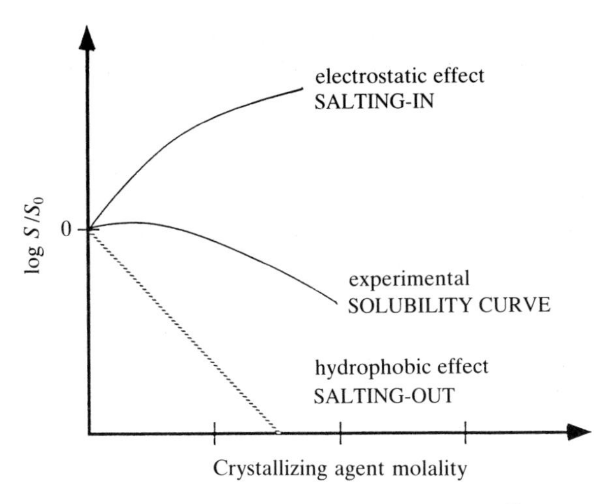

Figure 1. Contribution of electrostatic and hydrophobic effects on solubility S, normalized to that in pure water S_0. Redrawn from (7).

3. Methods for phase diagram determination

3.1 Description of a phase diagram

As the solubility of a biological macromolecule depends on various parameters (see Table 1 in Chapter 1), a two-dimensional solubility diagram is a representation of its solubility (mg/ml or mM macromolecule in solution) as a function of one parameter, all other parameters being constant. The diagram, represented in *Figure 2*, comprises the following zones:

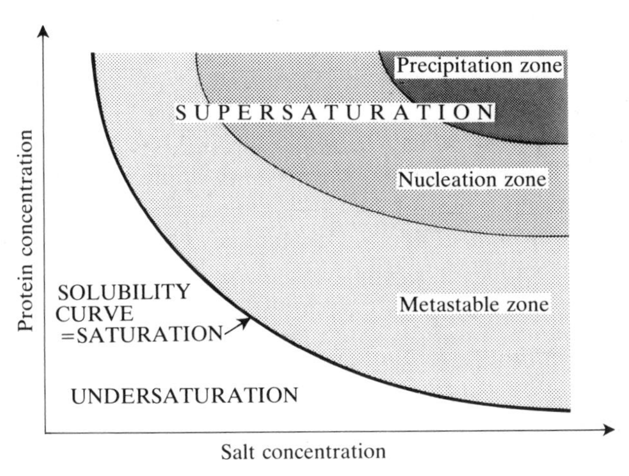

Figure 2. Schematic description of a two-dimensional solubility diagram, illustrated by the change of protein concentration with crystallizing agent concentration.

(a) The solubility curve S divides the undersaturated and supersaturated zones. In an experiment where crystallizing agent and biological macromolecule concentrations correspond to conditions of curve S, the saturated protein solution is in equilibrium with crystallized protein. This corresponds to the situation at the end of the process of crystal growth from a supersaturated solution, or of dissolution of crystals in an undersaturated solution; additional protein does not dissolve, but adding a solution of the same salt at the same concentration (e.g. reservoir) which contains no protein leads to dissolve the crystals.

(b) Under the solubility curve (S) the solution is undersaturated and the biological macromolecule will never crystallize.

(c) Above the solubility curve (S), the concentration of the biological macromolecule is higher than the concentration at equilibrium for a given salt concentration. This corresponds to the supersaturation zone. The level of supersaturation is defined as the ratio of biological macromolecule concentration over the solubility value. Depending on the kinetics to reach equilibrium and the level of supersaturation, this region may itself be subdivided into three zones. (i) The precipitation zone is where excess protein immediately separates from the solution in an amorphous state. (ii) The nucleation zone is where excess of biological macromolecule separates under a crystalline form. Near the precipitation zone, crystallization may occur as a shower of microcrystals which can be confounded with precipitate. (iii) In a metastable zone a supersaturated solution may not nucleate for a long period of time, unless the solution is mechanically shocked or a seed crystal introduced. This zone corresponds ideally to the growth of crystals without nucleation of new crystals.

3.2 Principle of phase diagram determination

The solubility curve can be determined either by crystallization of a supersaturated solution or by dissolution of crystals in an undersaturated solution. In both methods the protein concentration in the supernatant converges toward the same asymptotic value at equilibrium as shown in *Figure 3*. For double checking of solubility values, both crystallization and re-dissolution methods should be run in parallel, in order to ensure that equilibrium is reached. During the course of the experiments, all parameters which are not varied (such as pH, temperature, salt and buffer concentrations or nature, and so on) must be carefully kept constant and the biological macromolecule must be checked for stability.

For both crystallization and crystal dissolution, crystallization conditions must be previously defined. *Protocols 1* and *2* illustrate such a screening, depending whether an initial condition is known for a given crystallizing agent (11).

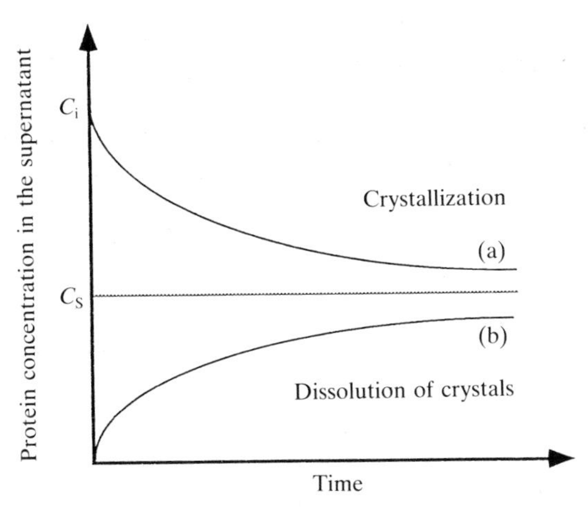

Figure 3. Variation of protein concentration in presence of crystals during equilibration process to reach the saturation value C_S: (a) by crystallization of a supersaturated solution or (b) by crystal dissolution.

Protocol 1. Screening for crystallization conditions of HEW lysozyme in different concentrations of NaCl (11)

1. Lysozyme is known to crystallize at 20–30 mg/ml in NaCl (0.7 M) in NaAcO (50mM) at pH 4.5 and 18 °C.
2. Solubility is to be determined in 0.3, 0.4, 0.7, 1.0, and 1.5 M NaCl.
3. Prepare a protocol for hanging drops (4+4 μl) in a Linbro box (see Chapter 4 for details):
   - use rows labelled A, B, C, and D for the different protein concentrations;
   - use columns labelled 1,2,3,4, and 5 for respectively 0.3, 0.4, 0.7, 1.0, and 1.5 M NaCl;
   - start with 0.7 M NaCl (row 3): use 10, 20, 30, and 40 mg/ml lysozyme in drops A3, B3, C3, and D3, respectively;
   - increase protein concentration when decreasing salt concentration and *vice versa*, e.g. for 1.0 M NaCl choose 5, 10, 20, and 30 mg/ml, whereas for 0.4 M NaCl choose 30, 40, 50, and 75 mg/ml lysozyme.
4. Prepare the experiments with stock solution of protein and salt in NaOAc (50 mM, pH 4.5) as described in Chapter 4 and store the experiments at 18 °C.
5. Observe the experiments once a day for a week.
6. Keep the conditions where crystals appear within one week (even in a large number, because the aim is not to grow large single crystals) and discard the conditions where no crystals occur as well as those where only precipitation is observed.

Checking for crystallization conditions when changing a parameter (new salt, other pH, or temperature values) can be done by hanging drop or dialysis method, starting from standard conditions as shown in *Protocol 2*. Dialysis offers the advantage of changing the nature of buffer or salt with a same protein sample.

Protocol 2. Screening for crystallizing conditions of HEW lysozyme in a new crystallizing agent

1. HEW lysozyme is known to crystallize at 20–30 mg/ml in 0.7 M NaCl in NaAcO (50mM) at pH 4.5 and 18 °C. Crystallization conditions of lysozyme have to be defined in new crystallizing agents [KSCN, $NaNO_3$, KCl, NH_4Cl, $MgCl_2$, NH_4 citrate, NH_4AcO, $(NH_4)H_2PO_4$].
2. Prepare control experiments with standard crystallization conditions, e.g. 20 mg/ml lysozyme with 0.6–0.8 M NaCl in NaAcO buffer (50mM) at pH 4.5 and 18 °C.
3. Prepare similar experiments, but substitute partially NaCl by the new crystallizing agent to be tested, e.g. 0.5–0.7 M NaCl when 0.1 M new crystallizing agent is added.
4. Analyse the results:
   - if fewer or no crystals appear, the crystallizing agent is less effective for crystallizing HEW lysozyme than NaCl (e.g. NH_4 citrate, NH_4AcO, and so on). Greater amounts of crystallizing agent are required for crystallization at the same protein concentration.
   - if microcrystals or precipitates are observed, the crystallizing agent is more effective for crystallizing HEW lysozyme than NaCl (e.g. $NaNO_3$, KSCN).
5. Follow *Protocol 1* to define crystallization conditions at different concentrations of the new salt.

3.2.1 Determination of solubility by crystallization

Once crystallization conditions have been defined by screening experiments (*Protocols 1* and *2*), solubility values are determined as in the following example of the influence of salt concentration on HEW lysozyme solubility:

- define the parameter to vary (i.e. salt concentration), keeping all other parameters strictly constant (i.e. pH, temperature, nature of salt and buffer);
- choose at least four values of the variable (different salt concentrations over a large range), because solubility curve usually do not fit with linear curves;

- prepare two or three different initial protein concentrations for a given parameter value as shown in *Figure 4*. The protein concentration in the supernatants will converge to the same constant solubility value during the crystallization process.

An example of experimental procedure to determine HEW lysozyme solubility as a function of NaCl concentration is given in *Protocol 4*. Such a curve has to be measured for each salt, if the influence of the salt nature on protein solubility is studied. The same method is applied for solubility variation due to pH or temperature changes. Crystal space group must be determined as polymorphism may occur.

3.2.2 Dissolution

To check the solubility value by crystal dissolution, a batch of previously grown protein crystals is required. The crystals are added to an undersatured solution, where they will re-dissolve until saturation is reached.

This method is mostly appropriate for the study of the influence of temperature. To determine protein solubility at different concentrations of a given salt, crystals can be prepared from the same batch, but must then be equilibrated carefully at respective salt concentrations. When solubility is checked in different salts, a batch of crystals is prepared in each appropriate salt. An alternative method to determine solubility changes with temperature is described in *Protocol 3*.

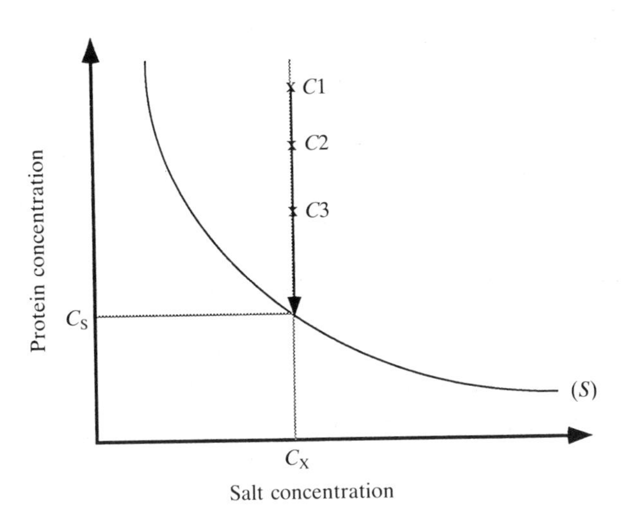

Figure 4. Determination of one point, C_S, of the protein solubility curve at C_x, a given crystallizing agent concentration, starting with three different initial protein concentrations $C1$, $C2$, $C3$.

Protocol 3. Solubility determination of HEW lysozyme at 18 °C by crystal dissolution

1. Prepare experiments for solubility determination as in *Protocol 4* (steps 1–7), but with a lower protein concentration.
2. Store the experiments at 0 °C for a week. Crystallization occurs rapidly.
3. Transfer the experiments in an incubator at 18 °C. Crystals dissolve until equilibrium is reached.
4. Continue steps 8–10 of *Protocol 4*.

The time necessary to reach the equilibrium is roughly the same with both crystallization and crystal dissolution methods.

3.3 Kinetic aspects

In the previous paragraphs, when presenting the zones of a phase diagram, and the different methods of determining solubility curves, the kinetic aspects of the process to reach the equilibrium have not been taken into account. In fact, the kinetics for reaching a final state of equilibrium depends on the initial supersaturation level of the protein solutions (13, 16).

3.3.1 Protein solubility determination

The methods of unstirred batch crystallization and dissolution are slow processes, requiring 4–12 weeks to reach stable values; the delay was shown to be shorter with the column method (17). This is due to a better exchange between a much larger crystalline surface and a smaller solution (supernatant) volume, as illustrated in Section 3.4.3.

3.3.2 The different supersaturation zones

A supersatured protein solution contains an excess of protein which will appear as a solid phase until protein concentration reaches the solubility value in the solution (supernatant). The higher the supersaturation level, the faster this solid phase appears.

- Precipitation occurs at very high supersaturation (about 30–100 times solubility value for HEW lysozyme). Immediately the excess of protein separates in an amorphous state from the solution. If the solution is centrifuged, the supernatant is still supersaturated and crystallization may occur.
- Nucleation takes place at a lower supersaturation than precipitation. In the case of HEW lysozyme, the nucleation range lies between about 5 and 30 times the solubility value. Crystals appear faster and in higher amounts with increasing supersaturation. Near the precipitation zone a shower of

microcrystals can hardly be distinguished from a precipitate by observation (see Chapter 10).

3.4 Techniques

The value of solubility must be determined in conditions where every parameter is well defined. As a consequence, vapour diffusion techniques are not well suited for the following reasons.

(a) At the beginning of the experiment the concentration of crystallizing agent is lower in the drop than in the reservoir (usually 1/2), but at the end of equilibration crystallizing agent concentrations are supposed to be identical. This is difficult to verify when working with small drop volumes.

(b) All components in the drop will concentrate twice as well the crystallizing agent and protein as the buffer and additives (and impurities!). As a consequence more than one parameter is changed during the experiment.

Nevertheless, hanging, sitting, or sandwich drops systems can be used if the salt concentrations is identical in both the drop and the reservoir all over the process. In this case the role of the reservoir is only to keep vapour pressure constant during the experiment, and the drop systems are used for a *batch technique* in microvolumes.

3.4.1 Macromethods

When large quantities of biological macromolecule are available, solubility determination can be achieved in large volumes, allowing the withdrawal of several aliquots. In a pioneer work, Ataka and Tanaka (18) used volumes of 8 ml for each experiment of a given protein and salt concentration. It is helpful to follow the protein concentration decrease in the supernatant until equilibrium value is reached, and the kinetics of the process. The main disadvantage of this technique lies in the requirement of large protein quantities. If only a few sample of aliquots are required, this can be scaled-down by using smaller test tubes (1 ml).

Each supersatured sample is prepared in a separate test tube which is then securely closed. As the three samples for the same salt concentration are in separate tubes, the refractive index of the mixtures should be measured to verify that they contain the same salt concentration.

3.4.2 Micromethods

To reduce the amount of protein required for solubility-curve determination, the experiments can be achieved in drops of small volumes (10–100 μl) in equilibrium with the reservoir (11, 12, 19). This method, although rather cumbersome, is well suited for the study of the influence of pH, temperature, or salts of various nature.

Protocol 4 shows how the HEW lysozyme solubility curve is produced using

NaCl as crystallizing agent and NaAcO (50 mM) as buffer (pH 4.5) at 18 °C with 100 μl sitting drops (11). Solubility of concanavalin A has been measured in 10 μl sitting drops (12).

Protocol 4. Determination of HEW lysozyme solubility curve at 18 °C as a function of NaCl (the crystallizing agent), in NaAcO (50mM) at pH 4.5.

1. Prepare the protocol with following previously defined crystallization conditions:

[NaCl] (M):	0.3	0.4	0.7	1.0	1.5
[lysozyme][a] (mg/ml):	120/100/80	100/70/50	25/20/10	20/10/5	10/8/6

   Use:
   - a reservoir of 15 ml in a plastic box or a sandwich box
   - 100 μl sitting drops in silanized pyrex plates (50 μl protein solution mixed with 50 μl salt solution).

   Duplicate one experiment for a given salt concentration: it will be opened periodically to check if the protein concentration of the supernatant has reached the equilibrium.
2. Prepare a 3M NaCl stock solution in NaAcO buffer (50mM) at pH 4.5 and use it to prepare 10 ml reservoirs of 0.6, 0.8, 1.4, 2.0, and 3.0 M by dilution with buffer. Measure the refractive index to verify salt concentrations.
3. Prepare HEW lysozyme[a] stock solution at 300 mg/ml in NaAcO buffer (50mM) at pH 4.5. Check protein concentration by optical density at 280 nm. Dilute to intermediate solutions at 100 and 20 mg/ml and verify by optical density measurements. Prepare the different protein solutions by dilution of the intermediate solution with buffer.
4. Filter the different solutions with 0.22 μm filter systems.
5. In five plastic boxes, pour 7.5 ml of the salt concentrations (0.6, 0.8, 1.4, 2.0, and 3.0 M), add 7.5 ml of buffer for final concentrations (0.3, 0.4, 0.7, 1.0, and 1.5 M NaCl). Measure the refractive index to verify salt concentrations.
6. On the three depressions of a silanized pyrex plates, prepare the 100 μl sitting drops, as in the following example for 0.3 M NaCl:
   - 50 μl 0.6 M NaCl mixed with 50 μl 240 mg/ml lysozyme
   - 50 μl 0.6 M NaCl mixed with 50 μl 200 mg/ml lysozyme
   - 50 μl 0.6 M NaCl mixed with 50 μl 160 mg/ml lysozyme

Protocol 4. *Continued*

7. Seal the box cover with grease, and store in the incubator at 18 °C.
8. After one month, open the *control* box and withdraw 10 μl of the supernatants to measure the protein concentration. Repeat this every two weeks until equilibrium is reached.
9. At equilibrium, the protein concentrations of the three drops in a box have reached the same value (± 10%) and remain constant with time.
10. When the equilibrium is reached (after 6–12 weeks after the crystals begin to appear), proceed for the solubility measurements.
    - Withdraw 50 μl supernatant of each depression in the 'measurement boxes' under binocular observation to avoid sucking of microcrystals.
    - Filter them by centrifugation over a 0.22 μm filter. This is achieved with a system of an Eppendorf tube (1.5 ml) equipped with a filter, commercially available from Millipore (Ultrafree-MC ref UFC3 OGV00), or from Costar (Spin-X). Alternatively, for very small quantities (≤ 50 μl), a cheaper system consists to introduce the solution on a Millex-GV_4 filter (Millipore), which is placed in a 0.5 ml Eppendorf tube for centrifugation.
    - Dilute for appropriate concentration measurements (optical density measurements at 280 nm or colourimetric measurements with a reagent, see Chapter 4).

[a] HEW lysozyme is previously dialysed in buffer solution (20 times the volume).

10–20 μl hanging drops (19) or sandwich drops (10–40 μl) may also be convenient for the determination of solubility curves (for practical advices, see Chapter 4). Protocols are the same as described above but volumes have to be adjusted. Greatest care must be given to prepare the drops and to seal the coverslips as the volume of the drops are too small for salt concentration measurement by refractive index. The protein concentrations are measured by colourimetric methods (12, 19).

3.4.3 Column method

A simple method was developed by Pusey and Gernert (17) as an alternative for the determination of solubility curves. This method is based on maximization of the exchange between the available crystalline surface area and minimal free solution volume to reach equilibrium. It overcomes the problem of prolongated equilibrium time, as equilibrium is reached within 1–5 days. Two columns are run in parallel: one for crystallization (supersaturated state), one for dissolution (undersaturated state). An example is given in *Protocol 5* for HEW lysozyme solubility determination in NaCl at different concentrations and temperatures.

Protocol 5. Solubility measurements of HEW lysozyme by column method (17)

1. Dialyse lysozyme in buffer solution (20 × volume).
2. Centrifuge the solution for clarification and pass through a 0.22 μm filter.
3. Adjust protein concentration to 80 mg/ml.
4. Stir gently the protein solution (80 mg/ml) and add an equal volume of NaCl (1.7 M or 10% w/v).
5. After 1 h stirring, allow formed microcrystals to settle.
6. Decante supernatant and add 2 × crystal slurry volume of 0.85 M (5% w/v) NaCl in buffer to resuspend the crystals.
7. Pour the suspension into a 10 ml column, plugged with glass wool, to reach 5 ml volume of packed crystals.
8. Plug outlet and inlet of the columns to solution collector rack and source reservoir respectively.
9. Place columns in a water bath to control temperature, with the crystalline bed below water level.
10. Pass three column volumes of 0.85 M (5% w/v) NaCl through the column to estimate saturation concentration.
11. Pass an additional column volume of solution at appropriate pH and salt concentration to equilibrate.
12. Pour supersaturated solution on crystallization column and under-saturated solution on dissolution column and let equilibrate for 8–12 h.
13. Collect 0.3 ml of eluent and measure protein concentration. Solubility value is reached when protein concentration of crystallization and dissolution columns are within a range of 3%.
14. Set temperature to a new value if variation of protein solubility is to be determined as a function of temperature variation. Repeat steps 12 and 13. Equilibration is reached within 10–24 h with temperature variations of 0.5–1.5 °C respectively.

The major disadvantage of the column method is the requirement of gram quantities of protein. This has been overcome recently by using columns of 75–1000 μl (20). The results fit with previous ones obtained with larger column volumes.

4. Influence of parameter changes on protein solubility

Generally, an aqueous solution containing various additives is used as solvent, its ability to dissolve biological macromolecules varies depending on pH, ionic strength, and dielectric constant, in addition to temperature. This is linked to:

(a) nature and amount of buffer

(b) nature and amount of neutral salts

(c) addition of organic solvent.

Recently, the influence of these parameters on the solubility of HEW lysozyme has been studied (11, 13, 14, 17). Results are presented in the following paragraphs, but the comparison is sometimes difficult because experimental conditions (buffer systems, and so on) often slightly differ from one author to another.

4.1 Nature of salts

In 1888, Hofmeister ranked various ions (Hofmeister or *lyotropic* series) toward their ability to precipitate HEW proteins (21) by adding increasing amount of salts to a mixture of HEW proteins. More recently, it has been shown that ions act in reinforcing or denaturating biological macromolecule structures according to the same series (4):

- cations: $Li^+ > Na^+ > K^+ > NH_4^+ > Mg^{2+}$
- anions: sulphate^{2-} > phosphate^{2-} > acetate$^-$ > citrate^{3-} > tartrate^{2-} > bicarbonate$^-$ > chromate^{2-} > chloride$^-$ > nitrate$^-$ >> chlorate$^-$ > thiocyanate$^-$

Sulphate ions are called *lyotropic* (stabilizing the native form) whereas ClO_3^- and SCN^- are *chaotropic*. Chaotropic agents have a disruptive effect on water structure, and decrease the stability of the native conformation of macromolecules (helix breaker) promoting denaturation at high concentration. These series have been tested toward crystallization and solubility determination of HEW lysozyme (11). The experimental conditions are described in *Protocol 4* for solubility determination in NaCl and were achieved in a similar way for the various salts. The resulting solubility curves are shown in *Figure 5*. The results are as follows.

(a) The solubility of HEW lysozyme is more affected by anions than cations. This can be attributed to the basicity of HEW lysozyme (pI = 11) which has a highly positive net charge at pH 4.5.

(b) Cations follow the Hofmeister series. As for the anions, if they follow the Hofmeister series they are definitely in the reverse order. These results

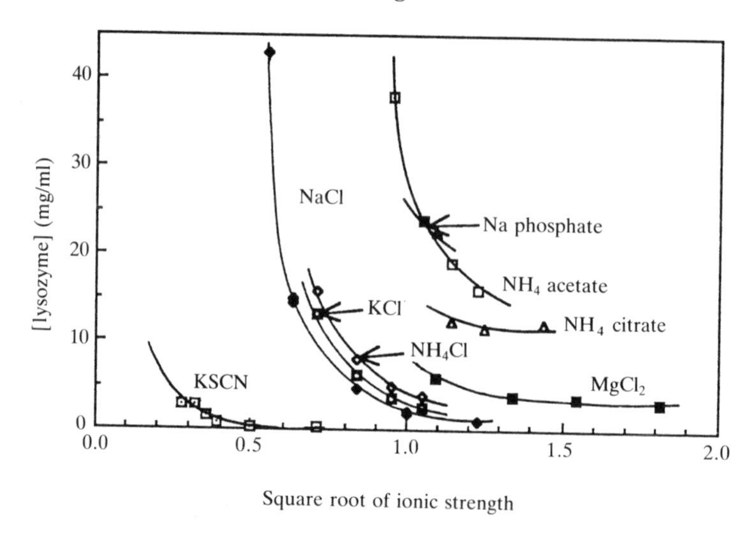

Figure 5. Phase diagrams of HEW lysozyme solubility (*S* in mg/ml) as a function of the square root of ionic strength, at 18 °C and pH 4.5 (NaAcO 50 mM). Data taken from (11).

do not agree with Hofmeister's observations, probably because (i) these series were tested against another target. It must be kept in mind that Hofmeister worked with a mixture of HEW proteins, which major component is known to be ovalbumin (acidic pI). (ii) pH could not be controlled in the study, as the concept of pH was only introduced in 1909.

(c) Thiocyanate, which is well known as a chaotropic agent at high concentration, appears to be very effective in crystallizing HEW lysozyme at low concentration. (i) The solubility curve indicates a decrease of HEW lysozyme solubility when increasing thiocyanate concentration, but at a low salt concentration (< 0.2 M) which normally corresponds to the salting-in range. The interaction of thiocyanate is probably due to binding with the protein, instead of exclusion. (ii) The same behaviour was observed with organic salts (Na *p*-toluenesulphonate, benzenesulphonate and benzoate) which successfully crystallized HEW lysozyme at low concentrations (22). (iii) The much higher effectiveness of KSCN, compared to NaCl, to crystallize HEW lysozyme at pH 4.5 is also observed with other proteins having a high pI: bovine pancreatic trypsin inhibitor and erabutoxin *b* (22).

(d) Lysozyme crystals are monoclinic when grown in the presence of SCN^-, but tetragonal with Cl^- (Na^+, K^+, NH_4^+, and Mg^{2+}), NH_4AcO, and NaH_2PO_4 (11). The tetragonal form is again obtained when crystallizing HEW lysozyme in presence of Na *p*-toluenesulphonate, benzenesulphonate, and benzoate at low salt concentrations (22).

4.2 Temperature

Protein solubility usually increases with temperature, but the reverse is also observed in some examples. As most proteins are stable between 0 and about 40 °C, temperature can only be varied within a limited range. The solubility diagrams of tetragonal HEW lysozyme at different temperatures given in *Figure 6* (data taken from references 11, 13, 14) show that:

(a) protein solubility is more sensitive to temperature variations at low salt concentrations;

(b) solubility at 20 °C is about 4–10 times higher than at 5 °C. This may be used to reduce nucleation rate by increasing the temperature, or by shifting from metastable zone to nucleation zone by lowering the temperature.

In their study, Pusey and Gernert (17) measured HEW lysozyme solubility between 10 and 32 °C with tetragonal and orthorhombic crystals grown at 20 °C and 40 °C, respectively. They observed the following.

- HEW lysozyme solubility is significantly lower with the tetragonal form compared with the data obtained with the orthorhombic crystals in the same range of temperature.

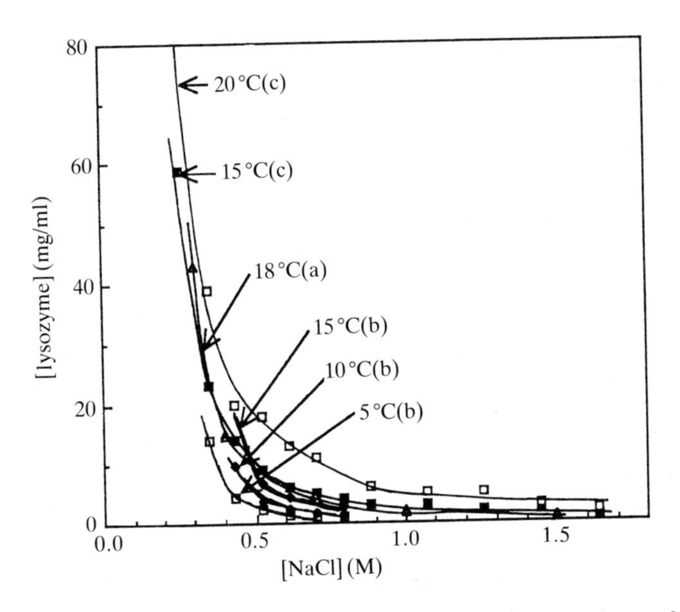

Figure 6. Solubility curves of tetragonal HEW lysozyme (*S* in mg/ml) as a function of NaCl concentration at different temperatures, pH 4.5 [NaAcO 50 mM (11, 14)], and adjusted with HCl (15). a, b, and c indicate that data are taken from (11), (13), and (14), respectively.

- Solubility data measured with tetragonal crystals are more scattered at high temperatures than at low temperatures. On the contrary, solubility values obtained with the high temperature orthorhombic form are more scattered at low temperatures than at high temperature.
- When a batch of orthorhombic crystals is kept for several days at low temperature, the population of tetragonal crystals increases, indicating a slow transformation. This was confirmed by the solubility measurements, as the solubility values measured with orthorhombic crystals decreased to approach the solubility value of lysozyme measured with tetragonal crystals.

These results illustrate the difficulty of comparing the behaviour of solubility within a large range of the variable especially when different crystal forms occur.

4.3 pH

Usually protein solubility is the lowest at the pI, as the net charge is equal to zero. There are only few data available in the literature to estimate the influence of pH on protein solubility over a large range, for the reasons given below.

- Most proteins are not stable over a large pH range.
- Checking protein solubility over a large pH range usually means changing the nature of the buffer. This hinders a true comparison of the data, as two parameters are modified at the same time.
- Crystals grown at different pH value can belong to different space groups (e.g. HEW lysozyme).

Figure 7 illustrates the variation of HEW lysozyme solubility between pH 4.0 and 5.5 (at 15 °C, 50 mM sodium acetate). These data taken from (14) agree with previous ones from (18), where pH was adjusted with HCl, showing very small dependence of HEW lysozyme solubility in this pH range. This is not universal, as in some other cases solubility can change drastically with a very small pH shift. In fact protein solubility varies mostly with its net charge, so when changing the pH:

(a) near the pK of most numerous charged residues, solubility varies very rapidly;

(b) out of the charged residues pK values, solubility changes smoothly.

Finally, the nature of the buffer must be emphasized, because at exactly the same pH the value of the protein solubility can be very different depending on the nature and the concentration of the buffer. By definition buffers are soft acids or bases and may present preferential binding, even of low affinity, with the biological material.

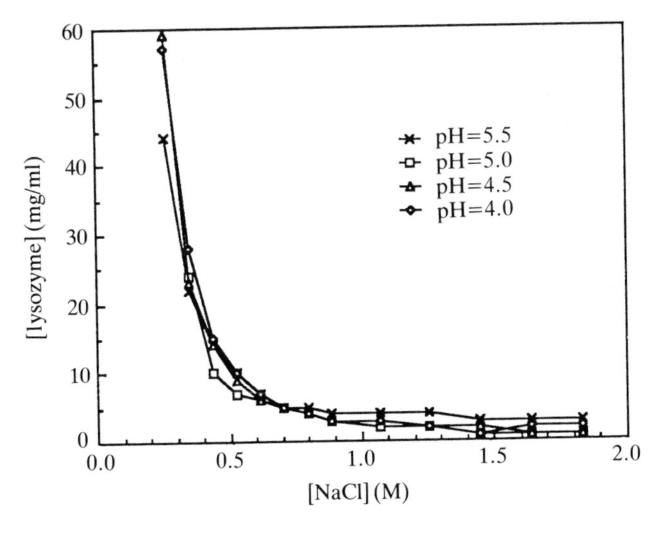

Figure 7. Solubility curves of tetragonal HEW lysozyme (*S* in mg/ml) as a function of NaCl concentration at different pH values (NaOAc 50 mM), and 15 °C. Data taken from (14).

4.4 Polymorphism

In the previous paragraphs, the example of HEW lysozyme crystals has been used to illustrate polymorphism when changing the nature of the salt (tetragonal, triclinic, and monoclinic), the temperature, and pH (tetragonal and orthorhombic). For a given value of each different crystallization parameters, only one crystal form appears, but different crystal forms can appear during the process of crystallization when:

(a) working at the borderline between two crystal forms;

(b) a parameter value varies during the crystallization process (e.g. pH and temperature shift).

5. Practical applications

Solubility diagrams are tools to determine the influence of crystallization parameters in order to define general rules. The knowledge of these rules defined with model proteins can be applied in practice without measuring all the diagrams for a new protein. One should try to visualize problems using a phase diagram, even though it cannot be determined experimentally when too small quantities of biological macromolecule are available.

- When the precipitation curve is very close to the solubility curve, the domain of crystallization is very narrow, which brings difficulties in defining the right conditions for growing large crystals. This is the case for the crystallization of HEW lysozyme (see *Figure 8*) with KSCN, where the

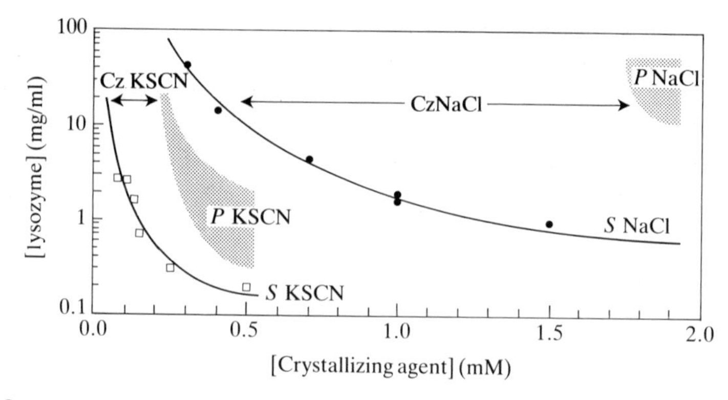

Figure 8. Comparison of crystallization zones, Cz, of HEW lysozyme in KSCN and NaCl, at 18 °C and pH 4.5 (NaOAc 50 mM); *S* being the solubility curves and *P*, the precipitation zones.

crystallization zone is limited to 100 mM KSCN concentration. It is well known how easily HEW lysozyme crystallizes in NaCl; in this case the crystallization zone extends over 1400 mM. It is thus worth checking crystallization conditions in different salts when dealing with narrow crystallization zones.

- Mounting of crystals. When a crystal is recovered from a drop, it is in a solution of given protein and crystallizing agent concentrations as shown in *Figure 9*. Very often reservoir solution is used to transfer the crystal. This can be done safely only when the remaining protein concentration in the drop is very low (B in *Figure 9*). When the solubility value is high (A in *Figure 9*), the crystal will start to dissolve as the reservoir contains no protein.

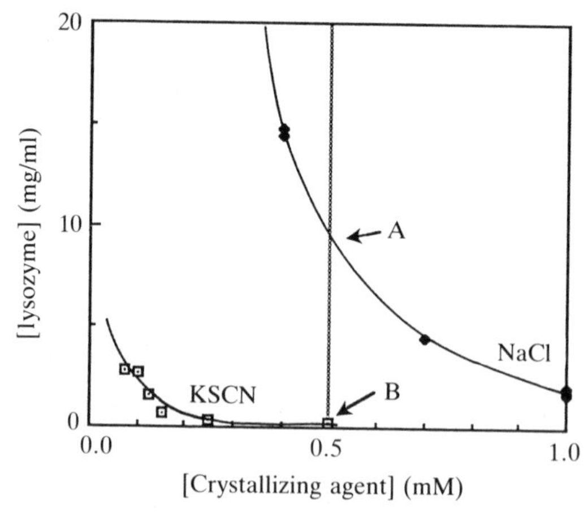

Figure 9. Comparison of HEW lysozyme solubility at 18 °C and pH 4.5 (NaOAc 50 mM), in (A) 0.5 M NaCl, and (B) 0.5 M KSCN.

- *Figure 9* points out that the solubility of HEW lysozyme in 0.5 M salt is 0.2 mg/ml in KSCN (B) but about 10 mg/ml in NaCl (A). This means that the uncrystallized amount of protein is 20 times higher in the second case at the end of the equilibrium process. The measurement of soluble protein at the end of a crystallization experience can be very helpful in order to (a) know the solubility in these conditions; (b) recover unused protein; (c) have an idea if the crystals can be mounted with the reservoir solution.
- Seeding. If protein is available in too small amounts for solubility diagram determination, the solubility can be approximatively determined by seeding of drops at different concentrations (of protein or salt) as shown in *Figure 10*. Crystals dissolve in undersaturated drops A, remain unchanged in saturated drops B, but they grow in slightly supersaturated drops C and nucleation occurs in more supersaturated conditions D.
- In a mixture of proteins, the first crystals contain the less soluble protein and this may not be the more concentrated one. Furthermore, when changing for example pH or temperature, the solubility of the major protein may be affected differently than the contaminant.
- With a programmable incubator, one can ideally start an experiment until nucleation occurs and then decrease the temperature as protein concentration decreases in the supernatant to grow crystals using the maximum biological material. This is obviously to avoid in case of polymorphism in the temperature range.

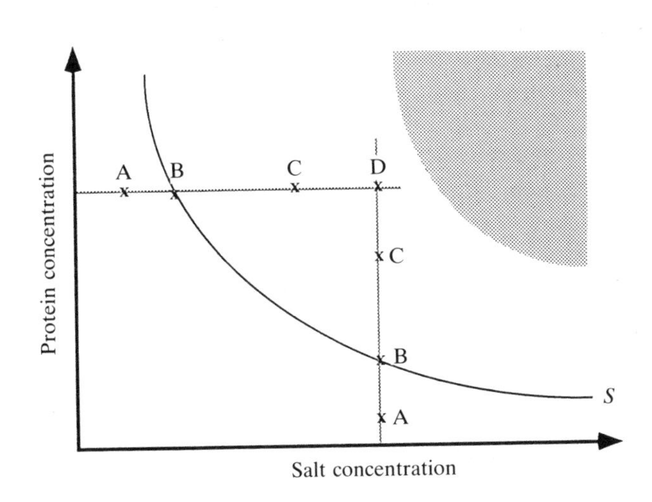

Figure 10. Theoretical phase diagram indicating the position of different saturation states: (A) undersaturation, (B) saturation, (C) supersaturation for crystal growth only, and (D) supersaturation for nucleation.

References

1. Boistelle, R. and Astier, J-P. (1988). *J. Crystal Growth*, **90**, 14.
2. Schein, C. H. (1990). *Biotechnology*, **8**, 308.
3. Arakawa, T. and Timasheff, S. N. (1985). *Methods in Enzymology*, **114**, 49.
4. von Hippel, P. H. and Schleich, T. (1969). In *Structure and stability of biological macromolecules* (ed. S. N. Timasheff and G. D. Fashman), pp. 417–574. Dekker, New York.
5. von Hippel, P. H. and Schleich, T. (1969). *Accounts of Chemical Research*, **2**, 257.
6. McPherson, A., Koszelak, S., Axelrod, H., Day, J., Williams R., Robinson, L., McGrath, M., and Cascio, D. (1986). *J. Biol. Chem.*, **261**, 1969.
7. Melander, W. and Horvath, C. (1977). *Arch. Biochem. Biophys.*, **183**, 200.
8. Green, A. A. (1932). *J. Biol. Chem.*, **95**, 47.
9. Cohn, E. J. (1925). *Physiol. Rev.*, **5**, 349.
10. Edsall, J. T. and Wyman, J. (1958). In *Biophysical chemistry*, (ed. J. T. Edsall and J. Wyman), Vol 1, pp. 241–322. Academic Press, NYC.
11. Riès-Kautt, M. and Ducruix, A. (1989). *J. Biol. Chem.*, **264**, 745.
12. Mikol, V. and Giegé, R. (1989). *J. Crystal Growth*, **97**, 324.
13. Ataka, M. and Asai, M. (1988). *J. Crystal Growth*, **90**, 86.
14. Howard, S. B., Twigg, P. J., Baird, J. K., and Meehan, E. J. (1988). *J. Crystal Growth*, **90**, 94.
15. Timasheff, S. N. and Arakawa, T. (1988). *J. Crystal Growth*, **90**, 39.
16. Feher, G. and Kam, Z. (1985). *Methods in Enzymology*, **114**, 77.
17. Pusey, M. L. and Gernert, K. (1988). *J. Crystal Growth*, **88**, 419.
18. Ataka, M. and Tanaka, S. (1986). *Biopolymers*, **25**, 337.
19. Chayen, N., Akins, J., Campbell-Smith, S., and Blow, D. M. (1988). *J. Crystal Growth*, **90**, 112.
20. Cacioppo, E., Munson, S., and Pusey, M. L. (1991). *J. Crystal. Growth*, **110**, 66.
21. Hofmeister, F. (1888). *Arch. Exp. Pathologie und Pharmakologie (Leipzig)*, **24**, 247.
22. Riès-Kautt, M. and Ducruix, A. (1991). *J. Crystal Growth*, **110**, 20.

10

The physical chemistry of protein crystallization

V. MIKOL and R. GIEGÉ

1. Introduction

Of all the different aspects of protein crystallization, the physics of crystal growth and its understanding have been the most neglected. Indeed, the growing of protein crystals is often believed to occur more through serendipity than as the result of a rational approach. This chapter will try to dispute this well-established belief. More knowledge on the fundamentals of crystal growth should make it possible to develop easy and convenient assays for defining optimal conditions for protein crystallization. Undoubtedly, the understanding of crystallization from a physicochemical point of view has received more attention recently and crystallization is slowly emerging from its veil of mystery. Using the physicochemical concepts of small molecule crystal growth (1), this chapter will go through all the steps of crystallization: nucleation, growth, cessation of growth, and characterization of crystals. It does not deal with the preliminary X-ray structure analysis reviewed in Chapter 12. Firstly, a brief summary of the basic physical concepts underlying each step is given. This provides a simple theoretical base for the biomacromolecule crystal grower. Then the main results obtained from physical studies on each stage and practical applications are presented.

2. Supersaturation/Nucleation

2.1 Supersaturation

2.1.1 Background

A saturated solution contains an amount of solute such that there cannot be either growth or dissolution if crystals are added to the solution. It corresponds to a thermodynamic equilibrium between two phases (liquid and solid). The chemical potential of each species 'i' is the same in both phases (crystal and solution):

$$\mu_{ic} = \mu_{is} = \mu_{i0} + RT \ln\gamma \, c_i \tag{1}$$

where μ_{i0} is the standard potential, γ the activity coefficient, and c_i the concentration of the species i. This is true whatever the species (protein, water, and so on). The concentration of solute in saturated solution is then defined as the solubility c_s (see Chapter 9). Supersaturation is reached when the chemical potential of the solute in solution is greater than that of the crystal. It can be achieved by varying any parameter affecting the chemical potential (temperature; protein concentration; activity coefficient, by changing the salt concentration; pressure, thus modifying the standard potential). Supersaturation is the driving force for crystallization; it is defined as the difference of the chemical potential between the solution and the crystal. However, the most common approximations are the ratios c/c_s and $(c - c_s)/c_s$ where c indicates the protein concentration. These quantities have the advantage of being dimensionless. It should be borne in mind that the same supersaturation value can be obtained by different ways (e.g. altering the concentration of solute, or raising or lowering appropriately the temperature). It must be noticed that the value of supersaturation (defined as the ratio of the protein concentration over the solubility c_s) currently encountered in protein crystallization experiment lies between 2 and 10. This is a much higher range than that observed in small-molecule crystal growth.

2.1.2 Experimental determination of supersaturation

Since supersaturation is the major physicochemical parameter to control in order to optimize crystallization, it should be emphasized that the determination of the solubility of the protein is an unavoidable prerequisite for any 'rational' experiments. The advent of DNA technology which allows the production of large amount of protein, and the development of micromethods to perform solubility measurements, make it possible to determine part of the phase diagram of the protein under study. The reader is referred to Chapter 9 of this book for experimental details.

2.2 Nucleation

2.2.1 Background

Nucleation of a new phase originates from energy fluctuations around the increased mean value of the free energy induced by the supersaturation. Nucleation is said to be *homogeneous* when it occurs in the bulk of the solution and *heterogeneous* on solid particles (dust, glass walls of the experimental set-up, and so on). Nucleation is called *primary* in all cases of nucleation in systems that do not contain crystalline matter. Sometimes nuclei are growing out of a parent one (e.g. branching); this is referred as to *secondary nucleation*.

Crystals will not grow out of all supersaturated solutions. To create a new phase, the system must overcome a certain energy barrier called activation free energy of germination ΔG_g. If supersaturation is too low, the amplitude of the energy fluctuations might not be high enough to jump over this energy barrier. One can divide supersaturation into two regions, a *metastable* region where nucleation cannot occur spontaneously but crystals grow and a *labile* region where stable nuclei can form and grow (*Figure 1*). The biochemist, having been involved in protein purification, is much more familiar with the precipitation curve. Noted here are two distinct types of precipitates which can occur.

(a) A crystal-like precipitate which is composed of very small crystals. In that case, the precipitation curve lies in the labile region. It corresponds to a crystallization kinetics curve characterized by a very short time-lag of crystal appearance.

(b) A gel-like precipitate or an amorphous precipitate or a flocculate (a more appropriated term from colloids chemistry) which results from non-specific van der Waals attraction. This is a new phase, non-crystalline, and the precipitation curve may cross any regions of the crystalline phase.

Following small molecule crystal growth theories (1), it can be derived that ΔG_g of a nucleus of radius r is composed of two terms:

$$\Delta G_g = [-kT(4\pi r^3)/V\ln\beta] + 4\pi\gamma r^2 \tag{2}$$

where V is the volume of a molecule inside the crystal, β is the supersaturation, k is the Boltzmann constant, and γ is the interfacial free energy between nucleus and solution.

ΔG_g is a function of a negative volume term and a positive surface term. Roughly, the latter corresponds to the energy one must provide to create a crystal surface unit and the volume term, proportional to supersaturation,

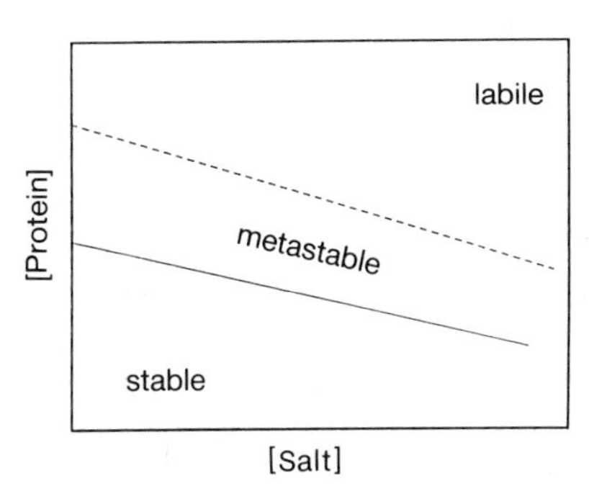

Figure 1. Stable, metastable and labile regions. The diagram is divided into three zones: the stable region (undersaturated) where crystallization is not possible, the metastable (supersaturated) region where nuclei cannot form but crystals can grow, and the unstable or labile (supersaturated) region where spontaneous crystallization can occur. The continuous line indicates the solubility curve and the dotted line the 'supersolubility' curve which is not so well defined as it may depend on, among other things, the agitation of the solution.

describes the energy gain resulting from the decrease of the free energy of the system. Without going into much detail of this derivation, it should be noticed that first ΔG_g increases with r until it reaches a maximum for a value of r called the critical radius (r_c) and then decreases as r tends to infinity. This means that a nucleus will be stable once it has grown up to the critical size r_c. A smaller crystalline aggregate will tend to dissolve, being unstable from a thermodynamic point of view. The formation of this new phase once realized will allow the system to reach thermodynamic equilibrium. Of particular interest is that ΔG_g decreases with supersaturation and increases with the interfacial crystal/solution free energy. Accordingly, a high supersaturation reduces the energy threshold to create a new phase and favours nucleation. The presence of foreign particles reduces the free interfacial energy and increases the frequency of nucleation. Thus a lower supersaturation is required to nucleate when dealing with heterogeous rather than homogeneous nucleation.

Up till now, we have been mainly concerned with the parameters which affect the energy threshold that must be overcome in order to form nuclei. A kinetic aspect must also be taken into account to draw a complete picture. Following small molecule crystal growth, the nucleation rate can be expressed as:

$$J_n = B_S \times \exp(-\Delta G_g/kT) \tag{3}$$

where the kinetic coefficient B_S is the product of solubility and a kinetic parameter. It appears that the greater the solubility, the faster the nucleation. Indeed, when there are many free molecules in solution, they are more likely to encounter and to associate with each other. In this case, nucleation occurs at a relative moderate supersaturation rate. When the molecule is sparingly soluble, however, nucleation will appear at a higher supersaturation with a larger time-lag, and a shower of crystals might come out. Thus, a high solubility will promote nucleation. Accordingly, to favour nucleation the take-home lesson for the first trials should be:

(a) to press the system as far as possible into the labile region of supersaturation (i.e. use protein concentration as high as possible), and this will favour fast growth as well;

(b) to choose conditions where the solubility of the protein is as high as possible (for instance far from pI). Warning: these recommendations are for optimizing nucleation and not optimal growth.)

2.2 Kinetics of supersaturation and nucleation

Not only does the nucleation rate depend on supersaturation but also on the way it is reached. Crystallization is often achieved by equilibrating a protein solution containing a precipitating agent with a reservoir containing a solution of different composition (see Chapter 4). The kinetics of equilibration depends upon the experimental arrangement and upon the method chosen (i.e. dialysis, vapour diffusion, see Chapter 4). The equilibration rate of the protein solution affects the kinetics of supersaturation and hence crystallization. Indeed, slowing down the equilibration rate seems to result in fewer and larger crystals. This was demonstrated during the crystallization of ribosomes (2) and lysozyme (3). This can be readily accounted for as follows: as equilibration proceeds, supersaturation develops, and once the system has reached the labile region nuclei may result (*Figure 1*). Let us suppose that nucleation has occurred right above the border between the two regions. The formation of small crystalline aggregates results in a decrease of the overall supersaturation and accordingly, as growth takes place, the system is pushed into the metastable region preventing additional nucleation. However, if equilibration is not achieved (e.g. of a hanging drop or a dialysis button), the system will be driven further into the labile region and further nuclei may appear. There is a tug-of-war between the growth of existing crystals and the nucleation of new ones, the goal being to decrease the free energy of the system as fast as possible. Thus, once a few crystals have appeared, if the equilibration rate is slow with respect to crystal growth then the system will remain in the metastable region and impede spurious nucleation. In order to control the supersaturation kinetics, not only must one know the solubility of the protein but also the equilibration kinetics.

In the case of dialysis, slowing the equilibration rate has been achieved (see Chapter 4) by incorporating a second dialysis membrane (4). The major parameters affecting the equilibration rate are temperature, osmotic gradient, dialysis membrane, and surface area of exchange. However, no study has yet thoroughly investigated the relative importance of each one.

In contrast, the influence of the major parameters of the vapour diffusion technique has been recently analysed and quantified (5). These include drop volume, dilution of the drop, and temperature. It was shown that the variations of the volume V of hanging drops in Linbro plates as a function of time can be described by an exponential function of the form:

$$V/V_0 = \delta + [(1 - \delta) \exp(-t/\tau)] \quad (4)$$

The term V_0 indicates the initial drop volume, δ the dilution of the drop (ratio of the precipitant concentration of the drop over that of the reservoir), and τ a time constant given by:

$$\tau = \eta \, V_0^{\varepsilon}/P_0 \quad (5)$$

where P_0 is the vapour pressure of pure water; η depends on the precipitant and is given in *Table 1* as well as ε. It should be noticed that this is a semi-empirical formula only valid for the units used i.e. V (μl), η (h), P_0 (Torr), and ε (see *Table 1*). This simple formula allows one to compute variations of supersaturation with time, provided that solubility is known, if water is the only volatile species. As an example, the kinetics of supersaturation of a

Table 1. Values of the parameters of the time constant (τ) for estimating the kinetics of water equilibration in hanging drops [from (5)]

precipitant	η (hours)	ε (unitless)
$(NH_4)_2SO_4$ 1.6 M 20 mM Bis Tris/HCl	32.1 ± 5.8	0.81 ± 0.06
PEG 6000 20% (w/v) 20 mM Bis Tris/HCl	149 ± 28	0.81 ± 0.06
MPD 20% (v/v) 20 mM Bis Tris/HCl	46.9 ± 7.7	0.81 ± 0.06

Note: When using another precipitant concentration, the value of η must be corrected. A reasonable estimation is that η varies as $1/c_p$ where c_p indicates the precipitant solution concentration. If another buffer at a different concentration is used, provided its concentration remains low, one can rely on the same values.

crystallization experiment by the vapour diffusion technique of concanavalin A is presented (*Figure 2*). In this case the equilibration of ammonia must also be taken into account to estimate properly the time-course of supersaturation.

2.2.3 Probing nucleation

When a system is supersaturated, phase transitions will result which drive back the system to equilibrium. It might be a second liquid phase (6) if the

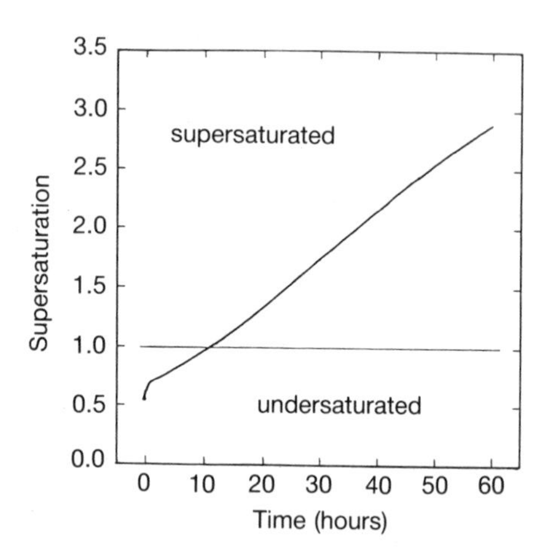

Figure 2. Variations of supersaturation of a concanavalin A crystallization experiment by the vapour diffusion technique. Supersaturation is estimated as the ratio of the protein concentration over its solubility. The initial drop (30 µl, 20 °C) composition is concanavalin A, 3 mg/ml, $(NH_4)_2SO_4$ 0.6 M, at pH 7.0; the reservoir contains 1.0 M $(NH_4)_2SO_4$ at pH 7.0. Two effects are taken into account, the equilibration of ammonia which is rapid and induces a quick change of supersaturation, and the equilibration of water. The reader is referred to reference (23) for a detailed description of this derivation.

protein concentration is very high, an amorphous phase, or a crystalline phase. The protein crystal grower is mainly interested in the latter. Thermodynamics can tell us which phase is the most stable but is not able to predict which one is reached first kinetically. Only visual examination with an optical microscope is fortuitous when analysing crystallization experiments. Probing the solution state of the molecule appears to help the experimenter to discriminate more rationally between solutions that lead to amorphous precipitates and to crystals.

A way of addressing this question was suggested by the work of Kam *et al.* (7). It allows the aggregation state of a molecule to be probed by light scattering. Crystallization is described as a co-operative step-by-step addition of monomers, whereas amorphous precipitation is an uncooperative polymerization process. This approach requires the analysis of the distribution of species within a supersaturated protein solution, (number of monomers, dimers, trimer, *n*-mer, and so on) which is extremely difficult to address experimentally. This is circumvented by introducing a theoretical model of the crystallization process. According to this model, in solutions leading to crystallization, an abrupt change in aggregate size is expected as a function of the protein concentration, whereas a nearly linear variation is predicted for precipitating solutions. This work was the first major approach to tackle protein nucleation and to provide the protein crystal grower with an easy assay. It requires:

- a basic laser light scattering set-up which consists of (*Figure 3a*):
  - (a) an illumination source: a laser with a visible wavelength He–Ne or Ar laser (e.g. Spectra Physics);
  - (b) optical components to focus the laser beam onto the cell which is maintained at a constant temperature;
  - (c) a photomultiplier to collect the scattered light;
  - (d) an autocorrelator to determine the autocorrelation function of the scattered light;
  - (e) a computer interfaced to the autocorrelator to adjust the autocorrelation function.

The cost of the set-up and the quality of the measurements will depend on the power of the laser (5–800 mW) and upon the sensitivity of the autocorrelator (number of channels, 64 is a good average).

- the knowledge of the solubility of the molecule
- measurement of the diffusion coefficient (obtained from the light-scattering experiments, and which is inversely proportional to the aggregate size by the Stokes-Einstein law[a]) of the protein solution as a function of the

[a] $r_h = kT/6\pi\eta D_t$, where r_h is the hydrodynamic radius, k is the Boltzmann constant, T the absolute temperature, η the viscosity of the solvent, and D_t the translational diffusion coefficient.

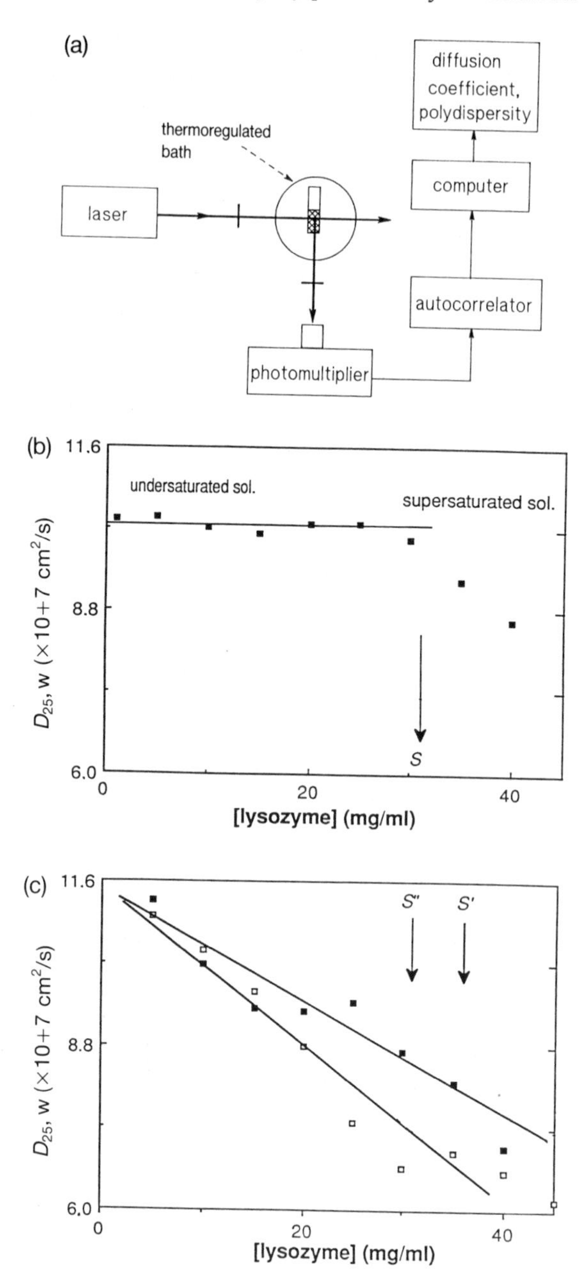

Figure 3. Crystallization and precipitation diagnostic of lysozyme by light scattering [adapted from (9)]. (a) Schematical drawing of a light-scattering set-up; (b) variations of the diffusion coefficient of lysozyme ($D_{25,w}$) as a function of protein concentration in 2% (w/v) NaCl, 40 mM NaAcO, pH 4.6. No significant variations are observed up to the solubility limit (S); crystallization can be expected when the solution is supersaturated; (c)

protein concentration in the supersaturation region. Before measurements, one should not forget to clean thoroughly the glass cell and filter and centrifuge all solutions, since light-scattering measurements are easily spoilt by any remaining dust.

However, the light scattering approach contains several flaws noted here.

(a) Experiments should be carried out in the supersaturated region, where the diffusion coefficient might depend on time.

(b) The analysis of the distribution is obtained through a semi-indirect approach.

(c) There are arguments coming from the study of small molecules in the supersaturated state (8) that the variations of the diffusion coefficient (abrupt change versus linear decrease) indicates rather the closeness of the system to the boundary between the metastable and labile regions than the ability to crystallize or to precipitate.

Along these lines, another approach was suggested recently (9), which aimed to understand the fundamentals of crystal growth at a simpler level. This provides a simple assay unbiased by theoretical considerations. This method is a somewhat negative assay as it predicts which solutions will certainly not yield crystals because aggregations are detected before supersaturation is reached. It gives information on the nature of the potential interactions between macromolecules. It cannot predict if large single crystals or crystal-like precipitates will be obtained. It requires a laser light-scattering set-up as well. The experiments should be carried out as noted here.

(a) The protein concentration is held constant and the diffusion coefficient of the scatterers (inversely proportional to the size by Stokes-Einstein law) is measured as a function of the precipitant concentration.

(b) The salt concentration is held constant and the diffusion coefficient of the scatterers is measured as a function of the protein concentration.

Values of the diffusion coefficient of the scatterers are then plotted (*Figures 3b* and *3c*) as a function of concentration (salt or protein). If any significant decrease of the diffusion coefficient is observed before reaching the solubility limit, it indicates that the protein will probably not crystallize in this precipitant solution (*Figure 3c*). These variations of aggregate size in the undersaturated region are interpreted in terms of non-specific interactions which will prevent the molecules forming a crystalline array when supersaturated.

---

variations of the diffusion coefficient of lysozyme ($D_{25,w}$) as a function of protein concentration in 40 mM NaAcO, pH 4.6, and $(NH_4)_2SO_4$, 0.8 M (■) and 1.0 M (□). The terms S' and S'' indicate the solubility in 0.8 M and 1.0 M, respectively. Nearly linear variations are observed suggesting that amorphous precipitation will occur when the solution is supersaturated.

Those two approaches have very similar conclusions. They correlate the concentration-dependence of the diffusion coefficient to the ability to form crystals. The former achieves this goal in the supersaturated region with the help of a theoretical model. The latter is mainly qualitative and deals within the undersaturated region. However, these methods have been applied to model systems, and still lack generality.

3. Mechanisms of crystal growth

3.1 Background

In crystal growth theory, one traditionally divides the process into two steps:

(a) mass transfer from the bulk of the solution to the crystal/solution interface;

(b) attachment of the molecule on the crystal (interface kinetics).

Mass transfer is governed by diffusion and convection due to density-driven gradients or by stirring (if any). The attachment of the molecules into the lattice can be accounted for by several mechanisms (1). Two of them, very detailed from a theoretical point of view, are briefly reviewed here, and seem to be relevant to protein crystallization. One refers to the growth of a perfect crystal and the second one to the growth out of a screw dislocation.

3.1.1 Two-dimensional nucleation growth

In the case of a perfect crystal face, a molecule impinging on the surface has little chance of being retained, because of the low number of bonds which can be established. Once the adsorbed molecules have formed a two-dimensional (2-D) cluster (nucleus) on the surface which exceeds a critical size, growth will proceed. This 2-D nucleus is then able to incorporate molecules colliding on the surface. This model is likely to apply for growth at high supersaturation.

3.1.2 Screw-dislocation growth

A screw dislocation (see Section 5.1) causes a kink in the crystal surface into which molecules can adsorb and thus incorporate into the lattice. The growth then continues in a spiral fashion around the centre of the dislocation. The steps expand by rotating around the emergence of a screw dislocation.

3.1.3 Cessation of growth

The cessation of growth is of great practical importance since a minimum size of the crystal is required to perform X-ray analysis. This phenomenon is poorly understood and it is often associated with poisoning of the crystal surface due to accumulation of impurities. The only direct experimental indication comes from electron micrographs (10) of lysozyme crystals surface. They reveal strong bunching of parallel step trains accounted for by impurities.

3.2 Experimental analysis of the growth mechanisms

3.2.1 Macroscopic analysis

Depending on the model used, the theoretical growth rate G of a face can be derived and then compared to experimental measurements using the following equations. The reader is referred to (1) for a detailed derivation.

- For diffusion as the rate controlling step, the growth rate will be:

$$G = DV\,(c - c_s)/l, \qquad (6)$$

  with D the diffusion coefficient, V the molecular volume, l the solute boundary layer.

- For 2-D nucleation growth:

$$G = k'(c/c_s)^{1/2} \exp\,(-\Delta G/kT) \qquad (7)$$

  where k' is a kinetic coefficient, ΔG the activation free energy of germination of a 2-D cluster.

- For growth on a screw dislocation:

$$G = (1/2)\,(D\,d^2\,k\,T\,c/\gamma)\,[\ln(c/c_s)]^2 \qquad (8)$$

  where γ is the interfacial free energy and d the molecular diameter. All other parameters were defined previously.

The equipment required to carry out such experiments is not complex. It consists of:

- a video camera to be mounted on the microscope;
- a time-lapse video recorder with a compressing time capability of at least 72-fold;
- a TV monitor to display the images;
- a crystallization cell with a temperature regulation device is highly recommended since heating of the sample due to prolonged transillumination can occur.

Working with the batch technique is simpler because it avoids taking into account the kinetics of equilibration. If drops are to be used, sandwich drops are recommended for obtaining better focusing.

Experiments must be performed carefully so as to provide stirring (to avoid diffusion limitation) if an attachment model is suspected. Similarly, a constant flowing of a fresh solution should be made. Otherwise one studies the growth rate as a function of supersaturation which itself depends on time. Most studies have been performed without maintaining a constant supersaturation and the results are not in agreement. However, the surface kinetics seems to be the rate-limiting steps.

Using electron microscopy (10) it was shown that growth proceeds by motion of steps, although spirals could not be visualized directly; a screw-dislocation mechanisms at low supersaturation was suggested. These results indicate that within the range of supersaturation investigated crystal growth occurs by motion of steps.

3.2.2 Optimizing crystal growth

At the first glance, the knowledge of the exact growth mechanism might not appear to be of great practical value for a protein crystallographer eager to solve the three dimensional structure of the protein under study. However, recording the growth sequence can bring useful information to optimize crystal growth and hence crystal quality. Indeed, it indicates:

- the time span for the first crystals to become visible (20/30 μm);
- an approximate growth rate;
- the origin of twinning and bunching (sedimentation or secondary nucleation).

It has been known for a long time in the field of small molecule crystal growth that the supersaturation conditions favouring an optimal nucleation are not the same as those promoting regular growth. Ideally, one should chose supersaturation conditions that induce the formation of a single nucleus just above the border between the metastable and labile regions. In doing this, the formation of the nucleus would drive the system back into the metastable region and regular growth would proceed. This ideal pathway can be partially achieved through seeding (see Chapter 5). The use of special crystallization set-ups to change the composition of the reservoir as a function of time in the vapour diffusion technique has proved useful and allows crystal growth to be optimized [see e.g. (3), (11)].

4. Characteristics of crystals

4.1 Background

The relationships between the morphology of a crystal and its internal structure have been extensively studied since the early days of mineralogy and crystallography. The *faces* of a crystal can be deduced from one face and the symmetry of the crystal. A face is indicated in brackets (*Figure 4*). A set of faces defines a *crystal form*. The *morphology* of a crystal is given by a set of forms which appear on the crystal. The morphology does not reflect the overall shape of the crystal which is given by the *crystal habit*. A crystal can be defined by cubic faces and have a needle or thin-plate habit. The habit takes into account the relative growth rate of the faces. It will depend on internal factors (structure bonds) and on external factors (supersaturation, impurities, and so on).

When the growth rate of each face depends only on their orientation (no poisoning by impurities) a polyhedron form results. In this case, the growth

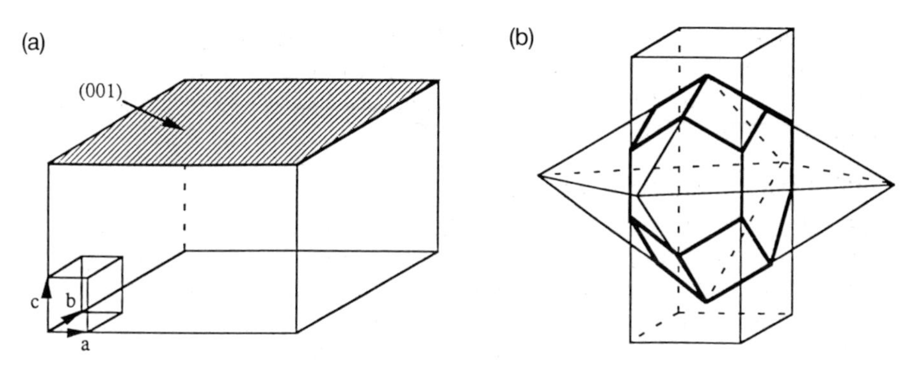

Figure 4. Face, form and morphology. (a) Face (001) of a tetragonal crystal form; (b) morphology resulting from the combination of two crystal forms both belonging to the quadratic system (a prism and a bipyramid).

rate is stationary and the growth form is simply determined by thermodynamic considerations. The smaller the interfacial energy, the smaller the growth rate and the greater the development of the face.

One sometimes relates the growth of thin plates or elongated needles to stronger binding of the molecule in the direction of the long axis which may imply better fitting and longer range order in this orientation. In order to modify the relative size of the crystal faces, one should try additives (see Section 4.2).

A slow growth rate is often an indication of low-defects content. Fast growth might result in an asymmetric crystal showing cusps, hollow interiors, or branches arising from solute depletion during rapid growth. Examples of habits are given in *Figure 5*.

Crystals sometimes exhibit some form of aggregation or intergrowth. It can result from random cluster or from symmetrical forms such as *twins* or *parallel growth crystals*. This latter type is the result of individuals growing on the top of each other in such a way that all corresponding faces and edges of the individuals are parallel. A *twin* or a *macle* is composed of intergrown individuals joined symmetrically about an axis or a plane. In some cases, a twin crystal might present a higher degree of symmetry than that of the individuals. As for large clusters or agglomerates, their formation is poorly understood but often related to a poor agitation of the solution.

4.2 Changing the habit or the crystal form

4.2.1 Supersaturation

Supersaturation variations can be used to change the habit or the crystal form. This can be achieved, for example, by changing the solute concentration. This was exemplified on the crystallization of carboxypeptidase where spherulites are grown at low supersaturation and large polyhedron

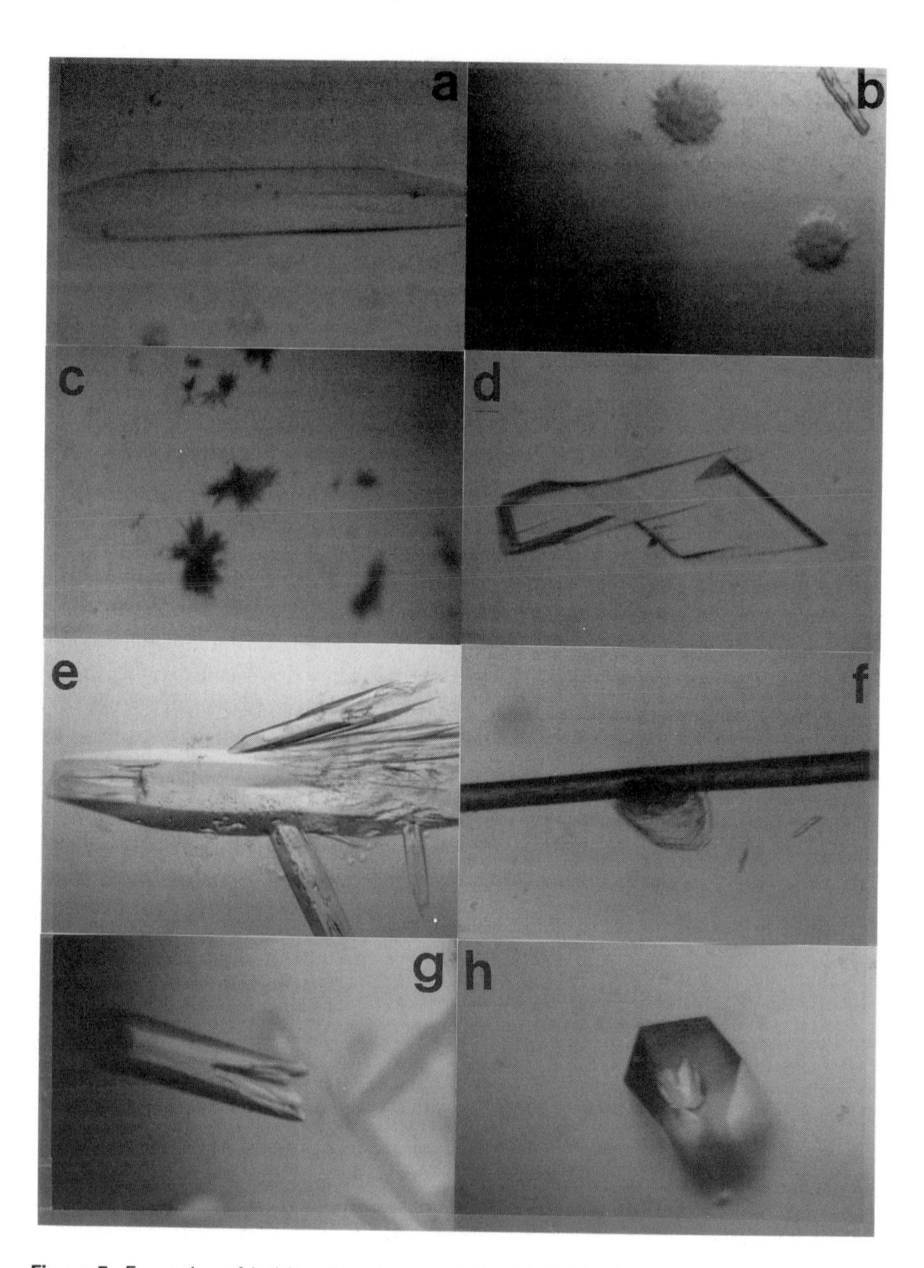

Figure 5. Examples of habits of protein crystals. (a) *Polyhedrons* are cystals limited by faces. (b) *Spherulites* are polycrystalline aggregates with an approximately radial symmetry. They are usually comprised of a radiating array of crystalline fibres. Successive generations of fibres repeatedly branch apparently at random. (c) *Dendrites* exhibit a smooth approximately parabolic tip, characterized by a tree-like structure. (d) Twin crystals. (e) Branching (secondary nucleation). (f) Crystal having a plate habit resulting from heterogeneous nucleation. (g) Crystal showing branches arising from solute depletion during rapid growth. (h) Hexagonal crystal showing an hollow interior resulting from a fast growth process.

crystals at higher supersaturation (12). With high supersaturation the growth unit incorporates rapidly into the faces where their adsorption energies are larger. Hence when there is an important anisotropy in the bond, a needle will grow more needle-like and a thin plate will turn out thinner and wider.

4.2.2 'Impurity'

Anything in the solution which is neither the solvent nor the solute (see Chapter 9) can be regarded as an impurity (e.g. buffer, additive). An impurity may have three effects, noted here.

(a) It can act as a habit modifier without affecting the crystal lattice (e.g. needles getting thicker). In this case the impurity is probably not incorporated into the lattice. Adding certain additives such as β-octyl glucoside with soluble proteins or dioxane presumably has this effect. In a recent report of the crystallization of Tumor Necrosis Factor (13) as long hexagonal rods, the addition of β-octyl glucoside is said to produce a dramatic improvement in the crystal size: the diameter of the hexagonal prism was increased approximately 5-fold while the length remained essentially unchanged. In this case β-octyl glucoside is very likely to have had some affinity for the face perpendicular to the 6-fold axis hindering the attachment of molecules hence slowing down its growth rate.

(b) It can affect the crystal form. The additive interacts specifically with the protein and so a new compound is crystallized composing of the complex of the additive and the protein. This is undoubtedly the case with phenol in the last reported crystal form of insulin (14).

(c) Finally, it may decrease the solubility of the protein and hence change supersaturation. However, the influence of such a tiny amount of additive on the solubility of any protein has never been quantified.

4.2.3 Temperature

Temperature was used in a few cases as an habit modifier but is more likely to change the crystal lattice; e.g. in the case of lysozyme (15), a tetragonal form is obtained in the presence of NaCl at an acidic pH and a temperature lower than 25 °C whereas an orthorhombic form is grown at higher temperatures. A part of the phase diagram of lysozyme has been established (16) and the solubility values clearly reveal this temperature-dependent phase transition.

4.2.4 Reducing the rate of mass transfer

As fast growth might result in asymmetric crystals with defects (see *Figure 5*), gel which reduces the depletion of solute and then favours a more homogeneous growth can be tried (see Chapter 6). Along this line, increasing the viscosity of the solution by adding 1–2% glycerol in the crystallization medium gets rid of the branches seen in *Figure 5g* and more round-shaped crystals could be grown.

5. Crystal defects

Although, for a protein crystallographer, the only defect protein crystals have is not diffracting X-rays to a higher resolution, defects can be classified according to the dimension of the perturbation they introduce. Two-dimensional defects are only briefly described here. The reader is referred to (1) for a complete presentation.

5.1 Planar defects or stacking faults

A crystal is seldom a perfect single crystal; it is considered as an ideally imperfect crystal composed of small mosaic blocks randomly misoriented. Each block is perfectly crystalline and is bound to its neighbour by dislocations. The mosaic spread of the crystal (*Figure 6*) is the extent of the angular misalignment of each block with respect to every other block, and accordingly reflects the content of dislocation of the crystals.

5.2 Experimental analysis of defects in protein crystals

X-ray diffraction analysis of protein crystals can reveal two types of defects:

- a short-distance one (internal order of the unit cell) estimated with the temperature factor;
- a long-distance one approximated by the rocking width of diffraction peaks (yielded by the mosaicity of the crystal) (*Figure 6*).

However, in the former case, it does not allow one to determine if the disorder is static or dynamic since X-ray diffraction only measures the average structure.

With a well-collimated beam mounted on a conventional source, it has been shown that the mosaic spread of lysozyme crystals was lower than 0.02 degree (17). Using synchrotron radiation (see *Figure 6* and reference 18) smaller values were obtained for other protein crystals. Thus the very small mosaic spread measured suggests almost perfect long-range order in the crystal and hence absence of dislocations. This view is compatible with the weak binding energies and mechanical softness of protein crystals.

Larger defects can be probed by the use of electron microscopy and they account for the poor diffraction quality of crystals. This was exemplified with crystals of alcohol oxidase, which have a nice polyhedron shape but do not diffract beyond 20 Å (19). Crystals were pre-fixed in phosphate buffer with glutaraldehyde, post-fixed in osmium tetroxide, and stained with uranyl acetate. Electron micrographs have revealed that the crystals have well-ordered regions interrupted by small holes. This account for the low diffracting power of these crystals.

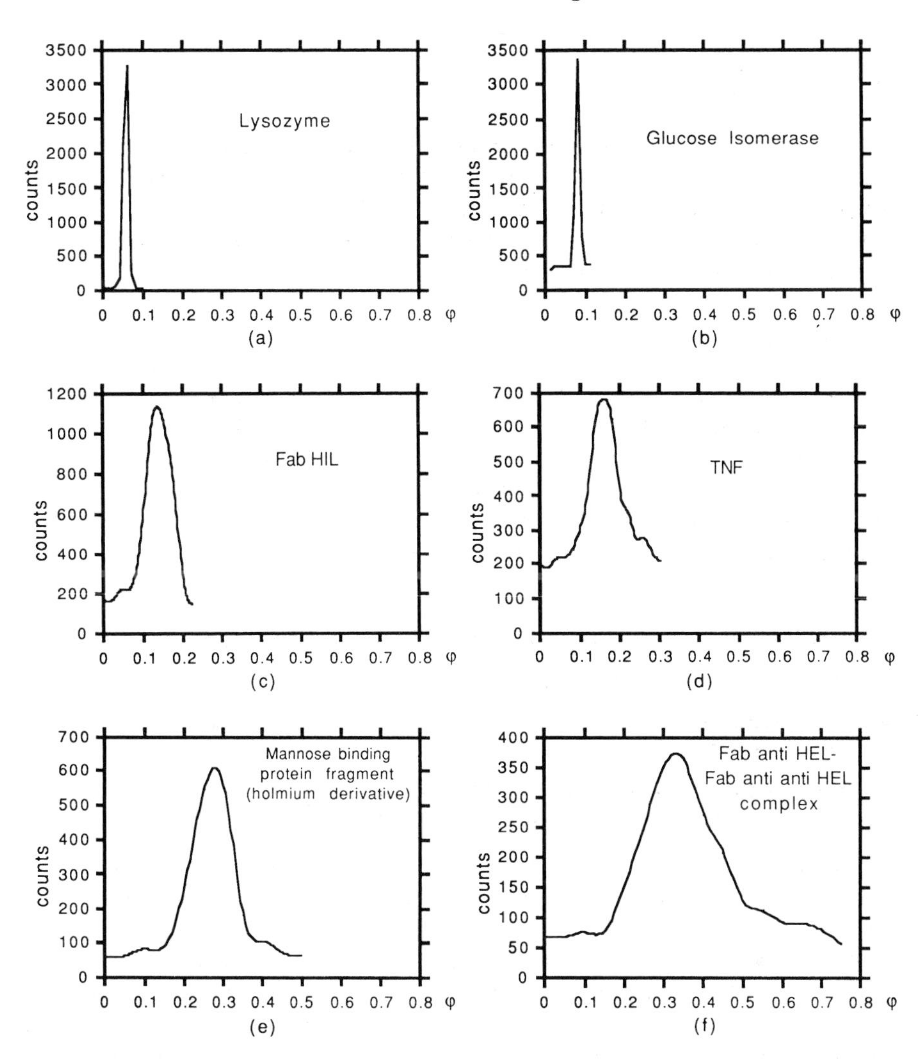

Figure 6. The angular rocking width of a specific (hkl) for various protein crystals, measured with a collimated synchrotron X-ray radiation at D23 LURE station, using MARK II detector. (a) HEW lysozyme; (b) glucose isomerase; (c) Fab HIL; (d) Tumor Necrosis Factor; (e) Mannose binding protein heavy fragment (holmium derivative); (f) Complex of Fab anti-lysozyme–Fab anti anti-lysozyme. Width (φ) is given in degrees. (Courtesy of R. Kahn, LURE Laboratory.)

6. Measuring crystal density

6.1 Background

The knowledge of the crystal density can be of great help in determining the number of protomers in the asymmetric unit (20). Wrong estimation of the solvent content of the unit cell has sometimes yielded erroneous structures. If crystals are available measuring protein density can be performed quickly and requires only basic equipment always available in a biochemistry laboratory.

Typically, 30–80% of the unit cell volume is occupied by solvent. This non-protein fraction of the crystal is composed of two parts; a 'free solvent' in equilibrium with the mother liquor and a 'bound solvent' associated with the protein. This latter solvent is essentially composed of water but ions or precipitating agent molecules are sometimes present. The number n of protomers per unit cell of volume V (cm^3) of a protein crystal can be derived as:

$$n = NV\,(\varrho_c - \varrho_s)/[M\,(1 - v_p\,\varrho_S)] \tag{9}$$

and the solvent fraction content Φ_S computed as:

$$\Phi_S = [(1/v_p) - \varrho_c]/[(1/v_p) - \varrho_s] = 1 - (nv_pM/NV) \tag{10}$$

where N is the Avogadro's number (6.02×10^{23}), ϱ_c and ϱ_s the density of the protein crystal and the solvent (free and bound) (g/cm^3), respectively, M the molar weight of one protomer (g/mol), and v_p the partial specific volume of the unsolvated protein (cm^3/g).

The volume of the unit cell is determined by X-ray measurements, the molar weight of a protomer is usually known, the partial specific volume of the protein can be either measured or computed with the amino acid composition (is often approximated as 0.737 cm^3/g). The densities of the protein crystal and of the solvent are determined experimentally. The number of protomers is then taken as the closest integer. Different methods exist to perform crystal density measurement, one of them is thoroughly described since it seems the most easy to perform with respect to the authors' experience.

6.2 The Ficoll method

This method was introduced by Westbrook (21). It consists of measuring crystal density in an aqueous density gradient where Ficoll (Mol. wt = 400 000) is the only solute. Ficoll is a large polymer of sucrose cross-linked with epichlorhydrin, currently used for separating cells. It can be bought from Pharmacia as Ficoll 400. One of its main advantages is that since the Ficoll solution remains essentially solute free, one can assume that the free solvent compartment of the crystal is only composed of water. Thus the density of the solvent (free and bound) can be approximated as 1 g/cm^3 and the expression of the number of protomers as:

$$n = NV(\varrho_c - 1)/M(1 - v_p) \tag{11}$$

and the solvent fraction content as:

$$\Phi_S = [(1/v_p) - \varrho_c]/[(1/v_p) - 1] \tag{12}$$

This method is simple to handle and has proved very useful. However, it suffers from one main pitfall. It was noticed (22) that in the case of loosely-packed crystals, an overestimation of the density was often made, presumably due to intrusion of Ficoll in the crystal lattice. Indeed, variations in the density were observed as a function of time and it seems that the exact value (calculated once the right X-ray structure was determined) can be obtained by extrapolation back to time zero. Accordingly, density should be always measured as a function of time.

Protocol 1. Experimental determination of crystal density

1. Prepare a series of Ficoll solution (from 30–60% w/w) by mixing appropriate amounts of Ficoll and water. Heat-up to 55 °C with gentle stirring, it takes about 1 h to dissolve.
2. Let them cool down, they turn out as yellowish viscous solutions.
3. Prepare the gradient in tubes of a relatively small inner diameter (about 3/4 mm) so that crystals can be easily spotted. Use a gradient maker if available. If not, layer fractions of decreasing density in a glass tube. Spin the tubes after placing each layer for 5 min at about 1000 *g*. Smooth the gradient by a longer centrifugation (1 h at 3000 *g* seems enough).
4. Calibrate the gradient with drops of a mixture of carbon tetrachloride (chloroform)/toluene (both of which had been saturated with water). For these organic mixtures, the density can be estimated by calibration with salt solutions (e.g. NaH_2PO_4), the density of which are readily measured by weight and volume or with a pycnometer.
5. Introduce the crystal (or two crystals if possible) and a tiny volume of mother liquor at the top of the gradient with a flame-narrowed glass capillary.
6. Centrifuge at 1000 *g* for a certain time (from 1–60 min) and measure the distance from the bottom of the tube to the crystal. Appropriate calibration drops allows one to obtain the value of the density.
7. After having measured the density take X-ray photographs to check that the cell lengths have not changed.

6.3 Organic solvent method

A gradient is made in a glass tube by combining various ratios of water-saturated xylene and carbon tetrachloride (or chloroform) (20). The gradient is calibrated with salt solutions. The crystal is then withdrawn from its mother liquor and placed in the density gradient with a volume of mother liquor as small as possible. Much care should be exercised so as to remove the excess of the mother liquor. This can turn out to be difficult and the crystals might fall apart.

In the case of fragile crystals which cannot stand handling cross-linking with gluteraldehyde might help manipulation. Soaking the crystal for a few hours in the precipitant solution containing 1–2% glutaraldehyde is enough to complete cross-linking. The second method requires a little more handling skills.

7. Concluding remarks

The biochemist tackling protein crystallization for the first time should keep in mind that there is no magic in growing protein crystals but only a great number of parameters to be aware of and to try to control. Most of them were not even recognized until recently. Even if routine assays for defining crystallization conditions are not yet available, we are now heading on the right track.

Acknowledgements

Salut Sally is gratefully acknowledged for her sound comments and M. Walkinshaw and M. Zurini for their constant support. R. Kahn kindly provided *Figure 6*.

References

1. Mullin, J. W. (1972). *Crystallization*. Butterworths, London.
2. Yonath, A., Mussig, J., and Wittmann, H. G. (1982). *J. Cell. Biochem.*, **19**, 145.
3. Gernert, K. M., Smith, R., and Carter, D. C. (1988). *Anal. Biochem.*, **168**, 141.
4. Thomas, D. H., Rob, A., and Rice, D. W. (1989). *Protein Eng.*, **2**, 489.
5. Mikol, V., Rodeau, J.-L., and Giegé, R. (1990). *Anal. Biochem.*, **186**, 332.
6. Thomson, J. A., Schurtenberger, P., Thurston, G. M., and Benedek, G. B. (1987). *Proc. Nat. Acad. Sci. USA*, **84**, 7079.
7. Kam, Z., Shore, H. B., and Feher, G. (1978). *J. Mol. Biol.*, **123**, 535.
8. Sorell, L. S. and Myerson, A. S. (1982). *AIChE Journal*, **28**, 772.
9. Mikol, V., Hirsch, E., and Giegé, R. (1990). *J. Mol. Biol.*, **213**, 187.
10. Durbin, S. D. and Feher, G. (1990). *J. Mol. Biol.*, **212**, 763.
11. Przybylska, M. (1989). *J. Appl. Cryst.*, **22**, 115.
12. Coleman, J. E., Allan, B. J., and Vallee, B. L. (1960). *Science*, **131**, 350.
13. Walker, N., Marcinowski, S., Hillen, H., Maechtle, W., Jones, Y., and Stuart, D. (1990). *J. Crystal Growth*, **100**, 168.
14. Derewenda, U., Derewenda, Z., Dodson, E. J., Reynolds, C. D., Smith, G. D., Sparks, C., and Swenson, D. (1989). *Nature*, **338**, 594.
15. Jolles, P. and Berthou, J. (1972). *FEBS Letters*, **23**, 21.
16. Ataka, M. and Tanaka, S. (1986). *Biopolymers*, **25**, 337.
17. Shaikevitch, A. and Kam, Z. (1981). *Acta Cryst.*, **A37**, 871.
18. Helliwell, J. R. (1988). *J. Crystal Growth*, **90**, 259.

19. Van der Klei, I. J., Lawson, C. I., Rozeboom, H., Dijkstra, B. W., Veenhuis, M., Harder, W., and Hol., W. G. J. (1989). *FEBS Lett*, **244**, 213.
20. Matthews, B. W. (1985). *Methods in Enzymology*, **114**, 176.
21. Westbrook, E. (1985). *Methods in Enzymology*, **114**, 187.
22. Bode, W. and Schirmer, T. (1985). *Biol Chem. Hopper-Seyler*, **366**, 287.
23. Rodeau, J.-L., Mikol, V., Giegé, R., and Lutun, P. (1991). *J. Appl. Cryst.*, **24**, 135.

11

Soaking of crystals

E. A. STURA and P. CHEN

1. Introduction

Once crystals of a macromolecule are obtained there are many circumstances where it is necessary to change the environment in which the macromolecule is bathed. By changing the environment of the crystals it is possible to bind inhibitors, activators, substrates, products, and heavy atoms to the macromolecule, which may have sufficient freedom to undergo some conformational changes in response to these effectors. In fact, macromolecular crystals typically have a high solvent content which ranges from 27–95% (1, 2). Part of this solvent (typically 10%) is tightly associated with the protein matrix as 'bound solvent' (3), which occupies well-defined positions in refined crystal structures and can be considered to be a hydration shell around the protein. This shell consists of both water molecules and other ions present in the solution. The remaining solvent is free to diffuse within the macromolecular reticulum and can be replaced.

The modification of crystals can be made using several techniques. The use and choice of the method will depend on the individual macromolecule. In this chapter we will consider the relative merits of the various methods for modifying crystals of macromolecules. It is important to consider some aspects of the nature of the lattice in which the macromolecule is locked to appreciate which changes can be made by soaking, and when it is necessary to co-crystallize in order to obtain suitable crystals.

2. General concepts

2.1 The crystal lattice

The size and configuration of the channels within the crystalline lattice will determine the maximum size of the solute molecules that may diffuse into the crystal. For most small molecules, solvent channels in the macromolecule crystal are sufficiently large to allow their diffusion to any part of the surface of the macromolecule accessible in solution except for the regions involved in crystal contacts. However, in some cases lattice forces will not permit reactions to occur which would involve conformational changes or rearrangements of the macromolecule in crystal. The disruption of the interactions responsible for intermolecular and crystal contacts may result in the cracking and dissolving of the crystals. In other cases, the forces that drive the conformational changes can be sufficient to overcome the constraints imposed by the crystalline lattice. These lattices may be more flexible and capable of accommodating limited conformational changes. In yet other case crystals may crack initially, then anneal into a new macromolecule conformation and retain their crystalline nature.

Macromolecules can also maintain their activity in the crystalline state. Glycogen phosphorylase, an allosteric enzyme that catalyses the degradation of glycogen into glucose-1-phosphate, is a particularly good example as crystals of this enzyme exhibit a wide range of behaviours when binding ligands. Phosphorylase *b*, crystallized in the presence of the weak activator inosine monophosphate (IMP) under low-salt conditions, is capable of binding substrates as large as maltoheptaose (seven glucose moieties in an α-1,4 linkage) with only local conformational changes (4). The strong activator adenosine monophosphate (AMP), much smaller in size, must be soaked-in slowly by increasing the concentration from 25–100 mM over a period of 2 days (5, 6). However, crystals of phosphorylase *b* crack and become unsuitable for X-ray work when soaked in NADH solutions. In order to achieve this complex in order to determine the binding site for this effector, crystals must first be soaked in a cross-linking reagent (6). Crystals grown in low salt conditions will not accommodate the changes in the transition to the fully active R-state. These consist of large changes in quaternary structure that involve a rotation of two enzyme subunits relative to one another, together with small changes in tertiary structure at the ligand binding sites and at the interface regions (7). In this case the two conformations of the enzyme (R and T) cannot exist in the same lattice, and crystals of these two enzyme states must be grown under different crystallization conditions (8–10).

2.2 Soaking of crystals versus co-crystallization

Complexes of macromolecules can be obtained either by co-crystallizing directly from solution or by soaking pre-formed crystals of the macromolecule in a ligand or reactant solution. Both methods have their own advantages and disadvantages. Soaking of crystals of a macromolecule whose structure has been determined in that crystal form, reduces the complexity of the crystallographic problem to that of determining the positions of the newly introduced atoms by difference Fourier methods. One must ensure that 'true' binding does occur as rearrangements in the macromolecule may be inhibited by the crystal lattice. In co-crystallization, the formation of the complex does not have to contend with lattice forces, but the solubility and the conformation of the complex may be sufficiently different from that of the native molecule to prevent the formation of isomorphous crystals. (For a more complete discussion of the crystallization of complexes see Chapter 5.) Co-crystallization of complexes can, however, yield crystals which are totally isomorphous with the native protein crystals; e.g. the complexes between the anti-progesterone Fab′DB3 and several steroids obtained by co-crystallization, rather than by soaking, are completely isomorphous with native crystals (11, 12). The limited solubility of the steroids was a major consideration in the choice of using co-crystallization of these complexes. The soaking of relatively hydrophobic compounds into crystals can in fact be performed successfully as for poliovirus. The drug binding studies were done by first making a saturated solution of the drug in 10% dimethyl sulfoxide and adding this to native crystals equilibrated in 25% ethylene glycol (which also serves as a cryo-protectant, see Section 3.7). The crystals were left to soak for 7–8 days, depending on the drug (Grant, Filman, and Hogle, personal communication). The long soaking times were necessary because the drug binding pocket in the various strains of poliovirus has been found to be occupied as observed in the crystal structure (13) and the compound in this pocket had to diffuse out before it could be replaced by the drug.

2.3 Changes of buffer

Soaking is not only used to introduce molecules into crystals, but also to change buffers, as may be required in order to change the pH to favour heavy-atom binding or to avoid conditions of incompatibility between certain heavy atoms and the crystallization buffer; e.g. ammonium sulphate is a poor mother liquor for heavy atoms binding at pH's above 6 because of the production of NH_3 which acts as a nucleophile. Ammonium sulphate can be replaced by Na and K phosphate, except when uranium and rare earth compounds are used as heavy atoms as these form insoluble phosphates. Because of their chelating properties, citrate buffers have also to be exchanged before soaking of heavy atoms. Change of water by heavy-water solvent will be required for neutron contrast variation measurements in neutron diffraction studies. Soaking crystals in a cryo-protectant may be necessary to attain the low temperatures required for cryo-crystallography. Crystals which have grown in one buffer system can often be transferred to a different buffer or cryo-protectant without major problems. It is important to remember that the solvent within the crystals does contribute to the diffraction at low resolution and therefore ‘native’ data should be re-collected for these crystals after the buffer change, to differentiate between those changes induced by the buffer and those caused either by low temperature or subsequent modifications to the crystals.

3. Soaking techniques

While in some cases it is possible to transfer crystals directly from the mother liquor in which they have grown to a fresh soak solution, more gradual changes in the crystal soaking, ending with the desired soak solution, can be effective in slowly annealing the crystal into the new conditions. Vapour diffusion can be used to add volatile organic solvents and to increase slowly the salt concentration in the crystals after they have grown, by adding such solvents or further salt to the reservoir. Flow cells, which have been used to change the mother liquor in which the crystals are bathed, such as for the introduction of a substrate, are well described elsewhere (14, 15).

3.1 Soaking of substrates, activators, and inhibitors

Many enzymes remain catalytically active in the crystal (e.g. 16 and 17) and the soaking of substrates will yield a mixture of substrates and products or only products. With the use of synchrotron radiation, the conversion of heptenitol to heptulose-2-phosphate in the presence of inorganic phosphate was followed crystallographically (18). Many analogues of inhibitors, activators, substrates, and products can be used for binding to the enzyme in the crystal for the determination of the binding sites and the mode of binding. Since it is often difficult to determine the binding of substrates, activators, and inhibitors to crystals by different Patterson methods, binding studies are often started only when reliable phases for the native crystals have been obtained. However, preliminary observations of the development of hair-line cracks on soaking crystals can be a good indication of binding as long as this is compared to a control experiment in which the native crystals are handled in the same fashion with the same buffer. Furthermore, it is important to develop a methodology for soaking that takes into account the possible dehydration of crystals while handling and avoids sudden shocks regarding pH and salt concentration. If soaking methods fail, co-crystallization of ligand and macromolecule may be the only option.

3.2 Soaking techniques for sitting drops

When sitting drops are used for crystallization (see Chapter 4 and reference 19), soaking experiments (see *Figure 1*) can follow *Protocol 1*. This protocol differs from that used for hanging drops since the evaporation from the reservoir solution is able to slow down sufficiently the evaporation from the protein drop containing the crystals while the procedure is carried out. The work is carried out under a dissecting microscope using a magnification of 10–100 fold.

Protocol 1. Soaking of crystals grown in sitting drops

1. Connect a glass or quartz capillary tube to a 1 ml glass syringe with a short piece of rubber tubing such as c-flex (Fisher #14–169–5c) which gives an excellent seal.

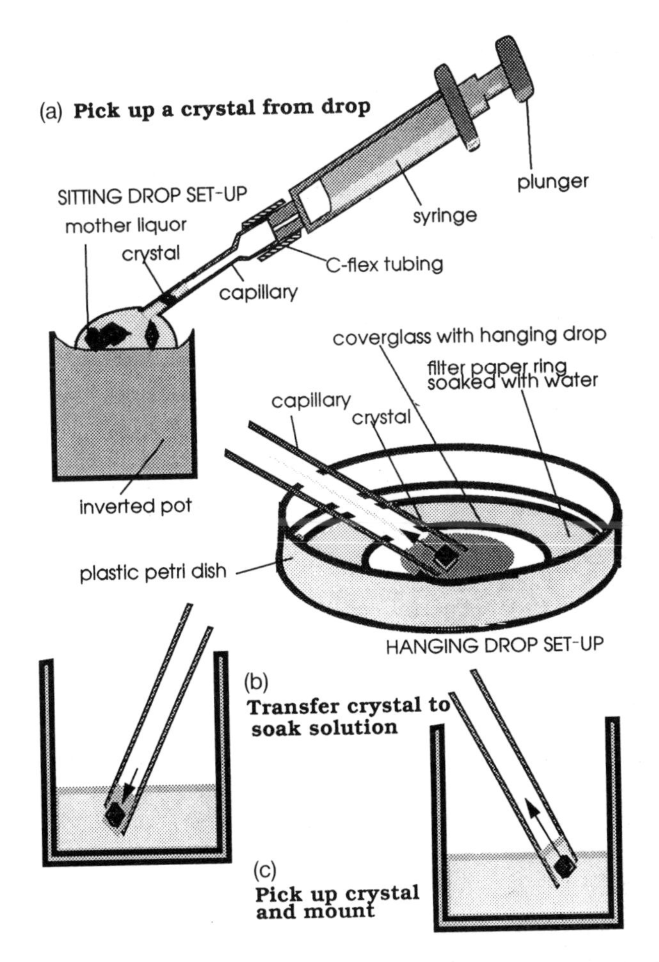

Figure 1. Schematic drawing of the transfer of crystals from either a sitting drop set-up, or a hanging drop set-up, to a well for soaking. (a) Crystals can be transferred directly from a sitting drop to a capillary because the vapour from the reservoir solution (not shown) protects the drop containing the crystal from dehydration. For a hanging drop vapour diffusion experiment the coverslip is placed at the centre of a plastic petri dish within a ring of filter paper soaked in water. Evaporation of water from the filter paper will ensure that the drop does not dry-out while the crystal is picked up into the capillary connected to the syringe. (b) The crystal is transferred into a small container containing the soak solution. (c) After soaking the crystal is removed from the soak solution for mounting. The walls of the container used for soaking the crystal should not be high, as this will restrict the angle at which the crystal can be picked up from the soak solution into the capillary, as further restrictions are also imposed by the dissecting microscope, which also limits the working angle.

Protocol 2. *Continued*

2. Snap open the end of the capillary with tweezers or scissors. The glass capillary may be siliconized if the experimental situation can benefit from a diminished adhesion of the solution to the glass wall such as when viscous solutions are handled. After siliconizing it should be extensively washed.
3. To increase the volume available for handling crystals, mother liquor (20–50 μl) can be added to the drop (some crystals require the mother liquor to contain protein for stability). If the crystals adhere to the well, withdraw liquid from the drop and gently eject it onto a chosen crystal. It is often possible to dislodge the crystal. Unfortunately, some crystals have severe adhesion problems and cannot be dislodged without breaking them. For such problematic crystals, glass pots with depressions coated with a thin film of Corning vacuum silicone grease should be used in the original crystallization set-up.
4. Pick up crystals into the capillary by pulling back the plunger of the syringe.
5. Transfer crystals from the syringe directly into a soak solution from which they are later picked up and mounted for X-ray studies.

3.3 Soaking techniques for hanging drops

Crystals grown by hanging drop (see Chapter 4) can be flushed with mother liquor from the coverslip into a larger container and then picked up as described in *Protocol 1* for sitting drops. Since it is difficult to find small crystals in a large container, the method described in *Protocol 2* may be preferable in such cases.

Protocol 2. Soaking crystals grown by hanging drops

1. Cut a circular piece of filter paper to fit a petri dish 4–5cm inside diameter.
2. Cut out a small circle from the centre of the filter paper such that the coverglass from the hanging drop can fit inside this without touching the paper.
3. Soak the filter paper with distilled water.
4. Place the coverglass with the hanging drop in the centre. Mother liquor is added to the drop (20–50 μl) and the crystals for soaking can be picked up and soaked as described in steps 2–6 of *Protocol 1*.

This set-up has been used for the stable transportation of crystals to synchrotron facilities, by soaking the filter paper with mother liquor instead of water.

3.4 Soaking of crystals in capillaries

Once the crystals have been introduced into the thin glass capillary, using either of the above two procedures, the mother liquor can be removed by allowing the crystal to adhere to the capillary wall and pushing the mother liquor out of the capillary onto a piece of absorbent paper, while the crystal remain *in situ* because of surface tension. Crystals that do not adhere to the capillary wall can be stopped from flowing with the mother liquor by wedging a hair against the crystal while the solution is removed (*Figure 2*).

Alternatively, the solution can be removed with a thin glass capillary tube (0.1–0.05 mm outside diameter) or a thin strip of filter paper. Once the mother liquor is removed, with the syringe still connected to the capillary tube, soaking is performed following *Protocol 3*.

Protocol 3. Soaking crystals in capillaries

1. Suck the new solution into the capillary, fully immersing the crystal.
2. Add paraffin oil to the open end of the capillary, leaving an air gap between the oil and the soak solution, for the duration of the soak.
3. After the soak period has elapsed, remove the oil and the solution.
4. Remove the excess solution around the crystal with filter paper.
5. Add either soak solution and/or oil to the open end of the capillary to maintain a moist environment for the crystal.
6. Seal with wax while still attached to the syringe. A wet strip of filter paper (5 mm wide) can be placed on the outside of the capillary to keep the crystal cool while the ends are sealed. The crystal is now mounted for X-ray diffraction work. Other techniques for mounting crystals can be found in Chapter 12 and reference (3).

This technique is particularly important for soaking compounds which are available only in limited quantities. Crystals soaked in capillaries were used in the work with phosphorylase *b* and heptenitol (20).

After data collection, capillaries may be opened and a soak solution added to the crystal with a Hamilton syringe, and the capillary sealed for the duration of the soak with paraffin oil. The oil and the soak solution are then removed and the crystal used for further X-ray studies such as for collecting an inhibitor complex data set after the native protein data have been measured, if the crystal has survived the damage from the first irradiation.

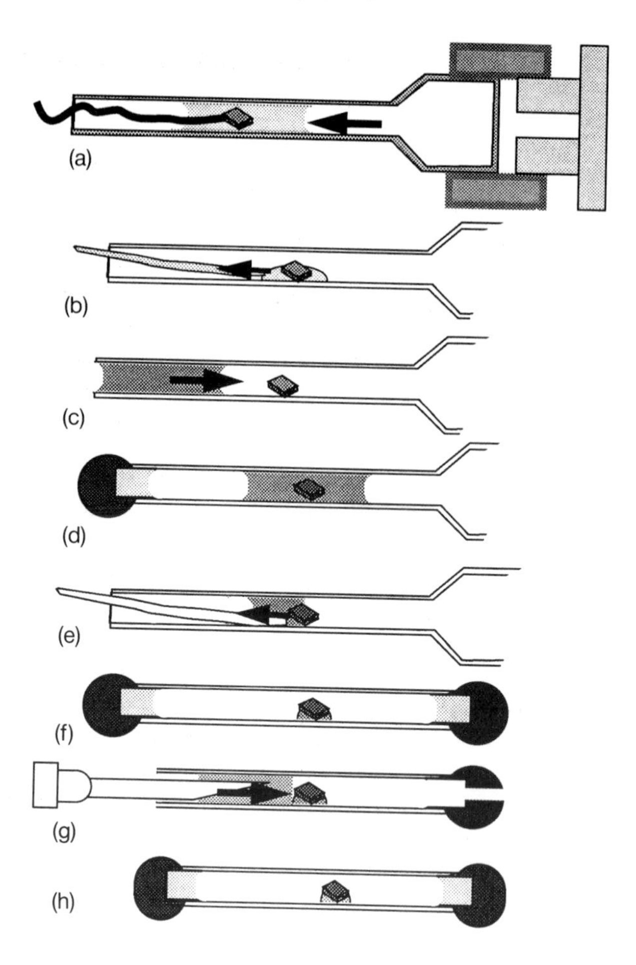

Figure 2. Schematic representation of the various stages involved in capillary soaks. (a) After the crystal is picked up into the capillary (*Figure 1a*) the mother liquor is removed from around the crystal by pushing the liquid out, while holding the crystal in position with a hair (in many cases the surface tension between the crystal and the capillary is sufficient to hold the crystal in place). (b) With a thin strip of filter paper, taking care not to touch the crystal, the excess liquid is removed. (c) The soak solution is drawn into the capillary to bathe the crystal. (d) Paraffin oil or buffer is added to the open end and the capillary is sealed with molten wax. (e) After soaking the capillary can be snapped open with thin-nosed forceps and the soak solution is removed using a thin piece of filter paper. (f) The capillary can now be sealed at both ends and the crystal used for X-ray diffraction studies. (g) Crystals that have been used for X-ray work can be soaked by snapping-off one end of the sealed capillary with forceps and opening the other end with a hot needle. A piece of wet filter paper is placed over the crystal to prevent the crystal from warming-up during this procedure. A solution is then introduced at the broken end of the capillary so that it bathes the crystal. Petroleum jelly is used to seal the experiment as it is easy to remove prior to resealing the thin-walled capillary tube with wax. (h) The soaked crystal can be used for X-ray diffraction analysis.

3.5 Soaking of crystals with cross-linking reagents

The use of gluteraldehyde to stabilize crystals is well documented from the early work on carboxypeptidase-A (21), where it was shown that cross-linked crystals were resilient to changes in mother liquor. The cell dimensions of the cross-linked crystals were shown to remain relatively constant, under a variety of low and high salt conditions as well as extremes of pH, from 5–11. Such cross-linked crystals also retained catalytic activity. Further examples, such as the cross-linking of phosphorylase *a*, with 0.03% gluteraldehyde for 1 h, also indicated that the reagent produces little change in the diffraction pattern of the cross-linked crystals to a resolution of 5.5 Å, while maintaining crystal integrity even after major conformational changes (17).

The use of bifunctional diimidates, which cross-link between the ε-amino groups of lysine residues (22, 23), provides a means of restricting the length of the cross-links that may be formed. Dombraãdi *et al.* (24) used diimidates of different length from the shortest, dimethyl malonic diimidate with an maximal effective length of 3.7 Å, to the longest, dimethyl dodecanoic diimidate spanning up to 14.5 Å. The use of such compound introduces a degree of selectability into cross-linking since the reacting groups must be within the maximal distance of the diimidate. Dimethyl malonic diimidate was used to stabilize crystals of glycogen phosphorylase against cracking in order to determine the mode of binding of the inhibitor glucose-6-phosphate (25). In the above study, a 2 mg/ml solution of dimethyl malonic diimidate in 0.1 M triethanolamine/HCl, 10 mM magnesium acetate, pH 7.8, was used to cross-link crystals for 2 h before the reaction was stopped by lowering the pH to 7.1.

3.6 Heavy-atom soaking

A common reason for the soaking of crystals of macromolecules is to pursue phase determination by the method of isomorphous replacement. This method has been central to X-ray analysis of protein crystals from the initial work on haemoglobin (26). The procedure is based on the introduction of a single or a limited number of heavy atoms per macromolecule, as an addition or replacement of an endogenous atom, without disrupting or significantly altering the crystal lattice. Because of the substantially greater atomic number of the heavy atom, this addition of electrons in the structure causes significant changes in X-ray recorded intensities. Such changes are then interpreted to estimated 'phases' which lead to the solution of the structure. Such modifications to crystals of macromolecules are normally carried out by soaking the native crystals in the mother liquor containing the heavy-atom compound. Although obtaining a good heavy-atom derivative is a trial and error process, there are general considerations which give the best chance of success.

It is clear that the preparation of isomorphous derivative crystals will depend on the pH, composition of the mother liquor, and temperature. Many successful pH values for heavy-atom soaking are 6–8. If the pH value is below 6, most reactive groups which could bind the metal atom will be protonated and blocked. Since many heavy-atom compounds are alkaline labile, at high pH they may form insoluble hydroxides. Except at low pH (i.e below 6), ammonium sulphate is a poor mother liquor for heavy-atom binding due to the production of the good nucleophile NH_3 (27). If possible, the crystals should be transferred to Mg or Na sulphate, or Na and K phosphate. However, an excess of phosphate is undesirable for uranium and rare earth binding. The temperature can change the rate of reaction and sometimes the degree of binding. In most cases the soaking temperature will be the same as crystallization temperature.

Other important considerations are the heavy-atom concentration and soaking time. The necessary concentration will depend on the solubility of heavy-metal compound. 1–2 mM could be chosen for the starting value. The soaking time can vary from several hours to months, but for the initial screening, 1–2 days is sufficient. If the crystal appearance is unchanged from the native crystal, and no changes are found from the diffraction pattern, increasing the soaking concentration and time may help. On the other hand, if the crystal cracks, the diffraction pattern changes too much, or diffraction resolution dramatically decreases, a lower concentration and shorter soaking time should be tried. In the Bowman-Birk inhibitor case (28), if the soaking concentration was any greater than 1 mM, the crystal showed no X-ray diffraction, even though the crystal remained intact. By lowering the concentration to 0.1 mM, the crystal diffracted as well as native crystals. Sometimes, back-soaking is necessary in order to reduce the cell changes to acceptable values (29). For back-soaking, after the initial soak the crystals are transferred to solutions with a lower concentration of heavy atoms, or to the original mother liquor, in order to reduce binding to the lower-affinity sites. Another useful criteria for evaluating the soaking conditions is the relative temperature factor of the derivative data compared to the native. When the derivative data have significantly larger temperature factors than the native it is likely that some disorder has been introduced into the derivative crystal. A reasonable temperature factor can often be achieved by lowering the concentration of the heavy-atom compound.

Since many heavy-atom compounds have a very vigorous photochemistry, soaking should be carried out under low power illumination or in the dark. Finally, freshly prepared soaking solutions should be used whenever possible. It is frequently the case that many heavy atoms (about 20–50) need to be tried before a good isomorphous derivative is found. There is no substitute for patience and hard work at this stage.

3.7 Soaking for cryo-crystallography

Cryo-crystallography provides a means of increasing the lifetime of some protein crystals by reducing radiation damage during data collection (30, 31) and allowing a complete data set to be collected from one crystal, often to the same resolution as crystals analysed at room temperature (32). It has been suggested that cooling may increase the internal order of parts of the protein which are mobile at room temperature (33), and also provide ways to observe enzyme–substrate complexes and unstable intermediates (34), with the possibility of collecting data on such intermediates with the use of Laue X-ray photography and synchrotron radiation (35).

The use of several cryo-protective solvents and combinations of such solvents is well described by Petsko (33). The most commonly used cryo-protectants are ethylene glycol, sometimes in combination with PEG to increase the stability of crystals, MPD (also commonly used in crystallization), and ethanol. Soaking of crystals for this application does not vary substantially from that used for soaking other compounds. The cryo-solvent is added to the mother liquor in small steps, slowly increasing its concentration to typically more than 50%. In some cases crystals are transferred to oil prior to freezing them (36). All such techniques are aimed at preserving crystal integrity by obtaining a transition from water to vitreous ice, preventing crystallization of the mother liquor.

Careful consideration must be given to the soaking of cryo-protectants and the freezing process, as it may affect crystals in many ways, leading to changes in cell dimensions and increase in mosaicity. The cryo-crystallographic work on complex co-crystals of an antipeptide Fab (17/9) with a nine-residue peptide, exemplifies some of the above considerations. The use of cryo-crystallography allowed the collection of data on extremely thin crystals, which would not have been possible otherwise. This was accomplished by immersing crystals of this Fab–peptide complex in 35% ethylene glycol solution for a short time (about 5 min) before flash freezing them under a stream of cold nitrogen at −150 °C (Rini, personal communication). Cryo-crystallography of macromolecules is a relatively new technique (36), and not yet routinely used. Problems still occur mainly because of the formation of ice around or near the crystal. Progress to reduce the moisture around the crystal can be facilitated such as by the use of an enclosure inside which the level of humidity can be reduced to less than 1% (Rini, Stura, and Wilson, unpublished results).

4. Conclusions

Soaking is the most common method used to obtain heavy-atom derivatives, although crystallization of previously modified proteins either chemically or biologically have also been used (37, 38). Soaking and co-crystallization are two different approaches to achieving complexes of macromolecules. The two procedures are both alternative and complementry to each other. By soaking effectors into pre-formed crystals it is possible to analyse the structure of complexes, only if crystal lattice constraints permit. The problem of cracking of the crystals, which may occur both with binding effectors (25) and heavy atoms (39), can be often resolved by the use of cross-linking agents. One must understand, however, that complexes obtained by soaking may differ from complexes obtained by co-crystallization (Ringe, personal communication). Flow cells, in which a constant supply of substrate is supplied to the enyme in the crystal and product is washed away, may answer the problem in cases where the rate of product formation is significantly slower than the rate of diffusion through the crystal (15, 40).

Acknowledgements

We would like to thank Dr Ian Wilson (I.A.W.) for helpful discussion and Drs James Rini, James Hogle, David Filman, and Robert Grant for unpublished results. E.A.S. would like to thank Prof Louise Johnson and Dr Keith Wilson for instruction in some of the techniques described here, and Dr Janos Hajdu for many discussions on methodology. This work was supported by National Institutes of Health Grants AI–23498, GM–38794, and GM–38419 (to I.A.W.). This is publication #6745-MB from The Scripps Research Institute.

References

1. Matthews, B. W. (1968). *J. Mol. Biol.*, **33**, 419.
2. Cohen, C., Caspar, D. L. D., Parry, D. A. D., and Lucas, R. M. (1971). *Cold Spring Harbor Symp. Quant. Biol.*, **36**, 205.
3. Blundell, T. L. and Johnson, L. N. (1976). *Protein crystallography*. Academic Press, New York.
4. Johnson, L. N., Stura, E. A., Sansom, M. S. P., and Babu, Y. S. (1983). *Biochem. Soc. Trans.*, **11**, 142.
5. Johnson, L. N., Stura, E. A., Wilson, K. S., Sansom, M. S. P., and Weber, I. T. (1979). *J. Mol. Biol.*, **134**, 639.
6. Stura, E. A., Zanotti, G., Babu, Y. S., Sansom, M. S. P., Stuart, D. I., Wilson, K. S., Johnson, L. N., and Van de Werve, L. (1983). *J. Mol. Biol.*, **170**, 529.
7. Barford, D. and Johnson, L. N. (1989). *Nature*, **340**, 609.
8. Johnson, L. N., Madsen, N. B., Mosley, J., and Wilson, K. S. (1974). *J. Mol. Biol.*, **90**, 703.
9. Fasold, H., Ortlandel, F., Huber, R., Bartels, K., and Schwager, P. (1972). *FEBS Lett.*, **21**, 229.

10. Madsen, N. B., Honikel, K. O., and James, M. N. G. (1972). In *Metabolic interconversion of enzymes*, (ed. Wieland, O., Heilmreich, E., and Holzer, H.), p. 448. Springer Verlag, Berlin.
11. Stura, E. A., Feinstein, A., and Wilson, I. A. (1987). *J. Mol. Biol.*, **193**, 229.
12. Stura, E. A., Arevalo, J. H., Feinstein, A., Heap, R. B., Taussig, M. J., and Wilson, I. A. (1987). *Immunology*, **62**, 511.
13. Filman, D. J., Syed, R., Chow, M., Macadam, A. J., Minor, P. D., and Hogle, J. M. (1989). *EMBO J.*, **8**, 1567.
14. Wyckoff, H. W., Doscher, M. S., Tsernoglou, D., Inagami, T., Johnson, L. N., Hardman, K. D., Allewell, N. M, Kelley, D. M., and Richards, F. M. (1967). *J. Mol. Biol.*, **127**, 563.
15. Petsko, G. A. (1985). *Methods in Enzymology*, **114**, 141.
16. Remington, S., Wiegand, G., and Huber, R. (1982). *J. Mol. Biol.*, **158**, 111.
17. Kasvinsky, P. J. and Madsen, N. B. (1977). *J. Biol Chem.*, **251**, 6852.
18. Hadju, J., Acharya, K. R., Stuart, D. I., McLaughlin, P. J., Barford, D., Oikonomakos, N. G., Klein, H. W., and Johnson, L. N. (1987). *EMBO J.*, **6**, 539.
19. McPherson, A. (1982). *Preparation and analysis of protein crystals*. Wiley, New York.
20. McLaughlin, P. J., Stuart, D. I., Klein, H. W., Oikonomakos, N. G. and Johnson, L. N. (1984). *Biochemistry*, **23**, 5862.
21. Quicho, F. A. and Richards, F. M. (1964). *Proc. Natl. Acad. Sci. USA*, **52**, 833.
22. Hunter, M. J. and Ludwig, M. L. (1962). *J. Am. Chem. Soc.*, **84**, 3491.
23. Browne, D. T. and Kent, S. B. H. (1975). *Biochem. Biophys. Res. Commun.*, **67**, 126.
24. Dombraădi, V., Hadju, J., Bot, G., and Friedrich, P. (1980). *Biochemistry*, **19**, 2295.
25. Lorek, A., Wilson, K. S., Sansom, M. S. P., Stuart, D. I., Stura, E. A., Jenkins, J. A., Zanotti, G., Hadju, J., and Johnson, L. N. (1984). *Biochem. J.*, **218**, 45.
26. Green, D. W., Ingram, V. M., and Perutz, M. F. (1954). *Proc. Roy. Soc.*, **225**, 287.
27. Sigler, P. B. and Blow, D. M. (1965). *J. Mol. Biol.*, **14**, 640.
28. Chen, P., Rose, J., Love, R., Wei, C. H., and Wang, B. C. (1991). submitted.
29. Mondragon, A., Wolberg, C., and Harrison, S. C. (1989). *J. Mol. Biol.*, **205**, 179.
30. Lipscomb, W. N., Hartsuch, J. A., Reeke, Jr., G. M., Quiocho, F. A., Bethge, P. H., Ludwig, M. L., Steitz, T. A., Muirhead, H., and Coppola, J. C. (1968). *Brookhaven Symp. Biol.*, **21**, 24.
31. Tsernoglou, D., Hill, E., and Banaszack, L. J. (1972). *J. Mol. Biol.*, **69**, 75.
32. Hope, H., Frolow, F., von Böhlen, K., Makowski, I., Kratky, C., Halfon, Y., Danz, H., Webster, P., Bartels, K. S., Wittmann, H. G., and Yonath, A. (1989). *Acta Cryst.*, **45**, 190.
33. Petsko, G. A. (1975). *J. Mol. Biol.*, **96**, 381.
34. Douzou, P., Hui Bon Hoa, G., and Petsko, G. A. (1975). *J. Mol. Biol.*, **96**, 367.
35. Hadju, J., Machin, P. A., Campbell, J. W., Greenhough, T. J., Clifton, I. J., Zurech, S., Gover, S., Johnson, L. N., and Elder, M. (1987). *Nature*, **329**, 176.
36. Hope, H. (1988). *Acta Cryst.*, **44**, 22.
37. Stura, E. A., Johnson, D. L., Inglese, J., Smith, J. M., Benkovic, S. J., and Wilson, I. A. (1989). *J. Biol. Chem.*, **264**, 9703.

38. Yang, W., Hendrickson, W. A., Crouch, R. J., and Satow, Y. (1990). *Science*, **249**, 1398.
39. Ringe, D., Petsko, G. A., Yakamura, F., Suzuki, K., and Ohmori D. (1983). *Proc. Natl. Acad. Sci. USA*, **80**, 3879.
40. Farber, G. K., Glasfeld, A., Tiraby, G., Ringe, D., and Petsko, G. A. (1989). *Biochemistry*, **28**, 7289.

12

X-ray analysis

L. SAWYER and M. A. TURNER

1. Introduction

This chapter covers the preliminary characterization of the crystals in order to determine if they are suitable for a full structure determination. Probably more frustrating than the failure to produce crystals at all, is the growth of beautiful crystals which do not diffract, which have very large unit cell dimensions, or which decay very rapidly in the X-ray beam.

It is impossible in one brief chapter to give more than a flavour of what the X-ray crystallographic technique entails, and it is assumed that the protein chemist growing the crystals will have contact with a protein crystallographer who will carry out the actual structure determination. However, preliminary characterization can often be carried out with little more than the equipment which is widely available in Chemistry or Physics Departments. If such is the case check that Cu radiation is available, and in the following discussion it is assumed that help from someone with crystallographic experience will be available. Thus, the crystal grower, remote from a protein crystallography laboratory, can readily monitor the success of his experiments. It must be remembered that X-rays are dangerous and the inexperienced should *not* try to X-ray protein crystals without help.

2. Background X-ray crystallography

It is necessary to provide an overview of X-ray crystallography, to put the preliminary characterization in context. For a general description of the technique the reader should refer to Glusker and Trueblood (1) or Stout and Jensen (2): for protein crystallography in particular, Volumes 114 and 115 of *Methods in Enzymology* (3) describe many of the advances since the seminal work of Blundell and Johnson (4). Amongst many excellent introductory articles, those by Bragg (5) and Harding (6) are particularly recommended.

2.1 X-rays

2.1.1 Why use X-rays

The scattering or diffraction of X-rays is an interference phenomenon and the interference between the X-rays scattered from the atoms in the structure produces significant changes in the diffraction observed in different directions. This variation in intensity with direction arises because the path differences taken by the scattered X-ray beams are of the same magnitude as the separation of the atoms in the molecule. Put another way, to 'see' the individual atoms in a structure, it is necessary to use radiation of a similar wavelength to the interatomic distances, typically 0.15 nm or 1.5 Å, and radiation of that wavelength lies in the X-ray region of the electromagnetic spectrum. It is also important to realize that it is the *electrons* which scatter the X-rays, and so what is in fact observed is the *electron density* of the sample. Because the electrons cluster round the atomic nuclei, regions of high electron density correspond to the atomic positions.

2.1.2 X-ray sources

X-rays are produced in the laboratory by accelerating a beam of electrons into an anode, the metal of which dictates what the wavelength of the resulting X-rays will be. Monochromatization is carried out either by using a thin metal foil which absorbs much of the unwanted radiation or, better, by using the intense low-order diffraction from a graphite crystal. To obtain a brighter source, the anode, which is water-cooled to prevent it melting, can be made to revolve in what is known as a rotating anode generator. For most work with proteins, the target is copper and the characteristic wavelength of the radiation is 0.1542 nm (1.542 Å).

An alternative source of X-radiation is obtained when a beam of electrons is bent by a magnet. This is the principle behind the synchrotron radiation sources which are capable of producing X-ray beams some thousand times more intense than a rotating anode generator (7). A consequence of this high-intensity radiation source is that data collection times have been drastically reduced, making kinetic crystallography feasible (8). A further advantage is that the X-ray spectrum is continuous from around 0.05–0.3 nm, dependent upon the particular machine, and this has distinct advantages for the crystallographer. The use of shorter wavelengths has usually been found to prolong the room temperature lifetime of a crystal in the X-ray beam. The main drawback is that synchrotrons are centralized facilities and consequently access is significantly less convenient, particularly for preliminary work.

2.2 What is a crystal?

A crystal is a regular, repeating array of atoms or molecules in three dimensions. It is convenient to describe such an object with the aid of a lattice, which is a geometric construction defined by 3 axes and the 3 angles between them. Along each axis direction a point will repeat at a distance referred to as the unit translation or unit cell repeat, and labelled a, b, and c, respectively. The angles between b and c, a and c, and a and b are α, β, and γ, respectively. The basic building block of a crystal, then, is a parallelepiped described by the dimensions a, b, and c, and α, β, and γ and called the *unit cell.*

There are 7 crystal systems which arise from the only possible combinations of these unit cell parameters. However, it is sometimes easier to consider a larger unit cell but with a simpler shape, e.g. with mutually perpendicular axes. This choice can be illustrated in the two-dimensional example shown in *Figure 1*. The choice of the basic building block containing a single 'molecule' can be made in a variety of ways because the lattice is no more than a geometrical construction affording a convenient description of the repeating figure. Crystallographers adopt the convention that the unit cell which is chosen is the one with angles nearest to 90 °. Such a cell with only one copy of the molecular structure is called *primitive* but, as noted above, a more convenient cell may have 2 or even 4 copies (see *Figure 1*, where the non-primitive, centred cell is at the right). There are 14 so-called Bravais lattices which can be constructed in 3 dimensions (there are 5 in 2-D). As an example of the limited number of lattices, construct a centred square lattice and it is evident that a smaller, primitive square lattice is also present.

Although the basic building block of a crystal is the unit cell and the lattice produced by its repetition has a characteristic symmetry (see *Table 1*), within the unit cell there may be further symmetry, e.g. the molecule itself may have symmetry about an axis which is either a proper rotation of 360 °, 180 °, 120 °, 90 °, or 60 ° only, or an improper one which involves 'inversion' through the point. Both of these can be illustrated with a molecule like methane. A 3-fold rotation axis (120 °) is evident when the molecule is viewed along an H–C

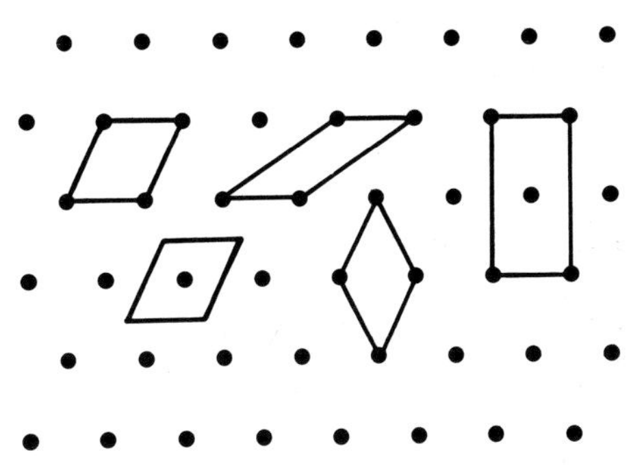

Figure 1. Each of the unit cells shown in this 2-dimensional example is a valid choice for the lattice of points. The cell on the right is a centred cell and has twice the contents of the others.

bond whereas a 4-fold improper rotation axis bisects an H–C–H angle in the plane of the other H–C–H so that a 90 ° rotation of one hydrogen about this axis brings it to a point on the opposite side of the C atom to an adjacent H atom. Proteins are made up of L-amino acids, and nucleic acids have a chiral ribose unit which precludes centres or mirrors. The combination of these symmetries and the crystal systems leads to the 32 *point groups* or *crystal classes*, of which only 11 can accommodate protein molecules. The rotations referred to above are the only ones allowed in the formation of a crystal, but of course other rotations about a point within the molecule are possible, as in the case of a spherical virus which has 532 point group symmetry. Only the 3-fold and 2-fold axes can be exploited in building up the crystal, leaving the 5-fold axis as a *non-crystallographic* symmetry element.

As well as the rotational symmetry possibly present in a unit cell, translational relationships between molecules exist. The spatial repetition of a crystal is such that a convenient packing involves axes which combine a rotation with a translation; e.g. if a rotation of 180 ° together with a translation of half a unit cell along the axis of rotation is applied twice, it will produce not the initial molecule (as with a pure rotation) but an equivalent one in the next cell. Such an axis is a *screw axis* and several consistent types exist.

It can be shown mathematically that there are only 230 combinations of these symmetry elements possible in 3-dimensions. Thus, any crystal must have a unit cell which conforms to one of these combinations, its *space group*. Further, the presence of symmetry elements within a unit cell means that there are at least two copies of the molecule which are related by an algebraic relationship: if there is an atom at position x, y, z in a cell with a 3-fold screw axis parallel to the b-axis, there must be an atom at $-x$, $1/2+y$, $-z$. The effect of this is to reduce the crystallographer's problem to one of locating the atoms in the *asymmetric unit*, rather than in the whole unit cell. Because all proteins and nucleic acid crystals comprise only one optical isomer, there are only 65 space groups available for such chiral molecules. *Table 1* shows the available crystal systems, classes, and space groups for a protein.

2.3 How do X-rays interact with crystals?

The explanation of how X-rays are scattered by crystals is largely the result of a beautiful simplification by Bragg, resulting in the law which bears his name. Consider a crystal lattice, represented in *Figure 2* by the rows of points A, B, C. For X-rays, X_2 scattered from row 2 to enhance those scattered from row 1, X_1, there must be an integral number of wavelengths difference. The relationship between the spacing of the rows, d, the wavelength, λ, and the angle at which the emergent ray is observed relative to the direction of the rows is:

$$n.\lambda = 2d.\sin\theta \tag{1}$$

Table 1. The crystal systems and related data for a chiral molecule

System	Necessary cell parameters	Bravais lattice	Class	Number	Available space groups	Multiplicity
Triclinic	$a,b,c,\ \alpha,\beta,\gamma$	P	1	1	P1	1
Monoclinic	$a,b,c,\ \beta$	P	2	3–4	P2, $P2_1$	2
	$(\alpha=\gamma=90\,°)$	C		5	C2	4
Orthorhombic	a,b,c	P	222	16–19	P222, $P222_1$, $P2_12_12_1$, $P2_12_12$	4
	$(\alpha=\beta=\gamma=90\,°)$	C		20–21	C222, $C222_1$	8
		F		22	F222	16
		I		23–24	I222, $I2_12_12_1$	8
Tetragonal	$a(=b),c$	P	4	75–78	P4, $P4_1$, $P4_2$, $P4_3$	4
	$(\alpha=\beta=\gamma=90\,°)$	I		79–80	I4, $I4_1$	8
		P	422	89–96	P422, $P42_12$, $P4_122$, $P4_12_12$, $P4_222$, $P4_22_12$, $P4_322$, $P4_32_12$	8
		I		97–98	I422, $I4_122$	16
Trigonal	$a(=b),\ c,\ \gamma=120\,°$	P	3	143–145	P3, $P3_1$, $P3_2$	3
	$(\alpha=\beta=90\,°)$	R	3	146	R3	3
		P	312	149,151,153	P312, $P3_112$, $P3_212$	6
		P	321	150,152,154	P321, $P3_121$, $P3_221$	6
	$a(=b=c),\ \alpha=\beta=\gamma\neq 90\,°$	R	32	155	R32	6
Hexagonal	$a(=b),\ c,\ \gamma=120\,°$	P	6	168–173	P6, $P6_1$, $P6_2$, $P6_3$, $P6_4$, $P6_5$	6
	$(\alpha=\beta=90\,°)$		622	177–182	P622, $P6_122$, $P6_222$, $P6_322$, $P6_422$, $P6_522$	12
Cubic	$a(=b=c)$	P	23	195, 198	P23, $P2_13$	12
	$(\alpha=\beta=\gamma=90\,°)$	F		196	F23	48
		I		197	I23	24
		P	432	207–8,212–3	P432, $P4_232$, $P4_332$, $P4_132$	24
		F		209–210	F432, $F4_132$	96
		I		211,214	I432, $I4_132$	48

The lattice types are P (= primitive), C (= C-face centred), F (= all faces centred), and I (= body centred). Alternative lattice types may occasionally be chosen. The symbols under Class refer to the rotational symmetry axes which are a characteristic of it. The Herman–Mauguin nomenclature for space groups gives the lattice type first, then the symmetry elements in an order which depends upon the crystal system. *Refer to International tables for X-ray crystallography*, Volume A, for a fuller explanation of these symbols. Number refers to the number in International Tables. Multiplicity gives the number of copies of the asymmetric unit in the unit cell.

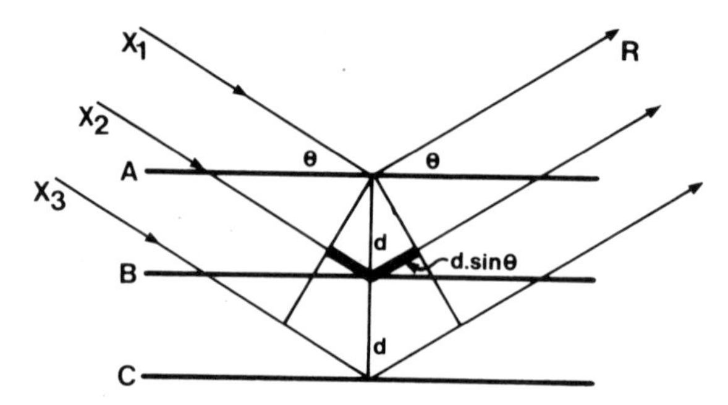

Figure 2. X-rays (X_1, X_2, X_3) reflected from lattice planes A, B, C. To observe a scattered beam of X-rays in direction R, the thickened path must equal a whole number of wavelengths. The ray from plane C travels twice as far as that from B, and so on.

Thus, as Bragg pointed out, X-ray diffraction can be regarded as the *reflection* of the beam of X-rays from the planes of points in the crystal lattice. Provided there are a large number of planes contributing to the interference, the position in space at which a given reflection is observed is highly defined. These positions are defined by the crystal lattice and since very few, if any, atoms actually lie on the lattice points, the scattered intensity is modulated by the atomic arrangement within the unit cell. To repeat, the direction of a diffracted ray is defined by the crystal lattice, the intensity of the ray depends upon the atomic arrangement within the unit cell. One further point concerns n, the order of diffraction, which is the number of wavelengths difference between the scattering from adjacent planes; the higher the order, the larger the angle of scattering. Alternatively, the scattering can be considered as arising from planes which are closer together: e.g. using the equation above, it can be seen that a reflection at θ can be considered either as the nth order from planes of spacing d, or the 1st order from planes of spacing d/n. Crystallographers generally adopt the latter approach.

A diffraction pattern for a protein crystal contains many reflections which must be appropriately indexed and the most convenient system is to use the order of diffraction with respect to each of the unit cell axes. The Miller indices as they are called, which were derived originally to label crystal faces for mineralogical studies, are illustrated in *Figure 3*. Each index along the a, b, and c axes, respectively, is derived by taking the reciprocal of the intercept that the first plane of the set, not passing through the origin, makes with each axis in turn. Thus, the 100 planes are the set which have a spacing of $a \times 1/1$ on the x-axis, $b \times 1/\infty$ on the y-axis, and $c \times 1/\infty$ on the z-axis. The 200 planes have a spacing $a \times 1/2$ on the x-axis, and so on. Notice that the planes $h00$ are all parallel to one another but the spacing decrease with increasing h. Hence the angle of diffraction increases with increasing h, consistent with Bragg's Law. The letters h, k, and l are used to refer to the indices in general terms.

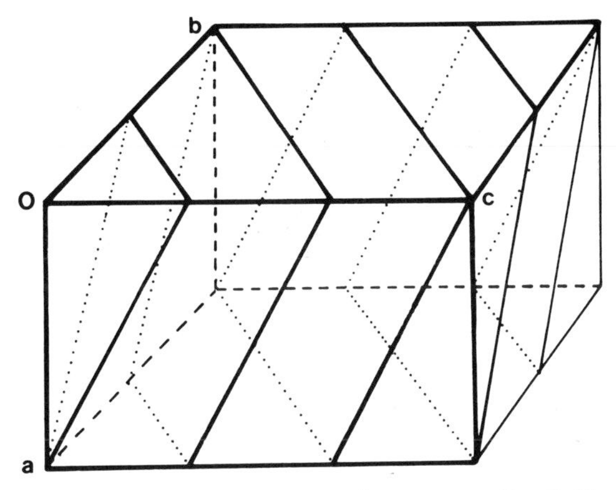

Figure 3. The set of planes 123 are shown as they cut a unit cell. The intercepts on *b* occur every 1/2 and on *c* every 1/3.

Each of the many sets of planes defined by the lattice gives rise to one reflection and *Figure 4* shows the relationship in 2-dimensions of the planes in the crystal (real space) to the points in diffraction space or *reciprocal space.* The points can be seen to make up another lattice (reciprocal lattice) whose axes and angles are derived from those of the crystal. This idea can be extended to 3-dimensions. It is important to realize that each reflection contains a contribution from every atom in the crystal and, conversely, each atom in the crystal contributes to every reflection. Thus, as the crystal is moved about in the X-ray beam, reflections flash out and can be recorded

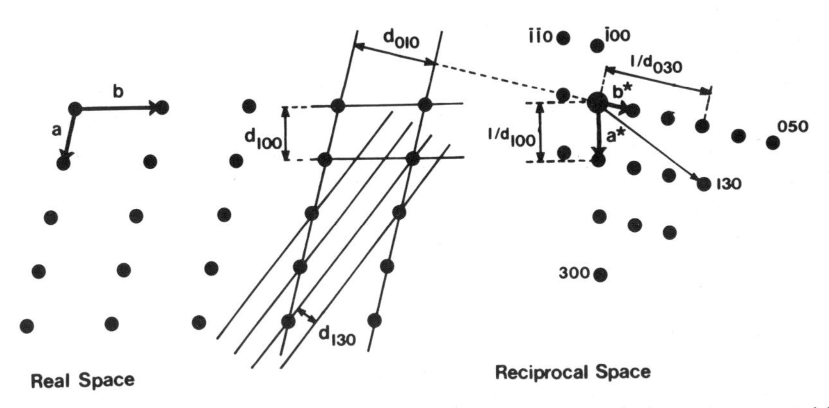

Figure 4. A diagram illustrating the relationship between sets of planes in a crystal in real or direct space and points representing a diffracted X-ray beam in reciprocal (or diffraction) space. Notice that the direction from the origin of reciprocal space (large point) to any point, e.g. 130, is perpendicular to the planes in the crystal and that the length is proportional to the reciprocal of the plane spacing.

when the geometrical arrangement of X-ray beam, crystal orientation, and detector satisfies Bragg's Law.

To help understand diffraction from a crystal, there is a construction introduced by Ewald and shown in *Figure 5*. As we move the crystal, the reciprocal lattice also moves about a fixed origin. With the crystal, X, as centre, a sphere is drawn of radius $1/\lambda$, and the origin, O, of the reciprocal lattice is taken as the point where the X-ray beam leaves the sphere after passing through the crystal. As the crystal is rotated about the z-axis (perpendicular to the page) the reciprocal lattice rotates until the point P lies on the surface of the sphere. The point P is the 410 reflection arising from the planes of spacing d_{410}. The angles at IX and XP, i.e. IXA and BXP, are equal to θ so that OXP = 2θ and OP is perpendicular to the crystal planes AXB. Now OP = $2 \times$ XO $\times \sin\theta = 2 \times (1/\lambda) \times \sin\theta$. However, OP = $1/d_{410}$ and so $1/d_{410} = (2/\lambda) \times \sin\theta$ which is Bragg's Law! Thus, the Ewald sphere gives a readily understandable way of relating the orientation of the crystal to the diffraction pattern observed. In order to collect a set of X-ray data, it is necessary to move the crystal (and detector) in such a way that every reciprocal lattice point passes through the sphere of reflection. There are various ways of achieving this, some of which are described in Section 4.

The space group in which a molecule crystallizes may impose certain conditions on the reflections which can be observed, so that by looking at the diffraction pattern of the crystal it is often possible to determine the space group unambiguously. Furthermore, the higher the symmetry of the crystal, the less data is actually required to be collected. A diffraction pattern has a centre of symmetry since reflections in opposite directions from the same

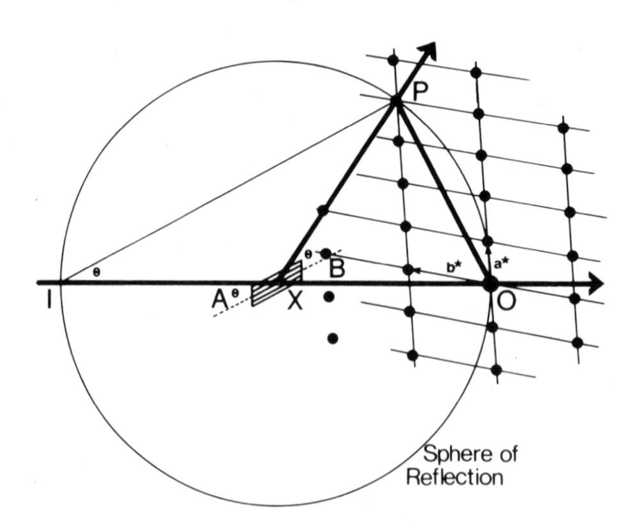

Figure 5. The Ewald construction. For clarity, this is shown as a planar diagram but IXO is the diameter of a sphere of radius $1/\lambda$.

planes must have the same intensity $I(hkl)$ = $I(\bar{h}\bar{k}\bar{l})$ is Friedel's Law (see *Figure 3*) (note: $\bar{h}$ (or $-h$) is pronounced '*h*-bar'). Thus the diffraction symmetry shown in *Table 2* has a centre of symmetry even though the space groups do not. The effects of the lattice type and symmetry elements upon the diffraction pattern are shown in *Table 3* and the effect can be explained with reference to *Figure 2*. If the beam X_3 scattered from row C is one wavelength behind X_1 scattered from row A, then X_2 scattered from row B is exactly half a wavelength behind and it will cancel out the reinforcing contributions from rows A and C. Thus, interposing planes midway between the planes separated by the unit cell repeat, as is the case for a centred lattice, leads to a systematic absence of reflections. Further, if a 2-fold screw axis is perpendicular to the planes, there will always be an identical (but rotated) set of scatterers to row A, on row B. Only when the index is even along the axial direction will constructive interference occur and the reflection be observed. Notice that simple rotation axes do not generate any systematic absences.

Table 2. Equivalent data for the chiral point groups

System	**Class**	**Laue Group**	**Equivalent reflections**[a]
Triclinic	1	$\bar{1}$	$I(hkl)$
Monoclinic	2	2/m	$I(hkl)$, **$I(\bar{h}k\bar{l})$**[b]
Orthorhombic	222	mmm	$I(hkl)$, **$I(\bar{h}\bar{k}l)$**, **$I(\bar{h}k\bar{l})$**, $I(h\bar{k}\bar{l})$
Tetragonal	4	4/m	$I(hkl)$, **$I(\bar{k}hl)$**, $I(\bar{h}\bar{k}l)$, $I(k\bar{h}l)$
	422	4/mmm	$I(hkl)$, **$I(\bar{k}hl)$**, $I(\bar{h}\bar{k}l)$, $I(k\bar{h}l)$,
			$I(h\bar{k}\bar{l})$, $I(\bar{k}\bar{h}\bar{l})$, $I(\bar{h}k\bar{l})$, $I(kh\bar{l})$
Trigonal[c]	3	$\bar{3}$	$I(hkl)$, **$I(kil)$**, $I(ihl)$[d]
	312	$\bar{3}$m1	$I(hkl)$, **$I(kil)$**, $I(ihl)$, **$I(\bar{k}\bar{h}\bar{l})$**, $I(\bar{i}\bar{k}\bar{l})$, $I(\bar{h}\bar{i}\bar{l})$
	321	$\bar{3}$1m	$I(hkl)$, **$I(kil)$**, $I(ihl)$, **$I(kh\bar{l})$**, $I(ik\bar{l})$, $I(hi\bar{l})$
Rhombohedral	3	$\bar{3}$	$I(hkl)$, **$I(klh)$**, $I(lhk)$
	32	$\bar{3}$m	$I(hkl)$, **$I(klh)$**, $I(lhk)$, **$I(\bar{k}\bar{h}\bar{l})$**, $I(\bar{l}\bar{k}\bar{h})$, $I(\bar{h}\bar{l}\bar{k})$
Hexagonal[c]	6	6/m	$I(hkl)$, **$I(kil)$**, $I(ihl)$, **$I(\bar{h}\bar{k}l)$**, $I(\bar{k}\bar{i}l)$, $I(\bar{i}\bar{h}l)$
	622	6/mmm	$I(hkl)$, **$I(kil)$**, $I(ihl)$, **$I(\bar{h}\bar{k}l)$**, $I(\bar{k}\bar{i}l)$, $I(\bar{i}\bar{h}l)$,
			$I(kh\bar{l})$, $I(\bar{i}\bar{k}\bar{l})$, $I(hi\bar{l})$, $I(\bar{h}\bar{i}\bar{l})$, $I(\bar{k}\bar{h}\bar{l})$, $I(ik\bar{l})$
Cubic	23	m$\bar{3}$	$I(hkl)$, **$I(\bar{h}\bar{k}l)$**, **$I(\bar{h}k\bar{l})$**, $I(h\bar{k}\bar{l})$, **$I(klh)$**, $I(lhk)$,
			$I(\bar{k}\bar{l}h)$, $I(k\bar{l}\bar{h})$, $I(\bar{k}l\bar{h})$, $I(\bar{l}\bar{h}k)$, $I(\bar{l}h\bar{k})$, $I(l\bar{h}\bar{k})$
	432	m$\bar{3}$m	$I(hkl)$, **$I(\bar{h}\bar{k}l)$**, **$I(\bar{h}k\bar{l})$**, $I(h\bar{k}\bar{l})$, **$I(klh)$**, $I(lhk)$,
			$I(\bar{k}\bar{l}h)$, $I(k\bar{l}\bar{h})$, $I(\bar{l}\bar{h}k)$, $I(\bar{l}h\bar{k})$, $I(l\bar{h}\bar{k})$, $I(\bar{k}l\bar{h})$,
			$I(\bar{k}hl)$, $I(\bar{h}lk)$, $I(\bar{l}kh)$, $I(h\bar{l}k)$, $I(l\bar{k}h)$, $I(k\bar{h}l)$,
			$I(hl\bar{k})$, $I(lk\bar{h})$, $I(kh\bar{l})$, $I(\bar{k}\bar{h}\bar{l})$, $I(\bar{l}\bar{k}\bar{h})$, $I(\bar{h}\bar{l}\bar{k})$

[a] The reflections listed here are identical. If Friedel's Law holds then $I(hkl) = I(\bar{h}\bar{k}\bar{l})$ and this generates an equal number of equivalent reflections. In protein crystallography, anomalous scattering which leads to a breakdown in Friedel's Law is used to help with phasing the reflections and so the two sets, equivalent to $I(hkl)$ and $I(\bar{h}\bar{k}\bar{l})$ must be kept separate.

[b] The reflections indicated in bold are those which are required to specify the Laue symmetry with the others being generated by repeated application of the symmetry elements.

[c] The axes in the trigonal and hexagonal systems referred to here are $a = b, c$, $\alpha = \beta = 90°$, $\gamma = 120°$

[d] When hexagonal axes are being used, $i = \bar{h}\bar{k}$

Table 3. Conditions affecting possible reflections

Element	**Symbol**	**Reflection observed for**	**Notes**
Primitive lattice	P		
Lattice centred on the C face	C	*hkl* with $h + k$ even	The C face is contained by *a* and *b*
Face centred lattice	F	*hkl* with *h*, *k*, and *l* all odd or all even	
Body-centred lattice	I	*hkl* with $h + k + l$ even	
Rhombohedral lattice	R	$-h + k + l = 3n$	'$= 3n$' means divisible by 3
Twofold screw axis ‖ c	2_1	00*l* with *l* even	For an axis along *a*, the row is *h*00
Threefold screw axes ‖ c	3_1, 3_2	00*l* with $l = 3n$	The two possible 3-fold axes have the same pitch but opposite hands
Fourfold screw axes ‖ c	4_1, 4_3	00*l* with $l = 4n$	
	4_2	00*l* with *l* even	cf. the 2-fold screw axis
Sixfold screw axes ‖ c	6_1, 6_5	00*l* with $l = 6n$	
	6_2, 6_4	00*l* with $l = 3n$	cf. the 3-fold screw axes
	6_3	00*l* with *l* even	cf. the 2-fold screw axis

2.4 How is a protein crystal structure solved?

The formation of a magnified image by a light microscope involves collecting all of the scattered light waves in the objective lens which recombines them in the correct way to produce the magnified image. But what is 'the correct way'? Associated with each wave is not only its amplitude but also its phase relative to the unscattered light. The focusing by the objective lens uses both amplitude and phase to produce the magnified image. In the case of X-rays, the crystal produces a diffraction pattern which needs to be recombined in the correct phase relationship, but in this case no lens exists which is able to perform the task, and in recording the pattern as one must, the vital phase information is lost. It must be calculated and this 'phase problem' is central to crystallography. Ironically, if the position of the atoms is known, then the phase for each reflection can be calculated! Whilst this problem may seem insuperable, if the positions of only a few heavy atoms are known their contribution can be calculated and this is generally sufficient to solve the phase problem for a protein. The preparation of heavy-metal derivatives of proteins has been dealt with in Chapter 11. If the derivative crystals have identical unit cell dimensions to (i.e. are isomorphous with) the native crystal, then the technique of multiple isomorphous replacement can be applied. This technique has been the basis of all new protein structure solutions, although it should be pointed out that molecular replacement (9) is applicable where a similar structure already exists.

A phase must be calculated for each reflection to be included in the calculation of the electron density map. The more X-ray reflections that are phased and included, the clearer the map will be and the better will be the resulting model of the protein. Thus the *resolution* of the data is usually reported and this refers to the minimum plane spacing included in the calculation: thus for a 3.5 Å map, all reflections with plane spacings greater than or equal to 3.5 Å will be included. The higher the resolution, the greater the amount of X-ray data which must be measured. Disregarding the symmetry of the reflection data, the total number of reflections is approximately $5V/d^3$ where V is the unit cell volume and d is the resolution.

2.5 Importance of preliminary characterization

There are a number of reasons why the preliminary characterization of a newly crystallized molecule is important. Most obviously, the first point to establish is that the crystal does diffract X-rays. As part of the process of checking that the crystal does diffract, some idea of the crystal lifetime in the X-ray beam will be obtained together with the resolution which can be achieved. Even when a crystal appears perfect, it should not be assumed that it will be suitable for X-ray work. Occasionally, some or all of the crystals in a batch give no discernable diffraction pattern. The reason for this is obscure but possible avenues to explore before abandoning the particular crystallization conditions used, are:

- try crystals from different drops, tubes, or preparations;
- search for crystals with a different morphology and X-ray them, sometimes different forms appear in the same tube;
- cool the crystal before and during the X-ray exposure, possibly down to liquid nitrogen temperatures (10);
- use synchrotron radiation with a shorter wavelength; it has been found to extend the lifetime of sensitive crystals (11);
- attempt to cross-link the molecules in the crystal with a bifunctional reagent such as glutaraldehyde (12).

These ideas may also be worth trying if the crystals produce feeble or rapidly fading diffraction, and the last point in particular may allow successful handling of crystals which are very fragile.

If the spots obtained on a *still* photograph (Section 4.2.2) are streaked, or the pattern looks like a small-angle precession photograph or rotation photograph (see *Figures 12* and *13*), then the crystal is likely to have a degree of disorder which may render successful structure determination impossible. The only recourse is then to re-examine the crystallization procedure.

The aim of this initial X-ray investigation should be to determine the unit cell dimensions and the space group. Not only must these be known to solve the crystal structure but also, with the crystal's X-ray lifetime, they dictate the strategy for efficient data collection; e.g. if a cell dimension is large then it may not be possible to resolve the spots satisfactorily in the laboratory. The amount of data to be collected is determined by the diffraction symmetry of the crystal and it is often possible to reduce the number of exposures by ensuring that the crystal is mounted in a particular way; e.g. it is best to mount a hexagonal crystal with the 6-fold axis parallel to the rotation axis of the instrument.

It is normal practice to determine the volume occupied per unit molecular weight (V_m) since this determines the number of molecules in the asymmetric unit. V_m has been found to be around 2.4 Å^3/dalton for globular protein crystals, although this value is subject to quite large fluctuations (13). It is obtained by dividing the unit cell volume by the product of the protein molecular weight and the number of equivalent positions (asymmetric units). Unfortunately, it is often found that a choice is possible and then some biological insight may help resolve any ambiguity. If it is possible to determine the crystal density and the weight loss on drying, the protein molecular weight can be calculated which, when compared with the known value, also gives the number of molecules per asymmetric unit. The approximate solvent content can also be calculated from the formula:

$$V_{sc}(\%) = 100(1 - 1.23/V_m). \quad (2)$$

One final point about the preliminaries is that biological information may emerge about the subunit structure. If it is found that the asymmetric unit contains half of the expected molecular weight, the protein must consist of an even number of subunits and it is probable that the molecular twofold axis coincides with a crystallographic one. This will be consistent with the space group which must possess such a symmetry element. Conversely, if a crystal is found to have 3 or 4 molecules in the asymmetric unit of a relatively low symmetry space group, then one should be alerted to the possibility of having missed a higher symmetry space group by not having collected data showing the principal axes.

3. Mounting crystals

Mounting a protein crystal is a procedure which requires a reasonable degree of manual dexterity. It is impossible to be dogmatic about the right and wrong way, and each person develops their own technique, modifying it as required from protein to protein depending on the size, strength, temperature behaviour, need to exclude oxygen, or toxicity. Although early workers did dry their crystals (14), 'flash cooling' is now seen as a way of greatly reducing radiation damage (15) as well as getting over the problem of the loss of water of crystallization of the salting-out solution within the protein crystal lattice.

3.1 Initial examination with a microscope

Well-formed protein crystals examined under the light microscope exhibit a symmetric arrangement of edges and faces which are related to the packing of the molecules. Thus, examination of crystal morphology may give a first glimpse of the symmetry of the unit cell. A stereo-zoom dissecting microscope, ideally fitted with a crossed polarizing attachment, with a magnification in the range ×10–×40 is best for such examination since crystals which cannot be seen readily with such an instrument are probably not going to diffract sufficient X-rays, even with synchrotron radiation. It is important to ensure that the illuminating light source does not heat the microscope stage lest undue evaporation and denaturation occurs. The use of crossed polarizers can indicate the direction of a principal axis. Rotating the crystal on the stage in the dark field (polarizer and analyser at 90 °), the crystal appears as a light colour until an optic axis lies along the direction of the polarizer whereupon extinction occurs, depending on the crystal system. During a full rotation, extinction occurs every 90 °. This effect will not be observed for cubic crystals, for tetragonal, trigonal, and hexagonal crystals viewed along their unique axis, or for non-crystalline material. Note that the crystal should not be contained in a plastic container (like a tissue culture plate) if polarized light is to be used because these containers affect the polarization, usually producing splendid colours! Salt crystals are usually highly coloured, even if they are small and, if a crystal is thought to be salt rather than protein, pressure with a fine probe will produce an audible 'plink' as the tip slips off the hard salt crystal. A protein, on the other hand, will shatter with very little pressure at all! It is well worth getting to know the crystal habit and its relationship to the axes since this saves considerable time in alignment in the X-ray beam.

3.1.1 Selection of a crystal for mounting

Protein crystals with dimensions of 0.2–0.5 mm are most suitable for use in an X-ray diffraction experiment. Use of smaller crystals is possible, however the diffraction pattern tends to be weaker, requiring longer exposure times and possibly poorer resolution. On the other hand, crystals much larger than 0.5 mm may not be uniformly bathed in the X-ray beam (depending on the size of beam collimator used) and generate their own problems associated with absorption of X-rays.

Suitable crystals for X-ray work should be single and should appear transparent (containing no cracks) with well-defined edges and faces. Birefringent single crystals, when observed under a polarizer, extinguish light sharply when rotated through 360 °. Less obviously twinned or multiple crystals may sometimes be detected if different sections of the crystal extinguish light at different rotations of the microscope stage. Crystals which have grown into one another, or have grown as clumps, may be carefully split using a fine probe or a fresh scalpel blade. Gently touch the crystal at the point where the extra piece joins the chosen crystal, keeping the blade parallel to the direction in which the crystals are to be separated. A gentle pressure is usually all that is required since crystals will generally cleave readily along the axial directions.

3.1.2 Selection of suitable handling buffer

In the average protein crystal, 30–60% of the volume is occupied by solvent, either hydrogen bonded to the surface of the protein molecule or free in the interstitial regions between molecules. In both cases it is required for maintaining the crystal lattice. Should the solvent content be changed (through evaporation or shifts in salt or buffer concentrations), the integrity of the crystal may be lost. Often, as in the case of the hanging drop, only very small (1–20 μl) quantities of mother liquor are available. For this reason it is necessary to find appropriate buffer conditions for convenient handling, storing, and subsequent soaking of the crystals. A good place to start is with the buffer and salt conditions under which the protein was crystallized. With the vapour diffusion technique, the drop and the well are equilibrated with protein present in the drop but not in the well. If crystallization has occurred without precipitant as a result of high concentration of protein, the test buffer should also contain protein to maintain equilibrium conditions. Otherwise, buffer from the well should be considered as first choice of a test solution to check the viability of the crystals. This is done in the following way:

- Check the supply of crystals for those which meet the criteria of:
  (a) transparency (this may be a problem for large metalloprotein crystals;
  (b) size (it is sensible to try changing buffer/mother liquor on the smaller crystals first);
  (c) being single, by extinction of polarized light.
- Depending on the system used for growth, it may be possible to draw only one or two crystals up into a pipette and transfer them to a drop (10–20 μl) of test buffer. This is more difficult when working with crystals from hanging drops as a significant volume of mother liquor will also be drawn up with the crystals. In the case of very small volumes, either add a drop of test buffer to the original drop or try lifting crystals out of the original drop with the aid of a probe and launching them quickly into a drop of test buffer on a nearby coverslip. Cover the test drops with an upturned vial, sealing the rim with silicone grease. Observation of the crystals over the period of a few hours will show whether no cracking or dissolution occurs as a result of buffer conditions. The extinction properties should also remain unchanged. Should the specimens lose their crystallinity, it will be necessary to modify the test buffer, e.g. by raising or lowering the precipitant concentration. Once a test buffer has been established, it may be possible to use it as a storage solution from which crystals can be withdrawn as required for mounting.

3.2 The basic techniques

Two related methods commonly used for mounting are described. Both take practice and a number of trials, preferably with a batch of old or non-precious crystals, before a decision can be made as to which steps are best suited to maintaining the crystal (and the sanity of the mounter). The first involves floating the crystal down through a volume of buffer in a Lindemann tube and the second drawing the crystal up into the Lindemann tube by a pipette attachment, allowing controlled movement of the crystal and buffer in the capillary.

3.2.1 Equipment for mounting a crystal

A list of equipment useful for mounting the crystal should contain the following items:

- glass slides and coverslips;
- Lindemann tubes (thin-walled glass or quartz capillary tubes, 1.0 or 0.7 mm diameter);
- Pasteur pipettes, some of which have been drawn in a flame to give a narrow bore and some cut off to increase the bore at the tip (for handling

larger crystals). It is necessary to have pipettes sufficiently small to fit inside a 1.0 mm Lindemann tube.

- small glass vials ('pots' made by cutting a vial to give a container perhaps 8 mm in diameter by 6 mm deep are ideal);
- sealing grease (silicone grease, petroleum jelly);
- modelling clay (plasticene);
- probe with a fine tip (like a sewing needle);
- scalpel with a new and pointed blade;
- spirit burner, low temperature soldering iron or Bunsen burner with which to melt wax;
- wax, forceps, supply of appropriate buffer;
- filter paper strips or cotton thread for soaking up excess mother liquor inside the Lindemann tubes;
- nylon, wire or glass fibres.

3.2.2 Method I

This procedure is summarized by *Protocol 1* and illustrated by *Figure 6*.

Protocol 1. Mounting crystals using method I

1. Apply plasticene to the closed end of a 1.0 mm Lindemann tube. This allows the tube to be left standing upright during the remainder of operations as well as providing a handle with which to manipulate the tube. You may choose to glue the capillary instead into a brass pin which fits into a goniometer head.
2. With a narrow pipette, fill the tube with buffer (in which the crystal is known to be stable) leaving the meniscus visible about 1-cm above the plasticene. There must be no air bubbles. Some space should also be left at the top to accommodate addition of more buffer when the crystal is dropped in.
3. Using a Pasteur pipette, carefully transfer one crystal in its buffer to the top of the Lindemann buffer column. It is wise to perform steps **1** and **2** before selecting the crystal.
4. Allow the crystal to sink to the lower meniscus.
5. Once the crystal has settled remove the excess buffer from the Lindemann tube with the fine pipette, leaving only a small amount surrounding the crystal.
6. Clip the tube between two wax blobs with a fine pair of forceps. Apply the wax at positions on the outside of the Lindemann tube approximately 5 mm up from the crystal and again 5 mm above the first blob.

Protocol 1. *Continued*

Application of wax prevents the tube from shattering when clipped. Wait until the wax is hard before clipping.

7. Dry the remaining buffer from around the crystal with the aid of a shred of filter paper inserted through the clipped end of the tube. A dry mount is preferred for two reasons. The faces are more easily visible when aligning the crystal on the X-ray camera and the absence of solvent may reduce the effects of crystal slippage. However, over-drying is to be avoided because solvent loss from the crystal can lead to a loss of the high-resolution X-ray data, thus the following steps should be performed as quickly as possible.
8. The crystal may be roughly aligned in the tube by manipulating it carefully with a glass fibre, nylon thread, or filter-paper strip. It may be useful to align the crystal so that extinction under polarizers occurs parallel and perpendicular to the capillary.
9. When the alignment is satisfactory, add a small plug of buffer to the open end of the tube, keeping it well away from the crystal.
10. Seal the tube with wax.

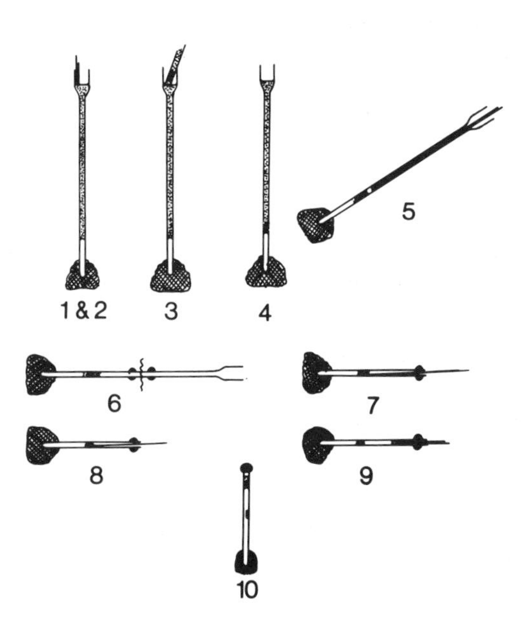

Figure 6. A diagram of the steps involved in mounting a crystal. The numbers refer to the steps described in *Protocol 10*.

Many variations are possible at the discretion of the mounter: e.g. it may be preferable to have two plugs of buffer in the Lindemann tube, one on either side of the crystal. This can be accomplished by adding a small amount of buffer at the bottom of the tube and leaving a gap of about 1 cm between it and the rest of the buffer column. The crystal will sink down as far as the first meniscus and the remaining steps are the same. It may also be found easier to manipulate the crystal, by tipping as well as by using a fibre, to its desired orientation before the crystal is dried-out completely with the small amount of buffer functioning as a lubricant. (Sometimes, however, wet crystals stick to the fibre with which they are being manipulated.) Final drying is done carefully after the crystal is positioned. If it is necessary to reposition the crystal, opening up the wax plug is done most easily with a heated needle.

3.2.3 Method II

The pipette arrangement of this method is shown in *Protocol 2* and illustrated in *Figure 7*.

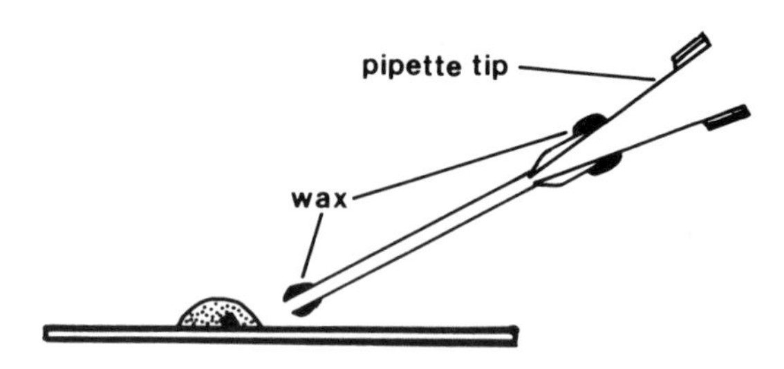

Figure 7. The construction of the Gilson pipette/Lindemann tube system described in *Protocol 2*, step 4.

Protocol 2. Mounting crystals using method II

1. Seal a plastic pipette tip into the wide end of a Lindemann tube with melted wax.
2. Clip off the closed end of the Lindemann tube with as even a break as possible. A glass tube may be cut smoothly by touching it, with a small amount of pressure, to a thin heated wire.
3. Fit the pipette tip onto a 2–20 μl Gilson pipette.
4. Manoeuvre a crystal to the edge of its drop with a glass fibre or probe. With the pipette set to 20 μl volume, a crystal is drawn up into the Lindemann tube along with about 5 μl of liquid by gradual suction controlled by the plunger of the pipette. Lift the end of the capillary out of the drop and continue drawing on the pipette until the plug of liquid advances about 15 mm up the tube.

Protocol 2. *Continued*

5. By alternately pushing and drawing, it may be possible to isolate the crystal from its plug of mother liquor if, for instance, the crystal settles onto the wall of the capillary. If this is unsuccessful, carefully remove the pipette tip from the pipette, being sure to compensate with pressure on the plunger for the suction effect caused as the tip is pulled off.
6. The plug of liquid containing the crystal is now free to move in the Lindemann tube. Its position can be optimized by gentle tilting of the tube until its bottom meniscus is approximately 1 cm from the bottom of the capillary. Seal the capillary with wax. Plasticene may be applied around the wax for easier handling.
7. Having sealed one end, the liquid plug will remain stationary. The remainder of the steps are now identical to steps 4–10 of *Protocol 1*.

Variations of this method are also possible. The less obvious variations are those related to drawing the crystal into the capillary. You might prefer to fix a tube (instead of a pipette tip) into the capillary such that the tube is long enough to reach your mouth. Then pressure and suction can be applied orally. Another option is to lead the tube not to your mouth but to a syringe which has been fixed to the Lindemann tube by a short (*c*. 3 cm) stretch of narrow-bore rubber hose. Finally, an unadorned, dry Lindemann tube will work on its own to draw up a crystal and the drop that contains it simply through capillary action.

3.2.4 Mounting for low temperatures

It is often necessary to maintain the crystal at temperatures lower than ambient: e.g. crystal stability may require working at around 0 °C. Working at low temperatures is probably best accomplished by housing the X-ray camera in a cold room but, since such a system is not always available, cheaper alternatives exist whereby a stream of cooled, dried air or nitrogen is allowed to flow over the crystal from a nozzle mounted on the X-ray instrument as close to the crystal as possible (16). Usually the stream is co-axial with the capillary tube and goniometer head (see *Figure 8*) and a plastic collar added to protect the instrument. Below room temperature but above the freezing point of the solution, the crystals should be mounted in the cold room, otherwise when the crystal is cooled on the camera or diffractometer the temperature gradient produced by the cooler will lead to water distilling along the tube and dissolving the crystal.

Recent developments in the cryo-crystallography of biological molecules have shown promise especially with regard to prevention of radiation damage. A general overview of these developments is given by Hope (10), and a novel thin-film procedure has recently been described by Teng (17). If it

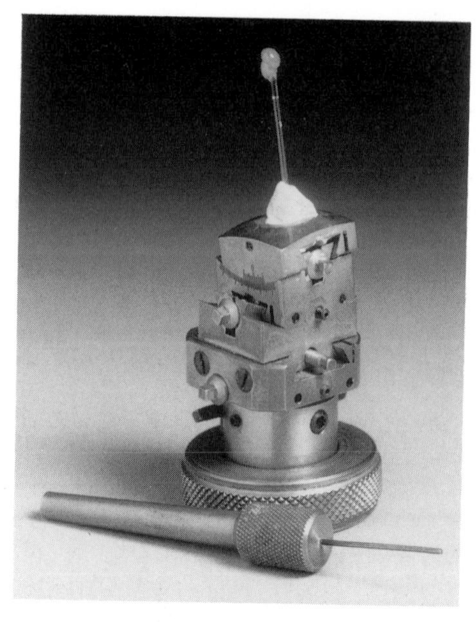

Figure 8. A typical goniometer head with the crystal sealed in a capillary fixed upon it. The key shown is for adjusting the slides and arcs. It has a fine Allen key at the other end for locking the arcs after adjustment. The knurled ring at the base will screw onto an X-ray camera or diffractometer.

becomes apparent that very low temperatures will be required (because conventionally mounted crystals have unworkably short lifetimes in the X-ray beam) a different mounting procedure must be applied.

4. Preliminary X-ray data

Particularly at the start of an investigation when nothing is known about the unit cell and space group, characterization is best performed with a precession camera (see below) which gives an undistorted picture of a reciprocal lattice layer. The drawback of precession photography is that the crystal must be accurately aligned with respect to the X-ray beam. Since the crystal may be sensitive to irradiation, the more rapidly it can be aligned the better, and this is accomplished through the series of steps described below. These lead to the taking of a precession photograph, but the procedure is the same up to step 4.2.2 for an oscillation photograph.

4.1 Optical alignment

Once mounted, the capillary tube containing the crystal is fixed to a goniometer head with plasticene, glue, or wax. It is convenient to align the tube in the goniometer head (see *Figure 8*) in such a way that obvious features of the crystal (i.e. a face) are positioned perpendicular or parallel to one of the graduated arcs and a crystal axis parallel to the spindle. This allows easier adjustment during alignment. In general, the axes of the unit cell are parallel or perpendicular to the faces, or the bisector of the angle between adjacent faces. Mount the goniometer head on an optical goniometer (attachments which allow the goniometer head to be rotated are available to fit onto the microscope stage) or the X-ray camera. With the graduated arcs, the orientation of the crystal can be changed by about ±20 °. Adjust the graduated arcs to align potential crystal axes with the equipment (e.g. a needle axis parallel to the spindle). Centring of the crystal is now carried out to ensure that it remains in the X-ray beam during rotation about the spindle. The two bottom sledges on the goniometer head are used to do this. First, rotate the crystal through 360 °, noting its position in the microscope crosshairs at 0 °, 90 °, 180 °, and 270 °. To centre the crystal, put one sledge perpendicular to the direction of view; this will correspond to either the 0/180 ° or the 90/270 ° positions. Move the sledge to the midpoint of the 0/180 ° (or 90/270 °) readings and repeat for the other sledge. The sledges are adjusted until the crystal (not the tube!) is stationary through a full rotation. Note that the crosshairs on the telescope may not define the centre of the rotation.

4.2 Photographic alignment

Because alignment photographs are essentially variations of the precession technique, it is necessary to have some understanding of the precession method in order to carry out the alignment effectively (18). The following section covers the precession method only briefly. Further reading of crystallographic texts is strongly recommended.

4.2.1 The precession method

The precession camera geometry is illustrated schematically in *Figure 9*. A crystal is mounted and aligned with the X-ray beam parallel to a real axis. This, from the definition of reciprocal space, means that a plane of reciprocal space is parallel to the film. The crystal is offset by a known angle μ and this offsets the reciprocal lattice similarly. The motion of the precession camera appears complicated but it is designed to allow the film holder to follow the motion of the crystal as the real axis precesses about the incident X-ray beam. As the crystal moves, the film holder follows at a constant crystal-to-film distance and the direct beam always hits the centre of the film when a zero-level photograph is being taken. The precessing motion rolls the plane OP around XO with O as a fixed point. All reciprocal lattice points within a circle of radius OP will at some time pass through the sphere of reflection and the diffracted rays like XP so produced will pass through the screen annulus to be recorded as a magnified image on the film at O′P′. Diffracted rays from all

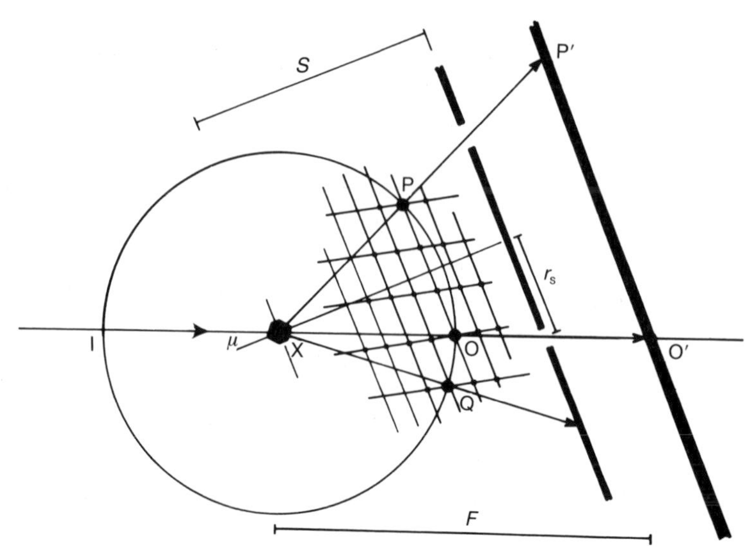

Figure 9. Schematic representation of the precession camera. The crystal is aligned so that the X-ray beam is perpendicular to reciprocal lattice plane OP. It is then offset by the precession angle μ. A screen with an annular hole of radius r_s is placed at distance S from the crystal. Screen, film (at O′P′, distance F from X), and reciprocal lattice plane OP are parallel. The crystal precesses causing the normal to OP to move in a circle around IXO which in turn makes all of the reciprocal lattice points on the plane OP within a radius OP of O, pass through the Ewald sphere and the annular screen to be recorded on the film.

other layers like XQ are blocked by the screen whose movement is linked directly to that of the crystal. The film, screen, and reciprocal lattice plane are always parallel to each other and perpendicular to the precessing axis. The great advantage of the precession method is its capacity to record one undistorted reciprocal lattice layer on each film. The magnification factor of this image of the reciprocal lattice is simply the crystal to film distance, F.

i. Still photographs

Still photographs are described in *Protocol 3.*

Protocol 3. Still photographs

1. Cut a small, asymmetric piece from the top right-hand corner of a 13 × 13 cm piece of X-ray film (e.g. Agfa-Gevaert Osray M3 or Kodak DEFS) and load it into a cassette (in the dark room). Mount the cassette on the camera. The clipped corner ensures you can relate the diffraction pattern to the camera.
2. Check the crystal-to-film distance is 75 mm. It should also be possible to set the crystal-to-film distance to 60 mm (commonly used with Mo

Protocol 3. *Continued*

radiation for small molecule crystallography) or 100 mm (useful for protein crystals with large unit cell dimensions).

3. Check the crystal centring and adjust the spindle of the camera so that the X-ray beam will pass through the chosen feature of the crystal at 90 °. Use the vernier scales to record the spindle and arc readings; usually the camera will have a log book in which to do this.

4. Take the still photograph by exposing the crystal to the unfiltered beam for a few minutes. The time of exposure varies with crystal size; longer exposure times are necessary for smaller crystals since their diffraction patterns are weaker. A convenient rule of thumb is to expose 0.5 mm crystals for 2–3 min, 0.25 mm crystals for 5–10 min, and smaller crystals for > 10 min on sealed tube or rotating anode X-ray sources.

5. Develop the film. A suggested procedure is given below:

(a) Open the film cassette in the darkroom (red safety-light). Put it in a holder.

(b) Develop the film for 4 min at 20 °C in developer solution (e.g. 0.7 l of Kodak LX24 developer diluted with 3.3 l of water).

(c) Move the film to the stop bath (1% acetic acid) for 15 s with agitation.

(d) Move the film to fixer (e.g. 0.81 of Kodak FX-40 fixer diluted with 3.2 l of water). The lights can be turned on once the film has been moved to the fixer solution. Leave the film in the fixer for 10 min. Note that if accurate intensity measurements are required, the film should be left for 10 min in the fixer with the lights off.

(e) Wash the film in running water for 30 min.

(f) Dry the film in an air oven at 60 °C.

This procedure is what should be done for films to be kept. For setting photographs, the measurements required (see below) can be made on the wet film often before it has properly cleared in the fixer.

The resulting photograph should show a series of concentric circles of diffraction spots corresponding to the outline of the Ewald sphere where it is cut by a series of planes in reciprocal space. *Figure 10* is a drawing illustrating this and *Figure 11a* is a still photograph of a lysozyme crystal. There are a number of observations you can make right away.

(a) Are there any spots at all?

If not, then is there a shadow of the backstop? Is the backstop misplaced (large black area in the film centre)? Adjust the backstop position if necessary. If there is no backstop shadow, are there X-rays when the

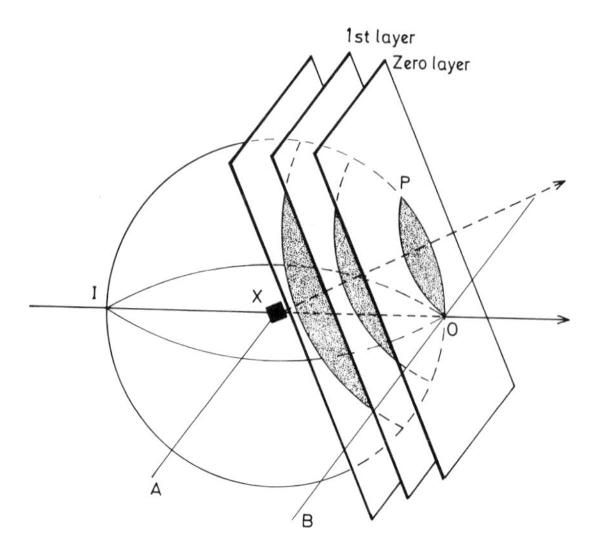

Figure 10. The Ewald sphere intersected by several reciprocal lattice layers. As the crystal is moved, the reciprocal lattice pivots about O but XA and OB remain parallel at all times. As shown, a film placed perpendicular to XO will record a series of concentric circles. As the crystal is rotated clockwise about AX, the circles will become smaller until the zero-layer circle becomes a point when a principal axis in the crystal is coincident with the X-ray beam. A counter-clockwise rotation will bring the layers $\bar{1}$, $\bar{2}$ and so on into the central region of the zero-layer circle.

shutter is open (seek help to check this)? Is the developer fresh? Is the beam hitting the crystal? If not, check first to make sure the crystal is still aligned in the crosshairs of the microscope eyepiece. If it is, the camera may need adjustment and help should be sought.

(b) How intense are the spots?
It may be necessary to adjust exposure times, particularly to increase the time if the diffraction pattern is weak.

(c) Do you see concentric circles of spots?
If concentric rings are not apparent in the array of spots, a main set of reciprocal lattice planes may be more than 20 ° away from being parallel to the film. Try rotating the spindle by 20 ° or until another obvious crystal feature (e.g. a face or the bisector between two faces) is parallel or

Figure 11. Setting photographs for the precession method. (a) Still. (b) 3 ° precession (mis-set). (c) Diagram showing the necessary measurements for correcting the mis-setting. (d) 3 ° precession (set); note the symmetry of the clipped spots and the fact that the backstop is not necessarily symmetrical with respect to the pattern. (e) A 'cloverleaf' to check the alignment of the screen. (f) A 15 ° precession photograph with filtered Cu radiation. Lysozyme was the crystal used for these photographs.

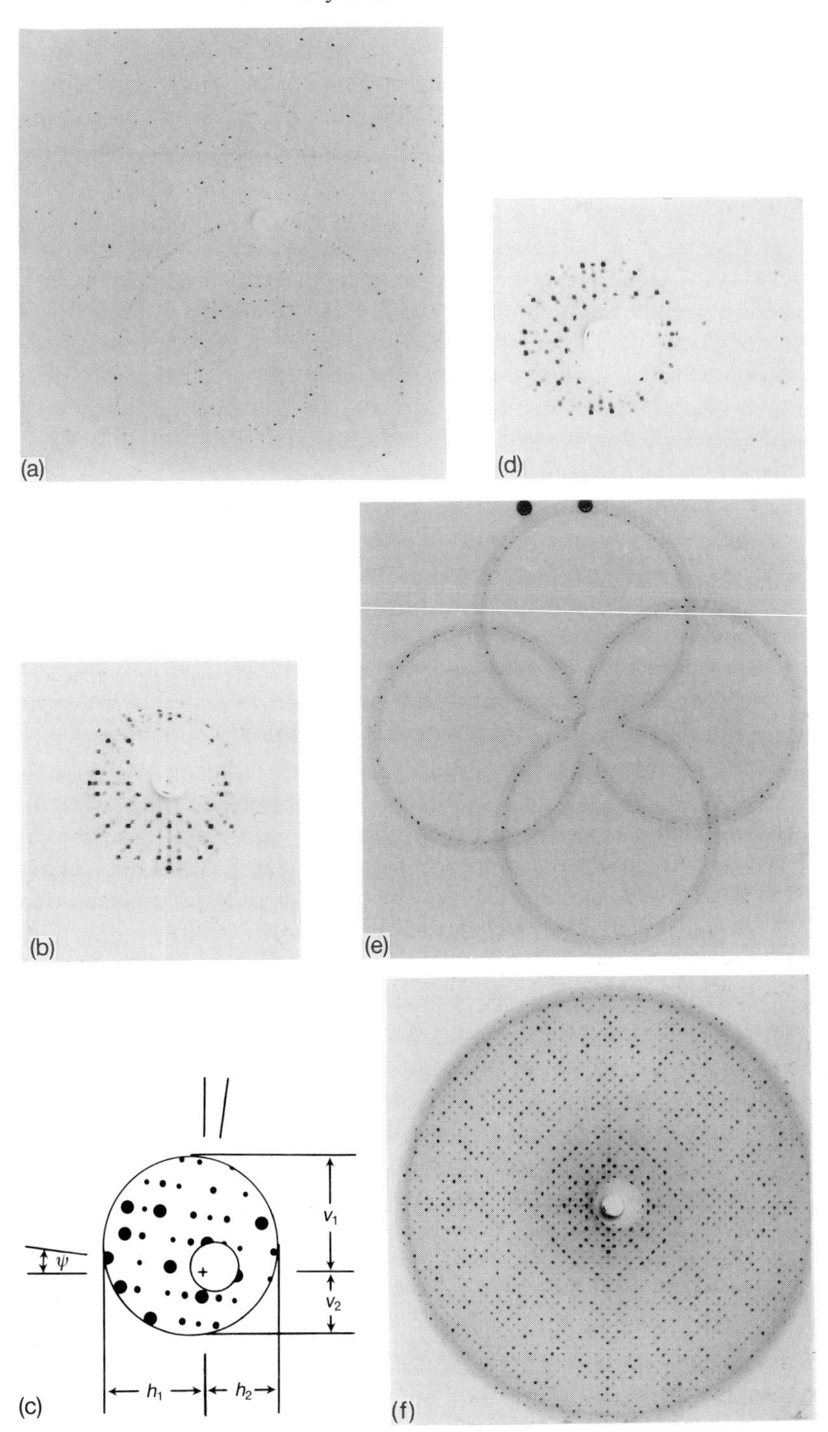
(a)
(d)
(e)
(b)
(c)
ψ
v_1
v_2
h_1
h_2
(f)

perpendicular to the beam. You may also have to adjust the horizontal arcs to achieve this. Note that every time an adjustment is made to the arcs, the crystal may need to be re-centred using the goniometer sledges. If the arc adjustments required exceed the angular units of the goniometer head, the crystal will have to be re-positioned.

(d) What if you see more than one set of concentric circles?
Initially, we are interested in the concentric circles arising from the reciprocal lattice layers perpendicular to the real crystal axes, the so-called *principal zones*. Diagonal (e.g. [011]) directions in the crystal have sets of planes associated with them too which will also show up as sets of concentric circles. Choose the set of circles which is most widely spaced and which tends to extend farthest out on the photograph. You will probably find that circles from a principal zone dominate other circles present on the same film.

(i) Especially with crystals in a highly symmetric space group, many sets of concentric circles may be observed. You must have enough photographs to compare these sets in order to decide which set represents a principal direction. Rotate the spindle 30 ° from the starting position and repeat steps 1–5. Repeat this twice more. This will give a set of still photographs from 0–90 °. It is important to cover a representative portion of reciprocal space systematically during your initial search for principal zones.

(ii) Choose the sets of concentric circles you want to align. The experienced eye will often pick out principal zones with little difficulty. Dominating sets of concentric circles 90 ° to one another may indicate monoclinic or higher symmetry. Two sets located at 60 ° to one another may indicate a hexagonal or trigonal system. Finally, if concentric circles are aligned at an arbitrary angle to each other, the cell may be monoclinic (with the b^* axis parallel to the spindle) or triclinic.

(iii) Measure the displacement (vertical and horizontal) of the centre of the circles from the centre of the primary beam. For a perfectly aligned crystal, the centre of the circles will coincide with the point where the main beam hits the film. Beware; the centre of the backstop shadow is not necessarily the centre of primary beam. The still photograph may be considered to be produced by the intersection of a concentric series of cones with the film (see *Figure 10*). The axis of these cones is the axis to be centred, and when perfectly set this axis is coincident with the X-ray beam and perpendicular to the plane of the film. The axis is always perpendicular to the reciprocal lattice planes. To correct the mis-setting, the crystal must be rotated to bring the cone axis coincident with the beam, thus bringing the reciprocal lattice planes parallel to the film.

(iv) Calculate the correction required to centre the concentric circles. The amount of correction necessary is calculated by applying the following formula:

$$\tan \varepsilon \approx \Delta / F \quad (3)$$

where ε is the angle of mis-setting, Δ is the displacement from the centre of beam position (mm), and F is the crystal-to-film distance (mm). For $F = 75$ mm, a mis-setting of 1 mm corresponds to an angular correction of about 0.8 °, for $F = 100$ mm, 1 mm is 0.5 °. Vertical missetting is corrected by adjustment of the spindle, sideways mis-setting is corrected by adjusting the horizontal arc on the goniometer head. The vertical arc rotates the pattern in the plane of the film. Marking the film as described in *Protocol 3*, step 1 ensures that the corrections are applied in the correct sense: stand behind the camera with the film as exposed to the X-rays and decide the movement to swing the centre to behind the backstop. Note that the horizontal movement will require corrections to be made to both arcs if neither is within 10 ° of horizontal. Remember to re-centre the crystal and to note the scale readings.

(v) If a large change has been required, repeat steps 4 and 8 of *Protocol 3* until the circles are aligned with the centre of the primary beam. Although it may be very close, the crystal will probably not be aligned perfectly until further adjustments are made.

ii. Small angle precession photographs

Fine adjustments of around 1.5 ° or less to the crystal, as aligned from a still photograph, are made in the following manner (*Protocol 4*).

Protocol 4. Small angle precession photographs

1. Mount a loaded film cassette on the camera. It is not usually necessary to use a full piece of film for small angle precession photographs.
2. Set the angle on the precession arm of the camera to 2–3 °.
3. Switch on the motor so that camera precesses.
4. Irradiate the crystal with unfiltered X-rays for an exposure time roughly 4 times that required for the still photographs.
5. Develop the film. The film should show a pattern similar to that illustrated in *Figure 11b*. A slightly misaligned crystal gives rise to an unsymmetrically filled circle like that shown. If the backstop shadow obscures a significant portion of the central circle, move the backstop towards the film.
6. As for the still photograph, it is essential to define the central point where the beam would strike the film. Some cassettes produce fiducial

Protocol 4. *Continued*

marks to help with this. With a sharp pair of dividers, score two lines on the film joining symmetrically related spots on either side of the X-ray beam. The intersection of these lines is the centre of the pattern and is not necessarily related to the centre of the backstop shadow.

7. Measure right and left horizontal distances (h_1 and h_2, *Figure 11c*) from the centre of the pattern to the edge of the circle and calculate the difference between them (Δ). The correction required to move the pattern to the central position is determined approximately by the following equation:

$$\Delta \approx 4 \times \varepsilon \times F \tag{4}$$

where F is the crystal-to-film distance, ε is the angular correction in radians. For $F = 75$ mm, $\varepsilon \approx 12$ minutes of arc; for $F = 100$ mm, $\varepsilon \approx 8.5$ minutes. Adjust the horizontal arcs of the goniometer head accordingly and record the values (see Section 4.2.2).

8. Repeat the correction for the vertical mis-setting (v_1 and v_2, *Figure 11c*) and apply the correction to the spindle. It is essential to return the precession arm to 0 ° exactly *before* changing the spindle setting. Watch the dial reading throughout the precession motion to see why this is important!

9. Take another small angle precession photograph following steps 1–9 of this protocol. Repeat as necessary until the crystal is perfectly aligned (see *Figure 11d*). Perfect alignment is characterized by a symmetrical circle of spots surrounding the centre of the pattern. The intensities of equivalent spots on opposite edges of this limiting circle must indicate an identical extent of 'clipping'.

10. Move the spindle to the other position at which a major zone was found. Perform *Protocol 4*, steps 1–9 at this new spindle setting. Note that the offset, Ψ, at 0 °, which can be measured directly with a protractor (and the correction directly applied to the vertical arc), becomes the *horizontal* mis-setting after a rotation of 90 °.

---

iii. Preliminary observations and measurements

From your sets of still and small-angle precession photographs, it should be possible to estimate the dimensions of the unit cell and to get some ideas about the symmetry. These estimates will not be accurate but may well be the limit that can be obtained from weakly diffracting crystals. Some qualitative observations may also be made.

(a) How close together are the spots?
If it appears that spots are overlapping each other, it will be necessary to

magnify the diffraction pattern by moving the film position farther away from the crystal. A smaller X-ray collimator will also help.

(b) How far out on the film does the diffraction pattern extend?
This gives a rough idea of the resolution. Resolution is discussed in more detail in Section 4.4. Well-diffracting crystals tend to produce patterns which extend to the film's edges. If the diffraction pattern is relatively feeble, it may be possible to move the film closer to the crystal keeping in mind that you want to avoid overlapping spots.

Calculations from stills: this gives a direct measure of real axis length.

(a) On a still which is aligned, measure the distances from the central beam position to each ring of diffraction spots, r_n, by halving the diameter.

(b) Perform the following calculation:
Put $\nu_n = \tan^{-1}(r_n/F)$ whereupon the axis dimension, d, is obtained from

$$d^* = 1/d = (1/\lambda)\ (\cos \nu_n - \cos \nu_{n+1}) \qquad (5)$$

where d^* is the reciprocal lattice spacing, λ is the radiation wavelength (1.542 Å for Cu), and F is the crystal-to-film distance. Average the results from several pairs of adjacent circles. For the most accurate result, take a still with filtered radiation after final setting as described in Section 4.2.3.

(c) Measure the other still photograph and make a similar calculation.

iv. Cloverleaf photographs

Before taking the full precession exposure, it is a good idea to check the position of the annular screen. To do this, 3 or 4 screened still photographs are taken on the same piece of film at various settings of the camera's precession arc as explained in *Protocol 5*.

Protocol 5. Cloverleaf photographs

1. Insert a loaded film cassette.
2. Insert a nickel filter between the generator tube and the collimator. This is generally a rotatable disc on the X-ray tube housing.
3. Set the screen position (56 mm for a 15 mm screen and $\mu = 15°$).
4. Set the arm to the desired precession angle. Angles ranging from $\mu = 9$–23° are typical in protein crystallography. The larger the angle the higher the resolution obtained but the longer will be the exposure. For feebly diffracting crystals, start with an angle of around 9°. Having set the screen and the precession angle, check that no collision occurs between any parts of camera during one complete revolution of the precession arm.

Protocol 5. *Continued*

5. Take four still photographs, one at each of 0 °, 90 °, 180 °, and 360 ° settings of the precession arm. The exposure time should be the same as that used for other still photographs.
6. Develop the film. With the screen in the proper position, each ring on the film should be populated with diffraction spots (see *Figure 11e*).

---

v. Full precession photographs

No further changes in crystal orientation or camera settings should be neccessary at this point. Zero-level photography is described in *Protocol 6.* An upper-level photograph requires the cassette to be moved toward the crystal by a distance of F × n × d* where F is the zero-layer crystal-to-film distance, n corresponds to the order of lattice layer at reciprocal spacing, d^*. For $n = 1$, d^* is $1/d$, the axis dimension calculated from the still photograph.

Protocol 6. Zero-level photograph

1. Insert the loaded film cassette. A full piece of film should be used.
2. Insert the annular screen (e.g. one of radius 15 mm). The crystal-to-screen distance, S, is related to the radius of the screen, r_s, and the precession angle, μ, by:

$$S_n = r_s \text{ cotg } \cos^{-1} (\cos \mu - nd^*) \tag{6}$$

   For zero layer, it reduces to $S_0 = r_s \text{ cotg } \mu$. All three parameters are variable, at least to the extent that camera geometry will allow.
3. Switch on the motor so that the camera precesses.
4. Expose the film overnight. A strongly-diffracting crystal will give a reasonable 15 ° precession photograph in 4–8 h. A feeble crystal may require 24 h or longer.
6. Develop the film. The resulting photograph should appear similar to that in *Figure 11f*.

Generally, the zero and first-layer photographs are taken at the first setting before the crystal is aligned along the second axis. It may not be possible to align the second axis if the twist of the rows on the first photograph is greater than about 20 °, unless the crystal is repositioned on the goniometer head.

4.3 Determination of cell dimensions and space group

The still photographs have already provided an estimate of unit cell dimensions. These can be used as a check for calculations made from full precession photographs where spacings of rows of spots on the film are related to the reciprocal axis lengths. The best way to measure the spacings is to measure the distance between as many consecutive spots as are fully recorded and divide by the number of spaces between them. When measuring spacing between spots, be aware that systematic absences may be present, particularly along the $h00$, $0k0$, and $00l$ rows (in other words, the reciprocal axes a^*, b^*, c^*). It may be better to measure spacings between spots along rows parallel to the axis directions to avoid confusion of, e.g. a reciprocal axis which appears twice as long as it is, with a reciprocal axis having every other reflection absent. Centring may also occur, giving the appearance of long reciprocal axes in adjacent parallel rows as well. The axial rows of the reciprocal lattice cross at the X-ray beam position on a zero-level photograph. If all of the angles between the rows are 90 °, then 'a^*' can be converted directly into 'a' since $a = 1/a^*$, and so on. However, in cases where one or more angles are not 90 °, the conversion is more complicated and requires using formulae found in the standard texts; e.g. in the monoclinic system $a^* = (1/a) \sin \beta$, $b^* = 1/b$, and $c^* = (1/c) \sin \beta$, with $\beta^* = 180° - \beta$ and conventionally β is chosen to be greater than 90 °.

Space group determination proceeds as follows:

(a) Observe the symmetry of the diffraction pattern of the zero-level precession photographs. This gives the diffraction (Laue) symmetry given in *Table 2* for the 11 relevant classes (and helps to ensure that principal zones have indeed been photographed). Diffraction symmetry always has an inversion centre; e.g. a triclinic cell, P1, has $\bar{1}$ diffraction symmetry. It is the appearance of extra symmetry which allows determination of crystal class. At this point the axes can be assigned as a, b, or c such that α, β, and γ are close or equal to 90 ° (unless a trigonal or hexagonal cell is suspected) and the cell is primitive (see *Table 2*). A zero-layer photograph by definition, arises from either the $hk0$, $h0l$ or $0kl$ sets of planes. These can be assigned arbitrarily in the case of certain space groups. The International Tables for X-ray Crystallography (19) will help with the task of assigning axes according to crystallographic convention. In general, the unique axis is b for monoclinic cells and c for cells of higher symmetry. The upper layer photographs, which contain no reciprocal axes, must also be assigned as hkn, hnl, or nkl where $n \geq 1$.

(b) Index the spots, h,k,l, on each film. Be aware that systematically absent reflections also require indexing. By keeping the film in a transparent envelope, axes and so on can be drawn on this without damaging the film.

(c) Analysis of systematically absent reflections in the diffraction pattern further pinpoints the space group. Determine the Bravais lattice first by considering the centring (I then F, B, or C according to *Table 2*), then any screw axes.

(d) Check that assignments of systematic absences are consistent with upper-level photographs as well. The upper layers also allow, e.g. distinction between a 6-fold and a 3-fold axis. (These look the same on a zero level photograph.)

(e) Identify, with use of the International Tables, a list of space groups compatible with the observed diffraction patterns. In some cases there is no ambiguity; e.g. $P2_12_12_1$, whilst in others no distinction is possible until the structure solution is under way, e.g. I222 and $I2_12_12_1$ have identical systematic absences as have the enantiomorphs $P3_121$ and $P3_221$ where only the hand of the screw axis differs. The convention '3_2' indicates a clockwise 3-fold (120 °) rotation and the subscript '2' indicates a translation of 2/3 along the unit cell. This is identical to an anti-clockwise rotation with a translation of 1/3.

(f) In general try to find as high a symmetry space group which is consistent with your observations and work to lower symmetries as need be.

(g) From the unit cell dimensions, the molecular weight of the molecule and V_m, determine the approximate number of molecules in the unit cell. Knowing the crystal system helps in this; e.g. if the crystals are orthorhombic, there must be a multiple of 4 molecules in the unit cell. (Note that a 'molecule' may also be some identically repeated portion of the protein or polynucleotide.)

4.4 Calculation of resolution

Resolution can be defined as the ability to distinguish individual parts of an object during its examination by radiation. Thus a diffraction pattern at 6 Å resolution means that data are available to interplanar spacings of 6 Å or larger. Similarly, diffraction to 1.8 Å resolution is expected to give much more information since the plane spacing is smaller and consequently the atomic positions are better defined.

- Use one of the still photographs for measuring maximum resolution. (Maximum resolution of a full precession photograph is limited by the chosen precession angle.) It is possible, however, that the diffraction pattern does not extend to the edge of the film, in which case the resolution is limited by the diffracting power of the crystal. Measure the distance from the centre to the limit of the diffraction pattern.
- Apply Bragg's equation:

$$d = \lambda/\{2 \sin [½.\tan^{-1} (r/F)]\} \quad (7)$$

where d is the minimum spacing which can be resolved (e.g. *resolution*), r is the distance from the centre to the edge of the diffraction pattern, and F is the crystal-to-film distance.

4.5 Other techniques for diffraction data collection

Precession photography gives an undistorted image of a reciprocal lattice plane. It does, however, require care and experience to align the crystal and it is fairly slow. Other methods commonly in use by protein crystallographers are diffractometry, the rotation/oscillation method, and the Laue technique. In combination with these methods is the growing availability of other 2-dimensional (or area) detectors such as the TV-type (ENRAF-Nonius FAST), the multi-wire (Siemens), and the image plate (Rigaku or Mar-research).

4.5.1 Diffractometry

Modern 4-circle diffractometers are the backbone of small-molecule crystal structure laboratories and have sophisticated control programs which allow cell dimensions to be obtained with little if any user intervention. However, most instruments use Mo radiation and the crystal-to-detector distance is often not large enough to allow easy resolution of spots with spacings typical of proteins. The instrument consists of a series of concentric circles, 3 forming an Eleurian cradle capable of rotating the crystal to (nearly) any angle relative to the X-ray beam, the fourth moving the detector in a horizontal plane. Considering the Ewald construction again (*Figure 5*), using the 3 circles of the goniometer, it is possible to orient the normal to any desired set of crystal planes in the horizontal plane in such a way that the normal bisects the angle between the incident X-ray beam and the detector. This satisfies Bragg's Law, and the reflection is observed by stepping the crystal from one side of the exact bisecting position to the other, thus moving the crystal through the reflecting position. A plot of detector counts versus angle will then show a peak as the reflection passes through the Ewald sphere. If such an instrument is available, preferably with a Cu tube and an extension on the detector arm fitted with a He path, it may be worth trying to determine the cell dimensions.

Whilst this approach may appear to be the simplest, and with good crystals it is very convenient, it is possible to find nothing in the initial scanning, and if no reflection is found in an hour the method is unlikely to be successful.

4.5.2 Rotation/oscillation photographs

The precession method gives a photograph containing just one level of the reciprocal lattice. The screen filters out all of the other reflections and so the method is rather inefficient. By rotating the crystal about an axis (AX is horizontal in *Figure 10*), the reciprocal lattice follows, cutting the Ewald sphere and producing extended circles as shown in *Figure 12* (*c. Figure 11a*). Provided the rotation angle is not too large, adjacent levels will not overlap and data for many layers are recorded on the single photograph. It is normal practice to oscillate the crystal many times during the exposure to minimize

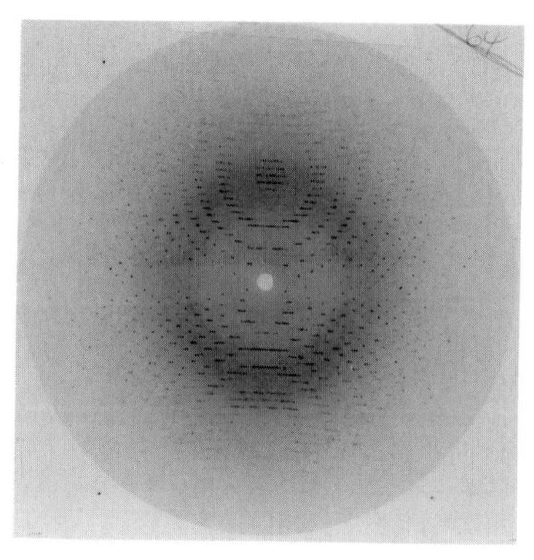

Figure 12. A 3 ° oscillation photograph of a hexagonal, cytochrome c_4 crystal taken with synchrotron radiation. Several major zones ([100], [1–10], and [010]) can be seen. The rotation axis was horizontal.

the effects of fluctuating X-ray intensity and crystal decay. When adjacent oscillation ranges are obtained, the partially recorded reflections on each film can be added to produce a full set. Carousels holding up to 8 film cassettes are available to make the process more automatic (20).

4.5.3 Area detectors (or position sensitive detectors)

Recently, position sensitive detectors have become commonplace and these combine the convenient digital output of the diffractometer with the high data capture rates of film. The mode of operation is essentially that of the rotation camera except that the oscillation angle is usually 0.1–0.5 ° and reflections are integrated as they pass through the Ewald sphere. The data are stored on magnetic tape or disc and further processed by computer in an automatic seach procedure. The unit cell dimensions are calculated. It may be necessary to plot out layers of data to ensure that the space group is correctly determined if no photographs have been taken. Since a medium resolution data set can be collected on such an instrument in the time taken to obtain a single precession photograph, the tendency with such an instrument is to shoot first and ask questions later!

5. Concluding remarks

The object of this brief excursion into X-ray crystallography has been to introduce the ideas and methods required to collect the information necessary for the first short publication on a new crystalline material. Papers should include not only the purification and crystallizing conditions, which should be reproducible, but also the techniques employed to obtain the X-ray diffraction data and the crystal lifetime in the X-ray beam and on the shelf. The unit cell dimensions and space group together with the resolution obtainable from a crystal have been the main concern of this chapter. V_m, the number of molecules in the asymmetric unit, the solvent content, and any comments about the subunit structure are also generally mentioned. Finally, crystallographers always enjoy talking about their subject and you will have discovered that the technique requires a modicum of dedication. Therefore do seek guidance in getting your project under way!

References

1. Glusker, J. P. and Trueblood, K. N. (1985). *Crystal structure analysis: a primer*, (2nd edn). Oxford University Press, New York.
2. Stout, G. H. and Jensen, L. H. (1989). *X-ray structure determination*, (2nd edn). John Wiley, New York.
3. Hirs, C. H. W., Timasheff, S. N., and Wyckoff, H. W. (ed.) (1985). *Methods in Enzymology*, **114**, 1; **115**, 1.
4. Blundell, T. L. and Johnson, L. N. (1976). *Protein crystallography*. Academic Press, London.
5. Bragg, W. L. (1968). *Sci. Amer.*, **219**, 58.
6. Harding, M. M. (1968). *Chem. Brit.*, **4**, 548.
7. Helliwell, J. R. (1987). *Protides Biol. Fluids*, **35**, 46.
8. Johnson, L. N. and Hajdu, J. (1990). *Eur. J. Biochem.*, **29**, 1669.
9. Rossmann, M. G. (1990). *Acta Cryst.*, **A46**, 73.
10. Hope, H. (1990). *Ann. Rev. Biophys. Biophys Chem.*, **26**, 107.
11. Acharya, R., Fry, E., Stuart, D., Fox, G., Rowlands, D., and Brown, F. (1989). *Nature*, **337**, 709.
12. Quiocho, F. A. and Richards, F. M. (1964). *Proc. Natl. Acad. Sci. USA*, **52**, 833.
13. Matthews, B. W. (1968). *J. Mol. Biol.*, **33**, 491.
14. Hodgkin, D. C. and Riley, D. P. (1968). In *Structural molecular biology*, (ed. A. Rich and N. Davidson), pp. 15–28. Freeman, San Francisco.
15. Henderson, R. (1990). *Proc. Roy. Soc. Lond.*, **B241**, 6.
16. Hajdu, J., McLaughlin, P. J., Helliwell, J. R., Sheldon, J., and Thompson, A. W. (1985). *J. Appl. Cryst.*, **18**, 528.
17. Teng, T.-Y. (1990). *J. Appl. Cryst.*, **23**, 387.
18. Buerger, M. J. (1964). *The precession method in X-ray crystallography*. John Wiley, New York.
19. Hahn, T. (ed) (1987). *International tables for X-ray crystallography*. D. Reidel Publishing Co., Dordrecht, Netherlands.
20. Arndt, U. W. and Wonacott, A. (ed) (1978). *The rotation method in crystallography*. North-Holland Publishers, Amsterdam.

13

Automating crystallization experiments

K. B. WARD, M. A. PEROZZO, and W. M. ZUK

1. Introduction

This book on the practical aspects of preparing single crystals of macromolecules suitable for diffraction analysis describes in detail the various activities required to find conditions suitable for optimum crystallization. These often involve the repetitive manipulation of small (1–10 μl) volumes of reagents to produce a solution which must then be inspected visually at intermittent times throughout a period ranging from several days to several months. This experimental protocol is ideally suited for the use of laboratory automation equipment. Consequently, several research groups have developed equipment in which one or more laborious, repetitive, and tedious operations have been automated.

In this chapter, we briefly review these systems. We aim to provide enough information so that the reader can evaluate which system might be most suitable for a particular application. Readers can consult the original literature to learn more about a particular system, and develop their own system based upon the experiences of others.

The initial projects which aimed to automate crystallization procedures were driven by one or more of the following goals:

- to reduce the drudgery associated with the repetitive manipulations involved, and thus to increase the efficiency of these operations;
- to increase safety when crystallizing highly toxic compounds;
- to increase the throughput of material tested for crystallization by increasing the speed with which the operations can be performed.

As experience was gained with these systems, it soon became clear that they could also:

- allow crystallization experiments to be conducted remotely, via, e.g. telerobotics;

- provide valuable insight into mechanisms of crystallization by controlling and monitoring the parameters of the experiment, and by recording video observations during the progress of the crystallization trials;
- provide a convenient interface between the experimentalist and the experiment. This will expand the user community, even to those with limited experience in the preparation of crystals.
- provide a convenient means to record in detail the results of experiments and to develop a data base of conditions used in crystallization trials;
- provide dynamic modification of crystallization parameters; i.e. based upon the signals received by the monitoring equipment, the crystallization parameters could be modified and optimized during the time course of the experiment.

An 'ideal' automation system would possess all the capabilities mentioned above and, no doubt, many more. Practical systems presently in use fall short of this ideal, but each incorporates one or more of these advantages over manual systems.

2. Automated fluid-handling systems

Because most crystallization procedures involve the manipulation of small volumes of reagents, a great gain in efficiency can be had by employing some means to handle these fluids automatically.

2.1 Automated syringes

2.1.1 Introduction

The crystallization scheme of Chayen *et al.* (1) utilizes a 96-well microtiter plate (Dynatech Laboratories) to prepare microbatch crystallization experiments in a semi-automated mode. Each experiment volume is less than 2 μl, making this technique ideal for screening a wide array of conditions. One plate containing 24 experiments can be prepared in 15 min or less. To set up a tray, the operator moves the customized microbore pipette tip from well to well at a prompt from the control computer, an IBM PC compatible, while the computer drives the motorized solution delivery syringes. Monitoring for crystal growth is performed conventionally via microscope.

2.1.2 Hardware

The solution delivery system consists of four separate liquid tubing lines, each including a stepper motor-driven syringe, a series of ground glass syringes, and a customized multibore microtip. On each line, a 100 μl gas-tight Hamilton syringe is driven by the 12 000-step stepper motor (Model DB3, Douglas Instruments Ltd.) and is connected in series to two three-way valves (Omnifit Ltd). One port of each of the three-way valves is connected to a ground-glass syringe that acts as a solution reservoir. The downline syringe contains the solution to be dispensed, while all lines and syringes upline from this syringe are filled with silicone oil (Silicone Fluid 200/1CS, Dow Corning). From the downline syringes on each line, the four lines converge at the multibore microtip, which is fabricated from and integral with the tubing lines.

2.1.3 Software

A PC computer is used for experiment design, preparation, and execution. The control software, called XSTEP, is written in Microsoft QuickBASIC. Included in the control software is on-line help, syringe calibration, syringe manipulation, and experiment design. The stepper motors are controlled by DOS device drivers written in machine code.

2.1.4 Experiment execution

i. Experiment design

To prepare a set of 24 experiments, the experimenter determines the types and concentrations of each component to be dispensed with the help of the XSTEP software. For a given protein, a buffer system and precipitating agent are chosen, and the pH and concentrations of each component in the drop are determined. A 6 × 4 matrix of experiments is prepared. The matrix is set up as a gradient in crystallizing agent concentration versus protein concentration, or varying pH values. The matrix is presented on the screen, and the experimenter inputs the desired concentration or pH values in the corners of the grid. The inner values are automatically determined by linear stepwise interpolation.

ii. System setup

The XSTEP software contains routines for system manipulation and calibration that are used in preparing the syringes. The syringe lines are initially primed with silicone oil, used as an incompressible and immiscible liquid to displace the desired volumes of each component of the crystallization volume. The appropriate solutions are then placed in the downline ground glass syringes, and the tubing lines from the downline syringe to the microtip are flushed with that solution. Typically, one line dispenses protein, one line dispenses water, and the other two dispense buffer and precipitating agent. Protein is aspirated from the multibore microtip into one line to conserve it; only enough protein to prepare the plate is used. Finally, the wells in the plate are filled manually with oil using a multiple pipettor (Alpha Laboratories).

iii. Experiment execution

The experimenter holds the multibore microtip below the surface of the oil in the well as the stepper motors drive the syringes and dispense the solutions.

The sub-microlitre crystallization volumes of protein, precipitant, and other components are dispensed into the oil in the well. This oil acts as a barrier against evaporation as well as a thermal, hydrophobic cushion for the cyrstallization volume. Droplets in 24 wells are dispensed in 45 s.

2.2 Computer-controlled liquid delivery

Kelders *et al.* (2) utilize a computer-controlled liquid delivery system to automate several labour intensive steps of the 'hanging drop' vapour diffusion method.

2.2.1 Apparatus

A Model 221/401 sample changer and dilutor from Gilson Medical Electronics is used to deliver reservoir solutions and droplet solutions to two multiwell plates developed by Omnilabo Holland BV. One plate holds the mother liquor, while the second plate holds the droplets. Both plates are situated side-by-side in the working area of the Gilson apparatus. Each well of the two plates can accommodate 300 μl of solution. The bottoms of the plates are flat and transparent, permitting observation of the experiments with a microscope.

The Gilson apparatus consists of two parts. A Model 221 sample changer positions a syringe needle to an accuracy of 0.1 mm in the x and y directions above the multiwell plates, and a Model 401 dilutor distributes fluid to the syringe. The Model 401 can deliver fluids either from a small vial located within reach or from a remote reservoir via tubing.

2.2.2 Experimental methods

In a typical experiment, the Model 401 dilutor dispensed PEG solutions from a reservoir to each individual well. The apparatus was rinsed thoroughly with double distilled water after all wells were filled. Droplets were made by first filling the sample changer needle with distilled water to prime the system, drawing a small amount of air to prevent mixing, drawing 5 μl of protein solution with the needle and dispensing the protein solution in the wells of the second plate, which had previously been siliconized. Next, 8 μl of reservoir solution were added to the protein droplets. No external mixing was performed. After droplet preparation, the plate was covered manually by a piece of transparent adhesive plastic foil (Alkor, made by Bijholt and Heuvelen BV). The top, non-adhesive side of the plastic foil was covered with silicone grease, and a small hole was punched in the foil above each droplet to allow vapour equilibrium to be reached between the droplet and the reservoir. Plate II was then inverted and sealed onto plate I. The two plates were taped securely and stored at 20 °C. Droplets were subsequently examined using a microscope.

2.3 Automated pipetting stations (Cox and Weber)

Cox and Weber (3, 4) have developed an automated fluid handling system to produce crystals with the hanging-drop method. The system is organized around a pipetting station and a multiport rotary valve, both of which are under computer control.

2.3.1 Apparatus

The Cox–Weber apparatus prepares buffer and reservoir solutions by diluting stock solutions, and drops are produced by mixing small amounts of reservoir and protein solutions. The fluid handling system utilizes a microcomputer-controlled pipetting station, manufactured by Micromedic Systems, Inc., which contains an *x–y* translating table and two syringes (2 ml and 40 ml). Solutions are dispensed by the pipetting station into individual wells of a Linbro depression plate positioned on the table. The table also holds a plexiglass template containing siliconized 22 mm diameter glass cover slips. During an experiment, the table moves the Linbro plate horizontally beneath the pipette tip. The tip itself can move in the vertical direction while dispensing liquids. Experimental parameters are chosen with menu-driven computer software and all solution concentrations are calculated accordingly during the course of an experiment.

2.3.2 Experimental methods

Sixteen individual crystallization experiments are carried out in a 4×4 array in each Linbro plate. The experiments are arranged so that solution pH is varied along columns of the plate and precipitant concentration along rows. Two or three component buffer systems are used to control pH. Up to six solutions may be dispensed by an eight-port rotary valve. These solutions included doubly distilled water, two solutions for controlling pH, precipitant solution, and two additive components.

One solution is dispensed into each well of the 4×4 array before another is added. All tubing is flushed twice with stock solution prior to dispensing to a well. Denser solutions are added after water and buffer solutions to promote mixing. Solutions are dispensed from the pipette tip as the *x–y* translating table moves each well into position. Once all wells have been filled, the transfer syringe draws 40 μl of well solution into the pipette tip, expels 3 μl onto the associated glass cover slip, and flushes and rinses the tip with water. The process is repeated for each well and cover slip. Enough protein solution for four drops is drawn into the pipette tip and then added to the droplets on the cover slips. Finally, cover slips are manually inverted and placed on the appropriate well and sealed by vacuum grease which has been previously applied to the well. A crystallization tray with five well components can be arranged in approximately 25 min. Droplets are examined periodically using a $20 \times$ microscope.

2.4 System II of ICN

A system based upon the design of Cox and Weber (3, 4) is available commercially from ICN Biomedicals, Inc. It is called the Robotic Protein Crystallization System II, and the system allows one to prepare either hanging-drop or sitting-drop experiments. A major improvement over the earlier designs is the ability of the System II to transfer automatically coverslips containing crystallization droplets to the test plate.

3. Robot-based automation systems

In addition to handling fluids, some systems can also manipulate crystal growth labware such as cover slips and crystallization plates. These systems are somewhat more elaborate (and expensive), and offer the possibility to include automated storage and visual inspection of crystallization experiments.

3.1 The NRL Zymate robotics system

3.1.1 Introduction

The automated crystal growth system of Ward *et al.* (5–9) at the Naval Research Laboratory (NRL), will prepare, monitor, and design experiments automatically. A customized Zymark laboratory robot prepares vapour diffusion crystallization experiments in 15 well CrystalPlates®, either as sitting or sandwich drops. Initial screening experiments test a wide array of conditions by varying precipitating agent type, concentration, buffer type, and pH. A high-resolution video camera captures images of the crystalline droplets and stores them on an optical disc recorder. The images may then be observed directly by the experimenter. Routines have been developed to allow the host computer to digitize, process, and analyze the images to detect crystal growth. Currently, one CrystalPlate® of 15 experiments is prepared in 90 min.

3.1.2 Hardware

i. CrystalPlate®

The vapour diffusion experiments prepared automatically by the robot are set up in the ACA CrystalPlate® (ICN Flow), a robot-friendly expendable crystallization tray designed by Noel Jones, Keith Ward, and Mary Ann Perozzo for the purpose of automated protein crystallization experiments. The plate accommodates 15 experiments in three rows and five columns. Experiments can be either sitting or sandwiched drop vapour diffusion experiments. A 500 μl reservoir is in vapour contact with a 5–50 μl drop positioned between two cover slips. Because drops larger than 10 μl are sandwiched between glass plates, the optical path is ideal for video imaging. The cover slips are positioned on lips above oil troughs filled with oil by the robot to seal the cover slips onto the plate.

ii. Zymark modules

The automated crystallization system uses a Zymark laboratory robot in conjunction with many customized components to automatically prepare crystallization experiments. The commercially available components of the Zymark laboratory robot include a robot controller, a robot arm, two banks of three automated syringes (Master Lab Stations), and two Power and Events controllers. The robot controller, with a BASIC-like proprietary software called EasyLab, communicates with each component of the system via RS232 protocol, and includes an electronic board for control of each module. Central to the system is the robot arm, which is located in the centre of the 5′ × 5′ table top. It has four degrees of freedom: reach, rotary, vertical, and wrist. Within its cylindrical envelope of motion, it has positional accuracy within 0.2 mm. The robot arm attaches to various end effector hands that are located in stands placed radially around the robot. The four customized hands perform the tasks of dispensing liquids to prepare the droplet and the reservoir solutions, manipulating cover slips, and moving CrystalPlates®. The placement of the table components is shown in *Figure 1*.

iii. Robot hands

- Forklift hand. The forklift hand is a two-pronged fork attached to a Zymate general-purpose hand that matches the slots along the side of the plate. The robot arm guides the forklift into the plate, and moves it from storage rack to work station.
- Cover slip handling hand. The cover slip handling hand manoeuvers cover slips with a vacuum suction device attached to the end of a pneumatic piston that is mounted vertically on a general-purpose hand. It is shown schematically in *Figure 2*. The piston is extended downward and a vacuum pump is activated by the Power and Events controller. A cover slip is retrieved when an in-line vacuum transducer, sensing a drop in pressure upon contact with the cover slip, retracts the piston by switching the piston solenoid valve. The cover slip is positioned above its intended location by the robot arm and is lowered onto the plate by extending the piston.
- Liquid delivery hand. The liquid delivery hand consists of two cannulas, positioned 90 ° apart and mounted on a general-purpose hand. It is shown schematically in *Figure 3*. Five tubing lines from the Master Lab Stations converge radially in a plane perpendicular to the output cannula. Solutions are dispensed simultaneously, and mixing of the solutions is achieved as they are expressed through the cannula into the reservoirs. By rotating the hand on the robot arm 90 °, the other cannula, plumbed directly to a syringe on the Master Lab Station, is positioned to dispense oil to the CrystalPlate®.
- Syringe hand. A Zymark syringe hand fitted with a 25 μl syringe has been

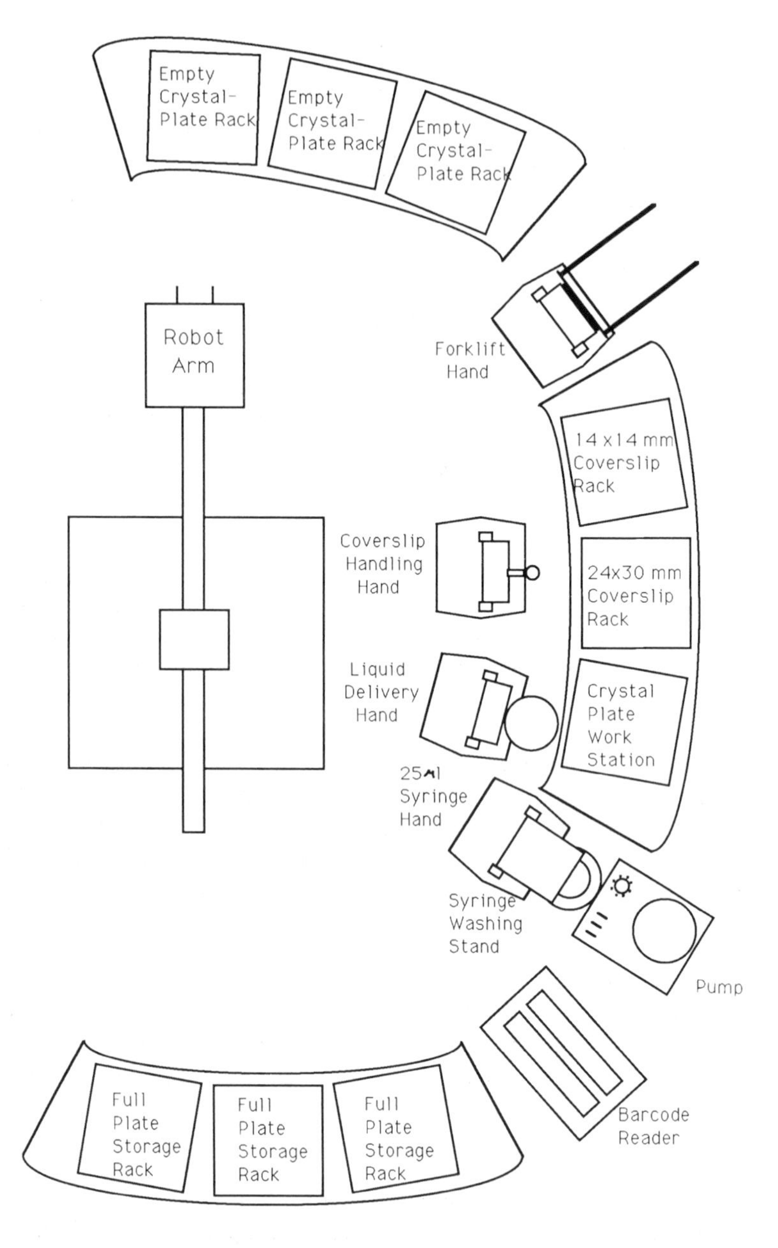

Figure 1. The radial layout of the robot hands and racks facilitates their access by the robot arm.

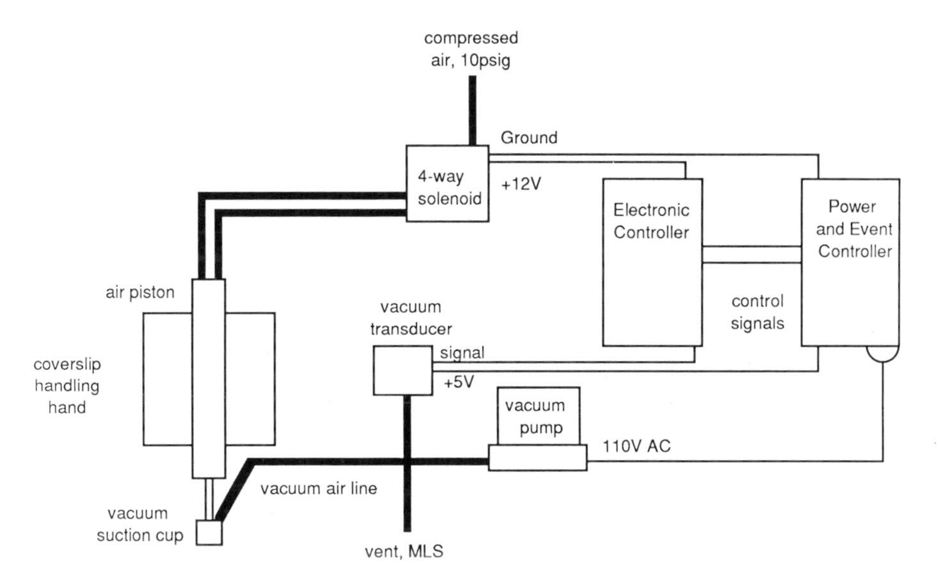

Figure 2. Schematic of the cover slip handling hand. The air piston is extended by switching a solenoid valve. A vacuum transducer senses when a cover slip is attached to the vacuum suction cup at the end of the piston, activating the solenoid, and retracting the air piston. The robot arm then positions the hand above the appropriate cover slip site, extends the piston, and vents the vacuum line to place the cover slip.

customized to run remotely through dedicated microprocessor control electronics. The syringe is activated to draw up solutions while it is attached to the arm. When it is parked, the syringe needle extends into a cylinder of water that is refreshed by a peristaltic pump. It is rinsed remotely by repeated activation of the syringe, while the robot arm performs other tasks.

iv. Motorized valves

Five motorized 8-port valves (Pharmacia LKB) allow selection of any five solutions, from 35 available, to prepare the reservoir solution for any row on the CrystalPlate®. The valves are plumbed to the syringes on the Master Lab Stations and are controlled by custom electronics.

v. Racks and stands

A total of thirty-three CrystalPlates® are stored in three racks prior to the experiment. The CrystalPlate® being prepared is positioned in a workstation. When the plate is completed, it is stored in one of three 8-slot experiment storage racks, where it can be accessed later for video monitoring. Cover slips of two sizes (14×14 mm^2 and 24×30 mm^2) are stored in separate racks. These plexiglas racks contain fifteen stacks of cover slips located in a

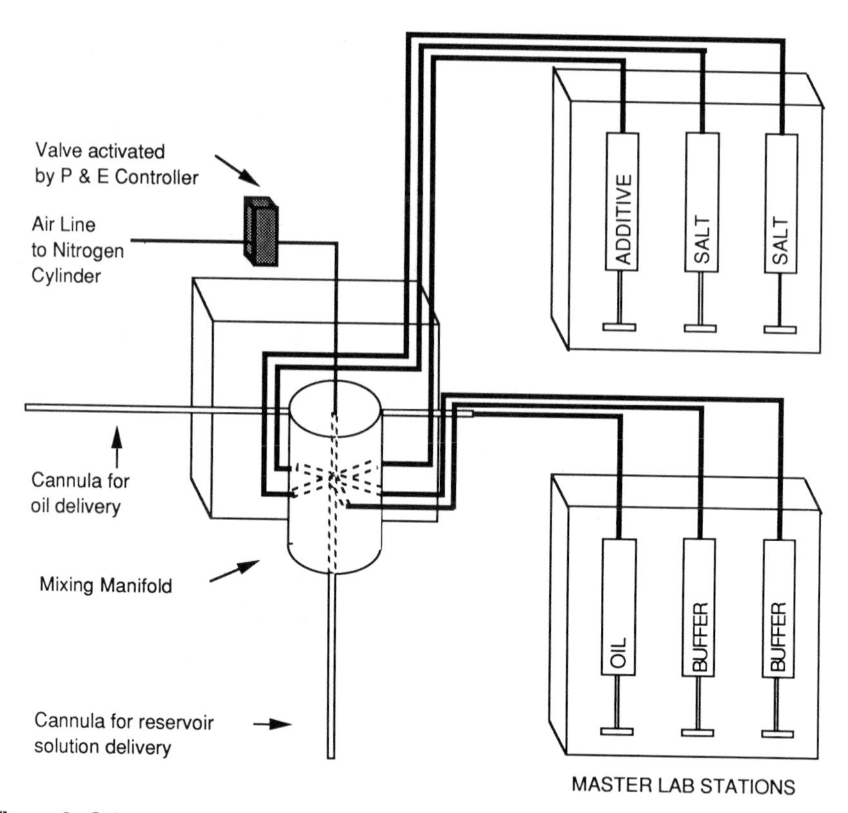

Figure 3. Schematic of the liquid delivery hand. The oil for sealing cover slips to the plate is dispensed from the horizontal cannula by rotating the hand 90 ° and driving the oil syringe of the Master Lab Station. The reservoir solutions are mixed and dispensed by driving the solution syringes simultaneously. The solution lines converge radially in the mixing manifold and the mixture is dispensed through the output cannula. The final drop is expelled by momentarily opening a solenoid valve on a pressurized nitrogen line.

3 × 5 array similar to that on the CrystalPlate® and are on stands located radially around the robot table.

vi. Computer control

A Standard 286, an IBM-AT compatible computer, controls the Zymate system via an RS232 interface. A second serial port serves a barcode label printer (Intermec #8625). The computer can also read input from a barcode reader (Symbol Technologies, LS6000A) through a keyboard wedge. *Figure 4* shows communication connectivity of the system.

3.1.3 Software

The automated protein crystallization system is controlled by a IBM-compatible 80286 computer, and Microsoft Windows 3.0 is used as the

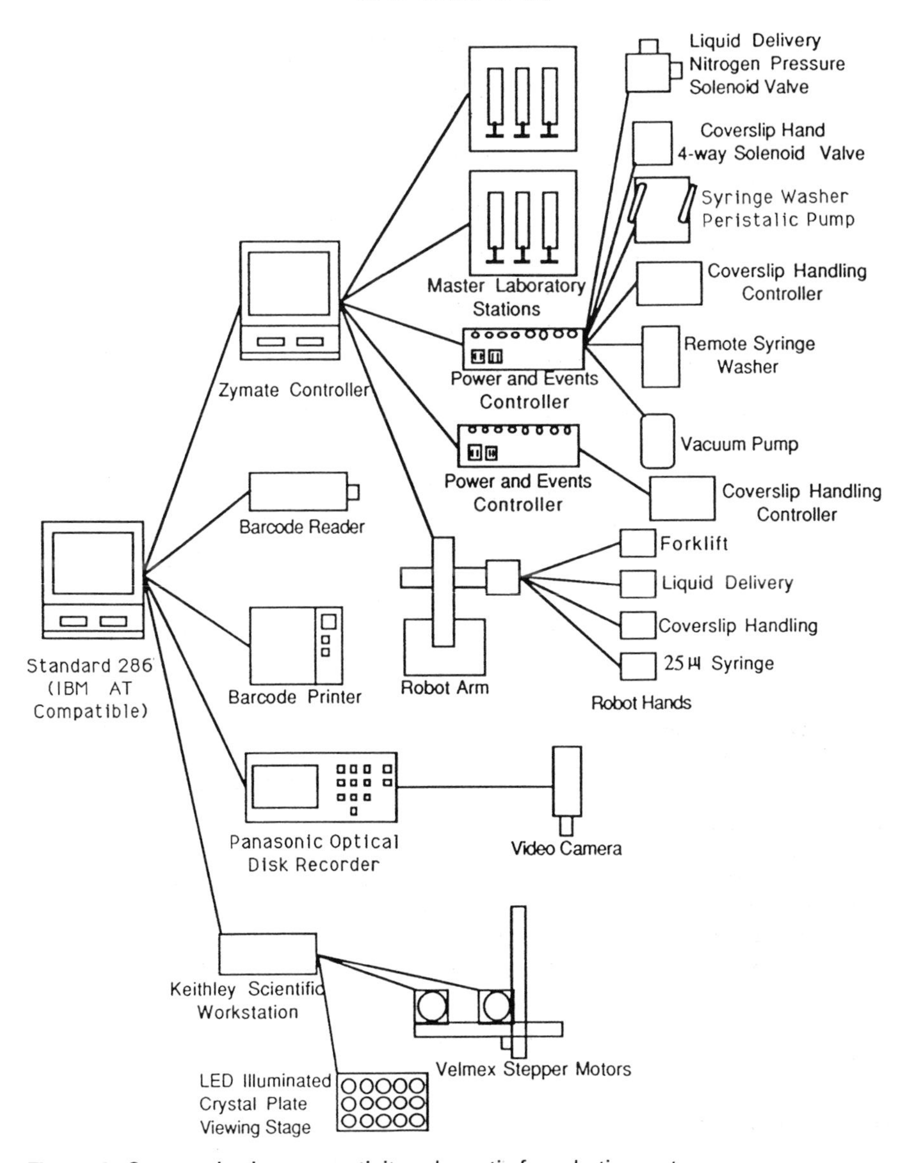

Figure 4. Communication connectivity schematic for robotics system.

graphical user interface. A Crosstalk terminal emulation package (DCA Communications) is used for generating barcode labels, and Microsoft Excel is used as experiment design software (M. Perozzo, in preparation). A text file containing all the experimental parameters is generated from an Excel template. Control software written in the C programming language downloads the parameters to the robot controller for one plate, writes a log file, records the barcode label, and initiates the experiment by starting a program on the robot controller. Simple do loops, goto, and if-then statements are used to structure commands that define the state of the robotic system components.

3.1.4 Experiment

i. Experiment design

Plate preparation begins with experiment design on a Microsoft Excel spreadsheet. The experimental parameters determined for the 3 × 5 well CrystalPlate® include buffer type, additive component type, pH, buffer concentration, and additive component concentration. Parameters that can vary along a row in droplet preparation are protein volume, reservoir volume, and seeding solution volume. Crystallizing agent concentration varies along a column.

A typical experiment will have a two-dimensional grid of varying concentration across each row, and varying pH down each column. One crystallizing agent, two buffer components, and water are usually dispensed into each well. A fifth additive ingredient, such as a detergent or counterion, may also be included. The spreadsheet presents the experimenter with lists of ingredients, concentration ranges, and pH ranges. The experimenter selects the lower and upper limits for each range, and the concentrations and pH of the intermediate wells are determined by linear stepwise interpolation. The pH values are achieved using two buffer solutions made with the same components, adjusted to the upper and lower pH limits for the buffer system. The volumes of each buffer to be dispensed for a given pH are calculated using the Henderson–Hasselbach equation.

ii. Set-up

After the experiment parameters are selected, an ASCII file of the values to be downloaded to the robot is generated. The textfile includes valve port numbers and solution volumes to be dispensed. A barcode label is generated with the format of protein/date/plate number/storage temperature.

iii. Execution

Under the control of a C language program, the computer sends commands to the robot controller to initialize the system, download experiment parameters, and initiate the plate preparation. A logfile of each plate prepared is generated. The robot is then commanded to prepare the plate.

The robot arm picks up the forklift hand and retrieves an empty CrystalPlate®. The barcode label, positioned manually on the edge of the plate prior to execution, is scanned by the laser barcode scanner. The plate is then positioned in the workstation by the robot arm. The robot arm parks the forklift hand and retrieves the liquid delivery hand. The oil delivery cannula is positioned over the first well of the plate and oil is delivered to an oil trough around each coverslip lip. When all wells have been prepared with oil, the robot arm rotates the hand 90 ° and positions the liquid mixing and delivery manifold over the first well. The syringes pump the appropriate volume of up to five components, mixed as described above, into the well.

The liquid delivery hand is parked and the cover slip handling hand is retrieved. The 14 × 14 mm^2 cover slips are retrieved from the cover slip rack, one at a time, and placed on the CrystalPlate®. The cover slip handling hand is parked, and the syringe hand is retrieved.

Up to this point, each operation is performed for each well in series on the plate. The next operation is delivering the drop onto the lower cover slip. Because the volume is very small, it is desirable to cover it as soon as possible in order to reduce evaporation, so only one drop at a time is prepared before retrieving the large cover slip and sealing the experiment. The syringe must also be rinsed before the next drop is prepared, and this is done while the cover slip is being positioned. The next series of operations is repeated for each drop.

The syringe hand is retrieved and a small aliquot of protein solution, between 3 and 10 μl, is obtained from a vial. The syringe then draws a small air bubble and then retrieves a small aliquot of the reservoir solution. The contents are dispensed onto the small cover slip on the CrystalPlate®. The syringe hand is parked and rinsed remotely as described above. The robot arm, meanwhile, retrieves the cover slip handling hand and places the large cover slip over the drop and reservoir. They are sealed off from the outer environment, but communicate by a vapour path with one another. The robot arm parks the cover slip handling hand and retrieves the syringe hand for the next droplet.

The finished CrystalPlate® is moved from the workstation to a storage rack, where it may be accessed for video-imaging analysis at any time.

3.1.5 Video monitoring of protein crystallization experiments

i. Video imaging equipment

A video acquisition system to monitor automatically the progress of crystallization experiments (5–7, 9) has been developed at NRL. The automated video monitoring system consists of an RCA TC 1005/01 high-resolution monochrome video camera with fixed optics mounted to an *x–y–z* stage driven by stepper motors, a Panasonic TQ-2025F Optical Disc Recorder (ODR), a PCVISION frame grabber installed in a Dell System 310 microcomputer, a Keithley System 501 Scientific Workstation interfaced to the Dell 310 and used for stepper motor control, and a Sony PVM-122 NTSC monochrome monitor. The system acquires images of individual crystallization droplets by moving the RCA video camera with the *x–y–z* stage above a stationary crystal plate. The *x–y–z* stage consists of three Unislide Motor Drive Assemblies (Velmex Inc.). Two Unislide B2509W1J slide assemblies serve as the *x* and *y* axes. Each of the B2509W1J slides has a resolution of approximately 13 microns per motor step. The *z* axis is provided by a Unislide

B2506BJ slide which has a resolution of approximately 7 μm. Images are focused by moving the camera up and down above the crystallization droplet using the high-resolution z assembly. Each slide is driven by a Bodine stepper motor. The motors are controlled by Velmex controllers interfaced to the Keithley Scientific Workstation to allow for direct computer control. The crystal plate is illuminated from below by three fibre optic strips attached to a single Xe light source.

Analog images of crystallization droplets for an entire crystal plate are stored on the Panasonic TQ-2025F ODR and are accessed at a later time for evaluation. The Panasonic ODR is a WORM device capable of recording 16 000 monochrome images on a single optical disc. The images may then be accessed randomly and quickly (1/30th s) and the recorder can display images for extended periods without synchronization problems or without damaging the medium, problems inherent in video-tape recording. The Panasonic ODR is interfaced to the Dell 310 microcomputer and is under software control. Experiments are monitored in the following way. The Zymark robot periodically retrieves and places completed crystallization plates on the imaging station. Images are acquired and stored for each droplet in the plate. After a tray is placed on the imaging stage, the light source is activated by the computer, the x–y–x stage moves the video camera sequentially to each well, and an image of each crystallization droplet is stored on optical disc. When the recording of an entire tray is completed, the plate is returned to the storage area by the robot and the background light is turned off. Images of an entire experiment stored on the optical disc may be scanned later by the experimenter quickly to ascertain if crystal growth has occurred in any of the droplets. In this way, experimenters never have to handle crystallization plates to monitor the course of an experiment.

ii. Digital processing and image analysis

The video-image acquisition system in our laboratory described above permits an investigator to quickly scan the progress of a crystallization experiment. However, in order to more fully automate crystallization methods, we are improving the system to allow the host computer to digitize, digitally process, and analyze droplet images to determine if crystal growth is occurring. Once an image has been processed, it may undergo simple analysis to detect its contents and determine the presence of particulate materials. This particulate matter may consist of protein crystals, precipitate, or impurities.

Once a droplet image has been acquired, it may be subsequently digitized with the PCVISION frame grabber. The digital image may then be processed to enhance features located therein. In order to locate objects in the image, specifically particulate matter, we apply an edge enhancement algorithm to accentuate object edges. Many complicated methods have been used to detect edges in a digital images (10). We have found that use of simple digital filters provides adequate results and performance in processing droplet images (9).

In our scheme of processing and analysis, a vertical Sobel filter (10) is applied to a droplet image, resulting edge features are located, and coordinates of edge points are stored. A horizontal Sobel filter is then applied to the original image and its edge points are recorded. The two edge-maps are then added, producing a fairly complete map of all object outlines in the original image. The final edge-map is then analyzed using straight-line finding techniques. Currently, using this processing and analysis procedure, the host computer is able to flag certain large features in a droplet image that may be crystals and differentiate to some degree between large crystals and precipitate. Refinement of analysis routines is necessary to properly locate and identify smaller crystals.

3.2 The Kodak robotics system

Rubin *et al.* (11) have described another version of an automated system which requires minimal manual intervention.

3.2.1 Design goals

The system was designed to minimize the amount of carryover fluids from previous experiments, to operate in a cold room (4 °C) environment, and to operate on as many as 15 consecutive Linbro plates (ICN Flow) with the minimum of manual intervention.

3.2.2 Hardware

The system uses a Hewlett-Packard Genechem robot (a Zymate robotic system with customized controller and software) and incorporates a unique greaser workstation to apply warm sealing oil to Linbro plates. A large slide is used to cover the entire working surface of the plate. The large plate greatly increases the speed with which a plate can be prepared, but prevents harvesting individual wells. The main purpose of the machine is screening, however, rather than the preparation of individual crystals for diffraction work.

3.2.2 Software

The system has a crystallization protocol design software module written in MDS command language or QuickBASIC. This screen-oriented, menu-driven system allows the user to select experimental parameters.

3.3 The Eli Lilly automated systems

Jones *et al.* (12–13) at the Eli Lilly Company in Indianapolis, Indiana, have described APOCALYPSE, an automated crystallization device similar in most respects to the NRL system described above. Indeed, the two groups collaborated on some aspects of the design. A major difference between the two systems is the nature of the software interface which allows the operator to design and control experiments.

3.3.1 The Asymtek system

More recently another system has been described (N. Jones, unpublished results) which Jones believes offers significant improvements over the former design. Unlike the former system which used a Zymate robot, the new system uses an Asymtek Automove Positioning Device, which is mounted in an inverted manner over an array of eight ACA CrystalPlates®.

Fluids are handled with a Hamilton Programmable Diluter/Dispenser. Fifteen 1.0 ml Hamilton syringes are used for fourteen stock solutions and one water supply. An additional 50 μl Hamilton syringe is used for preparing drops.

Automated operations are carried out by means of air-actuated tools which are located on a common arm. These tools are a cover slip pickup device, a high pressure line for dispensing petroleum jelly for sealing cover slips, a drop-making syringe, and a reservoir syringe.

4. Automated control of solvent vapour pressure

In most previous experimental designs, the relative humidity or partial vapour pressure of the solvent phase is either held constant or allowed to change over time as the crystallization experiment approaches equilibrium from its initial non-equilibrium state. Some workers have explored how this important parameter can be more carefully varied by direct experimental control.

4.1 Georgia Institute of Technology system

Wilson *et al.* (15, 16) have developed a novel system to regulate automatically the relative humidity in a sample chamber in order to control the equilibration rate between a hanging drop and a reservoir. The system utilizes computer regulation of N_2 gas flow over a hanging drop in a specially constructed crystallization cell to control the evaporation of solvent from the drop. During an experiment, the relative humidity in the cell is monitored using a simple humidity indicator (Thunder Scientific, Model PC-2101). A flat conductance cell with two miniature platinum electrodes is used to continuously monitor solution analyte concentrations. In addition, a thermal conductivity detector (Gow-Mac Instrument Co., Model 40) is used to measure the amount of water vapour removed from the cell.

Control of the system and data acquisition are accomplished with an Apple Macintosh II microcomputer using custom software developed with the LabVIEW laboratory program development environment (National Instruments, Austin, TX). The Macintosh II is interfaced to the crystallization cell through a National Instruments NB-MIO-16 multipurpose data acquisition board. Signals from the cell are first smoothed by National Instruments 5-B signal conditioning modules.

The Wilson–Suddath system is able to dynamically control the relative humidity in the crystallization cell. The protein droplet may be concentrated at a controlled rate by passing dry N_2 gas over the drop, while the droplet may be diluted by using a flow of N_2 gas which previously has been equilibrated with water. The authors have found that the control of relative humidity in this way provides a reliable method for dynamically controlling the growth of protein crystals. Studies using HEW lysozyme indicated that it was possible to increase the size of crystals obtained by closely controlling the vapour pressure of the crystallization system.

4.2 Automated control of reservoir

Another way in which control of the solvent partial pressure has been achieved is through controlling the composition of the equilibrium reservoir during the experiment, as described by Smith and DeLucas (17).

4.2.1 Hardware

A variable voltage DC power supply is used to control the flow of a peristaltic pump into a microdialysis or vapour-diffusion crystal-growth chamber. In this way, the time course of the concentration of crystallizing agent in the reservoir could be controlled and optimized.

5. Other systems

In addition to the more or less complete automated systems discussed above, there exist an increasing number of devices available which can be used to devise new systems or to serve as useful adjuncts to manual crystallization protocols.

5.1 Time-lapse video

The imaging system in our laboratory which was described above may also be used to produce very smooth time-lapse video recordings of individual crystallization experiments. The Panasonic Optical Disc Recorder, under software control, is utilized to record images of a single droplet at predetermined intervals. Before each image is acquired, the background light of the video stage is illuminated. After the image has been recorded and stored on optical disc, the background light is extinguished. When completed, the time-lapse video may be played back to observe the rates and other characteristics of crystal growth.

A video tape system has also been used by Koszelak and MacPherson (18) to produce time lapse microphotography of protein crystal growth. Images obtained with a Sanyo colour video camera mounted on a Swift microscope were recorded on a GYYR time-lapse VCR. The GYYR VCR has the capability of compressing real-time video from 1–240 times. At the maximum compression, 240 h could be recorded on a single 2 h VHS video tape. The authors principally used a compression of 72 times for their recordings. The VCR also provides a clock and calender display on the video monitor, permitting fairly accurate determination of crystal growth rate. The entire time-lapse system may be purchased from Resolution Technologies in San Juan Capistrano, CA. The authors used a polarizer between the sample and the light source to improve contrast and to permit observation of birefringence. The light source was cooled at the microscope stage through the use of a small fan. Using the time-lapse system, the authors initially studied lysozyme, canavalin, and the serine protease from penicillium cyclopium (18) and were able to, among other things, observe crystal growth from a visible nuclei stage, observe development of crystals with differing habits growing in the same drop, and measure and monitor overall crystal growth rates.

5.2 MG-1000 Microgripper

Using a piezoelectric polymer film, Microflex Technology, Inc. has developed an automated system which has the ability to manipulate small specimens, including single protein crystals (K. Ward, unpublished results). This apparatus, available from Bunton Instrument Company, Inc., or similar devices, may eventually be incorporated into more elaborate automation systems which will have provisions for harvesting and mounting crystals.

The device is called the MG-1000 Microgripper and can be used with commercially available micromanipulators. The Microgripper System consists of a Microgripper, interchangeable gripper tips (available in size from 5–1900 μm), and a Controller. The system is controlled either manually or via a standard RS-232 serial line, and offers continuous variation of gripper tip position over a 2.0 mm displacement to an accuracy of 1.0 μm. Forces exerted on the crystal being handled can range from 0–400 mg ($0–3.92 \times 10^{-3}$N).

5.3 Dynamic control

As more and more results are obtained from using automated systems, it is becoming clear that one should take more advantage of the ability to automatically monitor and control the progress of crystallization experiments.

5.3.1 Improved crystallization systems

As various groups collect data to describe the solubility phase diagram of purified protein samples, it is also becoming apparent that temperature can be a very useful variable to control the course of nucleation and crystal growth in a crystallization experiment. In addition, a number of workers have demonstrated that dynamic light scattering can provide useful clues as to the progress of nucleation kinetics, and that the presence of large (but still submicroscopic) crystal growth centres can be detected by the 'scintillation' signals produced from crystal facets by low-powered laser illumination (F. Rosenberger, in preparation).

The next step in automating crystallization systems will be to incorporate dynamic control, in which signals monitoring the course of the experiment will be evaluated and used to alter the experiment in progress. Several such systems using dynamic control are now under development. In these systems, the presence of crystal nuclei or submicroscopic crystals will be detected by light-scattering methods and the temperature of the crystal growth chamber will then be altered to achieve the most effective crystal growth.

5.3.2 Microgravity experiments

Much of the work on automation systems for protein crystal growth and the development of dynamic control systems has been driven by the desire to develop equipment which can be used to prepare crystals in the microgravity environment afforded on Space Shuttle or on present and future space stations. Future practical flight systems will be compact, miniaturized versions of the automated systems being developed for ground-based research. In addition, they will incorporate a good deal of telerobotic control, giving the experimenter on the ground the ability to direct and alter the course of the crystallization experiments.

6. Conclusion

Many advances have recently been made in developing automation systems for the preparation of single crystals of macromolecules. Systems developed have ranged from simple inexpensive automated syringe systems to elaborate systems having the ability to control all aspects of the design, execution, and analysis of the crystallization trial. All afford a marked improvement over manual manipulations.

Automation relieves the tedium and drudgery of manual manipulations of reagents and can therefore reduce the associated errors in experimental design and protocol. In the case where very large amounts of protein and many protein variants (e.g. expressed products of genetic recombinant experiments) must be routinely screened, the slow but round-the-clock ability of automation systems to accomplish these tasks is essential.

Note Added in Proof

A commercial system based upon the automated syringe system of Chayen *et al.* (1) (Section 2.1) has been announced by Douglas Instruments, Ltd. This

IMPAX 1 Protein Crystallization System is designed for microbatch experiments, but may be configured for vapour diffusion work. It is available from Cryschem in the U.S.

Oldfield *et al.* (19) recently described a modestly-priced system for protein crystallization trials comprising a Gilson Autosampler, two Gilson motor-driven syringes, and a three port valve. Software for controlling the system, written either in IBM QuickBasic or VAX Fortran, is available from the authors.

References

1. Chayen, N. E., Stewart, P. D., Maeder, D. L., and Blow, D. M. (1990). *J. Appl. Cryst.*, **23**, 297.
2. Kelders, H., Kalk, K., Gros, P., and Hol, W. (1987). *Protein Engineering*, **1**, 301.
3. Cox, M. J. and Weber, P. (1987). *J. Appl. Crys.*, **20**, 366.
4. Cox, M. J. and Weber, P. (1988). *J. Crystal Growth*, **90**, 318.
5. Ward, B., Zuk, W. M., and Perozzo, M. A. (1989). *Laboratory Robotics and Automation*, **1**, 157.
6. Ward, K. B., Perozzo, M. A., and Zuk, W. M. (1988). *J. Crystal Growth*, **90**, 325.
7. Zuk, W. M., Ward, K. B., and Perozzo, M. A. (1988). In: *Advances in laboratory automation robotics*, Vol. 4, (ed. J. R. Strimaitis and G. L. Hawk), pp. 217–34. Zymark Corp., Hopkinton, MA.
8. Ward, K. B., Perozzo, M. A., and Deschamps, J. R. (1987). In *Advances in laboratory automation robotics 1986*, (ed. J. R. Strimaitis and G. L. Hawk), pp. 413–33. Zymark Corp., Hopkinton, M.A.
9. Zuk, W. M. and Ward, K. B. (1991). *J. Crystal Growth*, **110**, 148.
10. Rosenfeld, A. and Kak, A. C. (1978). *Digital picture processing*, Vol. 2. Academic Press, New York.
11. Rubin, B., Talafous, J., and Larson, D. (1991). *J. Crystal Growth*, **110**, 156.
12. Jones, N. D., Deeter, J. B., Swartzendruber, J. K., and Landis, P. W. (1987). Abstract H4, p. 27. Abstracts of the American Crystallographic Association Annual Meeting. Austin Texas.
13. Jones, N. D. *et al.* (1987). *Acta Crysta.*, **A43**, C-275.
15. Wilson, L., Bray, T., and Suddath, F. (1991). *J. Crystal Growth*, **110** 142.
16. Wilson, L. (1990). Ph.D. Thesis, Georgia Institute of Technology.
17. Smith, H. W. and DeLucas, L. J. (1991). *J. Crystal Growth*, **110** 137.
18. Koszelak, S. and McPherson, A. (1987). *J. Crystal Growth*, **90**, 340.
19. Oldfield, T. J., Ceska, T. A., and Brady, R. L. (1991). *J. Appl. Cryst.*, **24**, 255.

14

Preparation of selenomethionyl protein crystals

S. DOUBLIÉ and C. W. CARTER, Jr.

1. Introduction

For many years multiple isomorphous replacement (MIR) has been the method of choice for phase determination. In most cases MIR is reliable, but occasionally it fails: heavy-atom binding may not occur, or if it does it may lead to loss of isomorphism. Use of Multiwavelength Anomalous Dispersion (MAD) has been introduced to circumvent these two problems (1, 2). On the one hand, the MAD method exploits the presence of anomalous scatterers, such as Cu or Fe in metalloproteins, or any of the heavy-metal derivatizing agents. On the other hand, all measurements relevant to determining a single phase can be made on the same crystal so isomorphism is exact, and electron density maps are of very high quality (3, 4). Thus, the MAD method will likely become a preferred method for phase determination.

Hendrickson has shown that selenium is a useful anomalous scatterer (3). Selenomethionine can totally replace methionine in *E. coli* (5). Substitution of methionine by selenomethionine offers a general method for introducing anomalous scatterers into cloned proteins. Preparation and crystallization of selenomethionine-substituted proteins are rather straightforward procedures that are relatively easy to perform (6). In brief, one needs to:

- express the cloned protein in a strain auxotrophic for methionine;
- ferment this strain in a medium in which methionine is replaced by selenomethionine;
- avoid oxidation during purification of the substituted protein;
- crystallize the substituted protein under conditions similar to those used with native protein.

2. Expression

2.1 Transformation

In most cases one can transform an existing methionine auxotroph with a plasmid producing the cloned protein. Auxotrophic (met^-) *E.coli* strains differ in their tolerance to selenomethionine. LeMaster studied the selenomethionine tolerance of several met^- strains and constructed a strain, DL41, which grows well on selenomethionine-containing media and therefore can be used as a general host for plasmid transformation (6)[a]. Plasmid transformation is carried out by standard procedures (7).

2.2 Transduction

In other cases, characteristics of a particular *E.coli* strain need to be preserved (8), and it is easier to introduce a met^- mutation by transduction, a mechanism of genetic exchange between donor and recipient cells via a bacteriophage that has incorporated donor genes into its genome. A convenient auxotrophic donor strain is one in which a gene from the methionine pathway has been inactivated by insertion of the transposon Tn10dCam[b]. Optimally, the inactivated met^- allele and the recipient strain will carry resistance to different antibiotics, so that cells selected on a medium supplemented with both antibiotics will be methionine auxotrophs and contain the plasmid of interest. Transduction is carried out as described in *Protocol 1* [adapted from reference (9)].

Protocol 1. Introduction of methionine auxotrophy into an *E.coli* strain via P1*vir* transduction[c]

1. Preparation of P1*vir* lysates of the donor auxotroph
   (a) Inoculate 5 ml of Luria-Bertani (LB) medium (7) with a single colony of the donor strain and rotate at 37 °C overnight.
   (b) Inoculate 0.05 ml of the overnight culture in 5 ml of LB medium containing 0.2% (w/v) glucose and 5 mM $CaCl_2$.[d]
   - Incubate for 30 minutes at 37 °C with aeration.
   - Add 0.1 ml of P1*vir* lysate (5×10^8 phages/ml).
   - Shake or rotate at 37 °C for 1.5–3 h until the cells lyse.
   - Add 0.1 ml of chloroform and vortex.
   - Spin down cell debris at 4500*g* for 10 min.

[a] The strain DL41 can be obtained from Dr Barbara Bachmann, *E.coli* Genetic Stock Center, Yale University School of Medicine, 333 Cedar Street, PO Box 3333, New Haven, CT 06510.
[b] These auxotrophic met^- donor strains can be obtained from George Weinstock, Department of Biochemistry and Molecular Biology, The University of Texas Health Science Center at Houston, PO Box 20708, 6431 Fannin Street, Houston, TX 77225.
[c] Use of P1*vir* is widespread in laboratories using *E.coli* genetics and should be readily available from such laboratories. Those unable to secure this reagent can contact G. Weinstock (see footnote [b]).

Protocol 1. *Continued*

- Transfer the supernatant to a sterile, screw-capped tube. Add 0.1 ml of chloroform and mix.
- Store the lysates at 4 °C.

2. Transduction to the recipient strain using P1*vir* lysate

(a) Inoculate 5 ml of LB medium with the recipient strain and shake at 37 °C overnight.

(b) Centrifuge the overnight culture at 1500*g* for 10 min and resuspend the cell pellet in 2.5 ml of 10 mM $MgSO_4$ containing 5mM $CaCl_2$.[d]

To five test tubes add recipient cells and P1*vir* grown on the donor strain as follows:

Test tube	Cells	Pl*vir* lysate
1	0.1 ml	none
2	0.1 ml	10 μl
3	0.1 ml	50 μl
4	0.1 ml	0.1 ml
5	none	0.1 ml

- Incubate the 5 test tubes for 30 min at 30 °C without shaking.
- Add equal volume of 1 M sodium citrate to each tube and mix.[e].
- Add 2.5 ml of appropriate molten (45 °C) top agar to each tube and plate on selective media.[f]

[d] P1 lysates are sensitive to chloroform if calcium is omitted. Calcium is also required for absorption of P1.

[e] Sodium citrate is added to chelate Ca^{2+} and prevent reinfection by P1 virus.

[f] Transductants should appear in plates 2, 3, and 4. 1 and 5 are controls for reversion and lysate contamination, respectively.

3. Fermentation

Although selenomethionine is recognized almost equally well with methionine by methionyl-tRNA synthetase [S. Doublié and C. W. Carter, unpublished observations; (10) and (11)], cells grow more slowly in selenomethionine and generally reach stationary phase at a lower final cell density. Furthermore, cells grown in selenomethionine tend to stay in stationary phase. A low percentage of LB in the starter culture will provide sufficient methionine to revive the cells from stationary phase, as well as thiamine and biotin. Hydrolyzed LB contains 5 mg/l of methionine (12), which will be incorporated preferentially to selenomethionine during fermentation (6). To minimize the amount of residual methionine in the purified protein, the dilution factor for the final inoculation must be adjusted according to the amount of LB used in the starter inoculum. This amount will be a compromise between a better growth rate and complete selenomethionine substitution. For each particular strain, one will have to determine the amount of rich medium in the starter culture, as well as the optimal concentration of selenomethionine throughout the fermentation (see *Protocol 2*).

Protocol 2. Cell growth in selenomethionine

1. Isolate single colonies by streaking an LB plate (7) supplemented with antibiotics with strain of interest. Incubate at 37 °C.

2. Ferment the cells in media containing all appropriate antibotics.
   (a) Inoculate 100 ml of starter medium (*Table 1*) with a single colony. Shake at 37 °C.
   (b) Inoculate a 10–20 litre fermenter[a] containing prewarmed fermentation medium (*Table 2*) with the 100 ml starter inoculum in mid-log phase.
   (c) Monitor cell growth in order to identify times for induction and harvest.

3. Induce specific cloned protein synthesis, if necessary.

4. Harvest the cells in mid- to late-log phase[b]. Resuspend the harvested cells in the appropriate lysis buffer and freeze with dry ice. Store at −80 °C.

[a] Cell growth should be done in a fermenter because regulated temperature and pH improve the yield.
[b] We have noticed cell lysis shortly after cells reached late log phase. Care must be taken to harvest cells as quickly as possible.

Table 1. Starter medium

Ingredients	**Concentration**
1. Minimal medium[a]	Minimal medium A[b] with carbon source at 5 g/l
2. All amino acids except methionine[a]	40 mg/l
3. Adenine, guanosine, thymine, and uracil[a]	0.5 g/l
4. Selenomethionine[c]	10–50 mg/l[d]
5. LB	*x*% (v/v) to be determined[e]

[a] One can also use LeMaster's medium (16) instead of 1, 2, and 3.
[b] Ausubel, F. M., Brent, R., Kingston, R. E., Moore, D. D., Seidman, J. G., Smith, J. A., and Struhl, K. (ed) (1987). *Current protocols in molecular biology.* Greene Publishing Associates and Wiley Interscience, New York, NY.
[c] Selenomethionine can be purchased from Sigma Chemical Company P.O. Box 14508, St. Louis, MO 63178 USA.
[d] Successful fermentations have been obtained with selenomethionine concentrations as low as 10 mg/ml (17), or as high as 40–50 mg/ml (6).
[e] As an example, Yang used a 100-ml starter medium containing 5% (v/v) LB for a 20 l fermenter (12).

Table 2. Fermentation medium

Ingredients	**Concentration**
1. Minimal medium[a]	Minimal medium A with carbon source 5 g/l
2. All amino acids except methionine[a]	40 mg/l
3. Adenine, guanosine, thymine, and uracil[a]	0.5 g/l
4. Selenomethionine	10–50 mg/l
5. Thiamine	2 mg/l
6. Biotin (if needed)	2 mg/l

[a] One can also use LeMaster's medium (16) instead of 1, 2, and 3.

4. Purification

Introduction of selenomethionine into proteins has two consequences that have an impact on purification. The altered chemistry of selenium makes substituted proteins more sensitive to oxidation than natural proteins. Moreover, if selenium atoms are exposed to solvent they can alter protein solubility and behaviour on chromatography resins. These properties require the following modifications to the normal purification, as shown in *Protocol 3*.

Protocol 3. Purification of selenomethionyl proteins

1. Purify as quickly as possible, with modifications to avoid oxidation.
   (a) De-gas all buffers by boiling or evacuation.
   (b) Include a reducing agent such as dithiothreitol (DTT) and a chelator such as Ethylene Diamine Tetraacetic Acid (EDTA) to remove traces of metals that could catalyse oxidation (13). Use 2–5 mM EDTA and 10–40 mM DTT.

2. Expect selenomethionyl proteins to be slightly less soluble than their natural counterparts.
   (a) Anticipate lower optimal ammonium sulphate concentrations in trituration protocols.
   (b) Anticipate increased retention in chromatography procedures.

3. Store purified protein in an oxygen-free environment, in 50% (v/v) glycerol at −80 °C, or better, as frozen droplets at −180 °C.

4. Undertake an amino-acid analysis to check the percentage of substitution. Selenomethionine is destroyed under the acid hydrolysis conditions used in amino-acid analysis (14), so that it is the disappearance of methionine that is monitored. Minimize the risk of mistaking contaminating methiomine for unsubstituted methionine by analysing a large (2–3 nM) sample.

5. Crystallization

Experience to date suggests that selenomethionyl proteins crystallize in conditions that are very similar to those used with native proteins (6), (12), (S. Doublié and C. W. Carter, in preparation). As a consequence of the lowered solubility of selenomethionyl proteins, either the protein or the precipitant concentration should be slightly reduced to achieve comparable degrees of supersaturation. Crystals should be maintained in a mother liquor containing DTT and EDTA, and stored in an oxygen-free environment such as an anaerobic chamber. Selenomethionine incorporation does not appear to alter diffraction limits and selenomethionyl protein crystals are isomorphous with native crystals. However, they are somewhat more sensitive to radiation damage.

6. Warning

Selenium is an essential element for most animal and bacterial life (15), but it is also a very *toxic* compound because of its ability to replace sulphur. In mammals (protein crystallographers), ingested methionine is a source of sulphur. As a result, selenomethionine can be *harmful* or even *fatal* if inhaled, swallowed, or absorbed through the skin. Experiments should always be done in a hood, and the experimenter should be sure to wear gloves.

References

1. Hendrickson, W. A. (1988). In *Crystallographic computing*, Vol 4, (ed. N. W. Isaacs and M. R. Taylor), pp. 97–108. Oxford University Press, Oxford, UK.
2. Hendrickson, W. A. (1985). *Trans. Am. Crystallogr. Assoc.*, **21**, 11.
3. Hendrickson, W. A., Pähler A., Smith, J. L., Satow, Y., Merrit, E. A., and Phizackerley, R. P. (1989). *Pro. Natl. Acad. Sci. USA*, **86**, 2190.
4. Yang, W., Hendrickson, W. A., Crouch, R. J., and Satow, Y. (1990). *Science*, **249**, 1398.
5. Cowie, D. B. and Cohen G. N. (1957). *Biochim. Biophys. Acta.*, **26**, 252.
6. Hendrickson, W. A., Horton, J. R., and LeMaster D. M. (1990). *EMBO J.*, **9**, 1665.
7. Maniatis, T., Fritsche, E. F., and Sambrook, J. (ed.) (1982). *Molecular cloning, a laboratory manual.* Cold Spring Harbor Press, Cold Spring Harbor, NY.
8. Carter,C. W. Jr. (1988). *J. Crystal Growth*, **90**, 168.
9. Silhavy, T. J., Berman, M. L., and Enquist, L. W. (1984). *Experiments with gene fusions*. Cold Spring Harbor Press, Cold Spring Harbor, NY.
10. Lawrence, F., Blanquet, S., Poiret, M., Robert-Gero, M., and Waller, J-P. (1973). *Eur. J. Biochem.*, **36**, 234.
11. Hoffman, J. L., McConnel, K. P., and Carpenter, D. R. (1970). *Biochim. Biophys. Acta*, **199**, 531.

12. Yang, W., Hendrickson, W. A., Kalman, E. T., and Crouch, R. J. (1990). *J. Biol. Chem.*, **265**, 13553.
13. Scopes R. K. (1982). *Protein purification.* Springer-Verlag, New York, NY.
14. Sheperd, L. and Huber, R. E. (1969). *Can. J. Biochem.*, **47**, 877.
15. Stadman, T. C. (1980). *Ann. Rev. Biochem.*, **49**, 93.
16. LeMaster, D. M. and Richards, F. M. (1985). *Biochemistry*, **24**, 7263.
17. Frank, P., Licht, A., Tullius, T. D., Hodgson, K. O., and Pecht I. (1985). *J. Biol. Chem.*, **260**, 5518.

A1

Suppliers of specialist items

Aldrich-Chemical Co., Inc., 1001 W. St Paul Avenue, PO Box 355, Milwaukee, WI 53201, USA. (chemicals)

Alpha Laboratories Ltd., Eastleigh, Hampshire, UK. (multiple liquid dispenser)

American Can Company, Greenwich, CT 06830, USA. (Parafilm® 'M', laboratory film)

Amicon Division, W. R. Grace and Co., 72 Cherry Hill Drive, Beverly, MA 01915, USA. (filters, membranes)

Applied Biosystems, Inc., 850 Lincoln Center Dr., Foster City, CA 94404, USA and Birchwood Science Park North, Warrington, Cheshire WA3 7PB, England. (biochemical instrumentation, chemicals)

Appligene, route du Rhin, BP 72, 67402 Illkirch Cedex, France. (biochemicals)

Bachem, Hauptstrasse 144, CH-4416 Bubendorf, Switzerland. (detergents)

BDH Limited, Broom Road, Poole, BH12 4NN, UK. (electrophoresis products)

Beckman, 4550 Noris Canyon Road, PO Box 5101, San Ramon, CA 94583, USA. (centrifugation, pipetting station)

Becton-Dickinson and Co., **Clay Adams Div.**, 299 Webro Road, Parsippany, NJ 07054, USA. (Falcon plasticware)

Bender and Hobein GmbH, D-8000 Munchen 2, Lindwurmstrasse 71, Germany. (free flow electrophoresis)

Bijhoelt and Heuvelen SV, The Netherlands. (transparent and adhesive plastic foils)

BioBlock Scientific, BP 111, F-67403 Illkirch Cedex, France. (scientific equipments)

BioRad, 1414 Harbour Way South, Richmond CA 94804, USA. (HPLC, IEF equipments, biochemicals)

Boehringer Mannheim, PO Box 310120, D-6800 Mannheim 31, Germany. (biochemicals)

Brookhaven Instrument Corp., 750 Blue Point Road, Holtsville, NY 11743, USA. (light scattering instrumentation)

Bunton Instrument Co., **Inc.**, 615 South Stonestreet Avenue, Rockville, MD 20850, USA. (microgrippers)

Calbiochem Behring Diagnostics, 10933 N. Torrey Pines Road, La Jolla, CA 923037, USA. (biochemicals, detergents)

Cambridge Repetition Engineers Ltd., Green's Road, Cambridge, CB4 3EQ, UK. (dialysis buttons for crystallization)

CEA verken AB, S-152 01 Strängnäs, Sweden. (X-ray films)

Charles Supper Company Inc., 15 Tech Circle, Natick, MA 07160, USA. (crystallographic equipment)

CJB Developments Limited, Airport Service Road, Porthmouth, Hampshire PO35PG, UK. (large-scale preparative electrophoretic apparatus)

Cole and Palmer Instrument Co., 7425 N. Oak Park Avenue, Chicago, IL 60648, USA. (scientific equipments)

Corning, Inc., **Science Products**, MP-21–5–8, Corning, NY 14831. USA. (glassware, pipettes)

Costar Nucleopore®, One Alewife Center, Cambridge, MA 02140, USA and Costar Europe, Ltd., PO Box 94, 1170 AB Badhoevedrop Sloterweg 305a, 1171 VC Badhoevedrop, The Netherlands. (titration and crystallization plates, pipettors).

Cryschem Inc., 5005 La Mart Drive, Riverside, CA 92507 USA; Moss Via Froutera, San Diego, CA 92127, USA. (crystallization boxes, protein crystallization system)

Dynatech Laboratories, Inc., 14340 Sullyfield Circle, Chantilly, VA 22021, USA. (titration plates for crystallization robots)

Douglas Instruments Ltd, 255 Thames House, 140 Battersea Park Road, London SW11 4NB, UK. (automatic batch crystallization system)

Dow Corning Corp., Dow Corning Center, Box 0994, Midland, MI 48686–0994, USA. (silicone oil, grease)

Dupont de Nemours and Co., Concord Plaza, Wilmington, DE 19898, USA. (chemicals, instrumentation)

Eastman-Kodak Co., 343 State St., Rochester, NY 14650, USA and Kodak House, Station Road, Hemel Hempstead, Herts. HP1 1JU, UK. (chemicals, films)

Enraf Nonius Delft, PO Box 483, 2600 AL Delft, The Netherlands. (diffractometry)

Euromedex, Produits de Recherche, 29 rue Herder, F-67000 Strasbourg, France. (chemicals, protease inhibitors)

Everett's Co., Park gate, Nr, South Hampton, UK. (vacuum wax, seals, and lubrifiants)

Fisher Scientific Company, 711 Forbes Avenue, Pittsburgh, PA 15219, USA. (biochemicals, scientific equipments)

Flow Laboratories International SA, via Lambro 23/25, I-20090 Opera (MI), Italy. (biochemical equipments, Linbro plate, CrystalPlate and coverslips)

Fluka Chemie AG, Industriestrasse 25, CH-9470 Buchs, Switzerland. (biochemicals, detergents)

Genzyme Corporation, 75 Kneeland Street, Boston, MA 02111, USA. (protease-free deglycosylation enzymes)

Gibco BRL, Bethesda Research Laboratories, Life Technologies, Inc. PO Box 6009, Gaithersburg, MD 20877, USA. (biochemicals, growth media)

Gilson Medical Electronics, Inc., 72 rue Gambetta, BP 45, F-95400 Villers-le-Bel, France and 3000 W. Beltine Hwy., PO Box 27, Middleton, WI 53562, USA. (sample changers)

Gow-Mac Inc., PO Box 32, Bound Brook, NJ 08805–0032, USA. (thermal conductivity detectors)

Heraeus Feinchemikalien und Forchungsbedarf GmbH, Alter Weinberg, D-7500 Karlsruhe 41-Ho., Germany. (chemicals, reagents for silanization;

Hewlett-Packard Co., Analytical Group, Mailstop 20B AE, Palo Alto, CA 94403, USA. (robotics)

Hamilton Co., PO Box 10030, Reno, NV 89520–0012, USA. (syringes)

Hilgenberg Glass Company, D-3509 Malsfeld, Germany (X-ray glass/quartz capillaries)

Huber Diffraktionstechnik Gmbh, D-8219 Rimsting, Germany (diffractometry)

IBF Biotechnics, 35 avenue Jean-Jaurès, 92290 Villeneuve-la-Garenne, France. (chromatographic matrices, biochemicals)

ICI Cambridge Research Chemicals, Gadbrook Park, Northwich, Cheshire CW9 7RA, UK. (chemicals)

ICN Biomedicals, Inc., Micromedica Systems Diagnostic Division, 102 Witmer Road, Horsham, PA 19044–2281, USA. (robotic protein crystallization system II, pipetting stations)

ICN Flow, 3300 Hyland Avenue, Costa Mesa, CA 92626, USA (biochemical equipments, Linbro plate, CrystalPlate® and coverslips)

Imaging Technology Inc., 600 West Cummings Park, Woburn, MA 01801, USA. (digitizers)

Intermec Corp., 4405 Russell Road, PO Box 360602, Lynnwood, WA 98046–9702, USA. (barcode printer)

Jouan SA, rue Bobby Sands, F-44800 Saint Herblain, France. (laboratory equipments)

Keithley Data Acquisition and Control, 28775 Aurora Road, Cleveland, OH 44139, USA. (instrument interfaces)

Kohyo Trading Company, Kyodo Bldg 4–1, 2 Chome, Iwando-cho, Chiyoda-ku, Tokyo, Japan. (detergents)

Leica SARL, see Wild-Leitz.

Marresearch, Grosse Theaterstrasse 42, Postfach 303670, 2000 Hamburg 36, Germany. (image plate)

Memmert GmbH and Co., Aeussere Ritterbacherstrasse 38, D–8540 Schwabach, Germany. (laboratory equipments, thermostated cabinets)

Merck, Frankfurterstrasse 250, D–6100 Darmstadt, Germany. (chemicals and biochemicals)

Micromedica System, Inc., (see ICN Biomedicals). (pipetting station)

Millipore Waters, PO Box 255, Bedford MA 01730, USA and Zone Industrielle, F-67120 Molsheim, France. (filtration, membranes, HPLC equipments)

Microflex Technology, Inc., The Millennium Centre, PO Box 31, Triadelphia, WV 26059, USA. (microgrippers)

National Institute of Standards and Technology, (Standard Reference Data) Bldg. 221/A323, Gaithersburg, MD 20899, USA. (Software with crystallization data bank)

National Instruments Corp., 12109 Technology Blvd, Austin, TX 78727–6204, USA. (instrument interfaces, laboratory software)

Neosystem Laboratories, Technopole du Rhin, 21 rue du la Rochelle, F-67100 Strasbourg, France. (peptides)

Nikon Corporation Instrument Div., Fuji Bldg 2–3, 3-Chome, Maranouchi, Chiyoda ku, Tokyo 100, Japan. (stereo microscopes)

Nikon Europe BV, Shipholm weg 321, 1171 AE Badhoevedorp, The Netherlands. (stereo microscopes)

Nunc Inc., 2000, North Aurora Road, Naperville, IL 60566, USA. (plastic tubes and plates)

Omnifit Ltd., 51 Norfolk Street, Cambridge CB1 2LE, UK and 2005 Park Street, Box 56, Atlantic Beach, NY 11509, USA. (valves)

Omnilabo Holland BV, Breda, The Netherlands. (multi-well plates)

Oxyl, Peter Henlein Strasse 11, D-8903 Bobingen, Germany. (detergents)

Panasonic Inc., One Panasonic Way, Secaucus, NJ 07094, USA. (optical disc recorder)

Pentapharm Ltd, Engelgasse 109, CH-4002 Basel, Switzerland. (protease inhibitors)
Peptide Institute, 476 Ina Miush-shi, Osaka 562, Japan (protease inhibitors)
Perpetual Systems Corporation, 2283 Lewis Avenue, Rockville, Maryland 20851, USA. (sitting-drop rods for crystallization)
Pfanstiel Laboratory, Inc., 1219 Glen Rock Avenue, Wavkega, IL 60085 0439, USA. (detergents)
Pharmacia-LKB Biotechnology, Inc., S-751 82 Uppsala, Sweden and 800 Centenial Drive, Piscataway, NJ 08854, USA. (chromatographic supports, biochemicals, motorized valves, peristatic pumps, and other instrumentation)
Phenomenex, 6100 Palos Verdes Drive S., Rancho Palos Verdes, CA 90274, USA. (HPLC columns for tRNA)
Pierce, PO Box 1512, 3260 BA Oud-Beijerland, The Netherlands. (laboratory supplies)
Polycrystal book service, PO Box 3439, Dayton, Ohio 45401, USA. (crystallography books)
PolyLabo Paul Block et Cie, BP 36, F-67023 Strasbourg Cedex, France. (scientific equipments)
Prolabo, 12 rue Pelée, F-7511 Paris, France. (chemicals, equipments)
Promega Corporation, 2800 Woods Hollow Road, Madison, WI 53711, USA. (RNAsin®-nuclease inhibitor)
Pye Unicam Ltd, York Street, Cambridge CB1 2PX, UK. (Philips X-ray generator)
Resolution Technology, 26000 Avenida Aeropuerto 22, San Juan Capistrano, CA 92675, USA. (time-lapse VCR)
Radiometer, A/S 49 Krogshojvej, DK 2880 Dagsvaerd, Denmark. (pH-meter, conductimeter)
Rainin Instrument Co. Inc., Mack Road, Woburn, MA 01801, USA. (filters)
Rigaku, Monschauer Strasse 7, D-4000 Düsseldorf-Heerdt, Germany & 3 Electronics Avenue, Danvers, MA 01923, USA. (X-ray generators, image plate)
Roucaire, BP 65, F-78143 Velizy-Villacoublay Cedex, France. (scientific equipments)
Seikagaku Kogyo Co, Ltd., 1–5, Nihonbashi-Honcho 2-Chome Chuo-ku, Tokyo, 103, Japan. (biochemicals, glycosylases)
Serva Feinbiochemica GmbH and Co., PO Box 105260, D-6900 Heidelberg, Germany. (biochemicals)
Setaram, 7 rue de l'Oratoire, BP. 34, F-69641 Caluire Cedex, France. (instrumentation, calorimeters)
Siemens AG, Mess., Pruf. und Prozesstechnik, Ostl. Rheinbrückenstrasse 50, D-7500 Karlsruhe 21, Germany. (diffractometry)
Sigma Chemical Company, PO Box 14508, St Louis, MO 63178, USA. (biochemicals)
Societe 3412, 65 avenue de Stalingrad, F-95104 Argenteuil, France. (crystallization boxes)
Sofranel, 59 rue Parmentier, 78500 Sartrouville, France. (X-ray glass/quartz capillaries)
Speciality Chemicals, PO Box 1466, Gainesville, FL 32602, USA. (Prosil® -28 reagent for silanization)
Spectra Physics, 333 N. First Street, San Jose, CA 93412, USA and Siemensstrasse 20, D-6100 Darmstadt Kranichstein, Germany. (HPLC equipments, lasers)
Spectrum Medical Industries, Inc., 8430 Santa Monica Blvd, Los Angeles, CA 90069, USA. (dialysis membranes -Spectrapore®)

Symbol Technologies, Inc., 1101 Lakeland Avenue, Bohemia, NY 11716-3300, USA. (barceode readers)

The Product Integrity Company, Enfield, CT 06082, USA. (programs for factorial analysis)

Tosohaas, 6th and Market Streets, Philadelphia, PA 19105, USA. (HPLC columns)

Transformation Research Inc., PO Box 241, Framington, MA 01701, USA. (protease inhibitors)

Vegatec S.A.R.L., 7 place des Onze Arpents, F-94800 Villejuif, France. (detergents)

Velmex, Inc., PO Box 38, E. Bloomfield, NY 14443, USA. (stepper motors, motorized slides)

Wild-Leitz (Leica SARL), 86, avenue du 18 juin 1940, F-92563 Rueil-Malmaison Cedex, France and CH-9435 Heerbrugg, Switzerland. (stereo microscopes)

Whatman Laboratory Sales Ltd, Unit 1, Colred Road, Parkwood, Maidstone, Kent, ME15 9XN, UK. (chromatography supports)

Wolfgang Müller, Reierallee 12, D-1000 Berlin 27, Germany. (X-ray glass/quartz capillaries)

Zymark, Zymark Center, Hopkinton, MA 01748, USA and ZAC Paris Nord II, 13 rue de la Perdrix, BP 40016, F-95911 Roissy Charles-de-Gaulle Cedex, France. (robotics)

Index

Index